William Saville Kent

The Naturalist in Australia

William Saville Kent

The Naturalist in Australia

ISBN/EAN: 9783337311636

Printed in Europe, USA, Canada, Australia, Japan

Cover: Foto ©berggeist007 / pixelio.de

More available books at **www.hansebooks.com**

THE NATURALIST IN AUSTRALIA,

BY

W. SAVILLE-KENT, F.L.S., F.Z.S., &c.,

PAST PRESIDENT ROYAL SOCIETY OF QUEENSLAND;
FORMERLY ASSISTANT IN THE NATURAL HISTORY DEPARTMENTS OF THE BRITISH MUSEUM;
LATE COMMISSIONER AND INSPECTOR OF FISHERIES AND FISHERIES EXPERT REPORTER TO THE
GOVERNMENTS OF TASMANIA, QUEENSLAND, VICTORIA, AND WESTERN AUSTRALIA

AUTHOR OF

"THE GREAT BARRIER REEF OF AUSTRALIA,"
"A MANUAL OF THE INFUSORIA," &c., &c.

Illustrated by 50 full-page Collotypes, 9 Coloured Plates by Keulemans and other Artists, and over one hundred Illustrations in the Text.

LONDON: CHAPMAN & HALL, LIMITED.
1897.

LONDON:
WATERLOW AND SONS LIMITED, PRINTERS,
LONDON WALL.

TO

THE HON. SIR JOHN FORREST, K.C.M.G.

PREMIER AND EARLY PIONEER

OF

WESTERN AUSTRALIA

WITH THE NATURAL HISTORY

OF WHICH COLONY

THE FOLLOWING PAGES MAINLY DEAL

This Volume

THE NATURALIST IN AUSTRALIA

IS DEDICATED AS A MARK OF HIGHEST ESTEEM

BY THE AUTHOR.

PREFACE.

HE endeavour is made in this volume to present to the English reading public a few glimpses of the Faunal and Floral products of that magnificent component of our Empire—the Island-Continent of Australia.

No attempt is made here to produce a systematic monograph or anything beyond a general compendium of any particular group or groups, the purpose being more in the direction of recording data concerning the life phenomena or peculiarities of biological types that specially attracted the author's attention during the period of close upon twelve years in which he acted as Commissioner of Fisheries, or specially engaged Fisheries expert to the greater number of the Australian colonies, and in the fulfilment of which professional engagements he extended his travels throughout the entire Australian coast-line.

The material here selected for description and illustration will, it is trusted, assist towards the promotion of a wider interest in the natural history wealth in sea and on shore possessed by the Australian peoples, and conduce towards its more intimate investigation both by dwellers on the land and by systematic explorers. Concerning many of the subjects here dealt with, as for example the form, habits, and architectural fabrications of the Termites and the varieties, distribution, and bizarre manners of many members of the Bird and Lizard races, a wide field is open both for original investigation and for inexhaustible recreation to every intelligent observer.

As was the case with the author's previously published volume, "The Great Barrier Reef of Australia," the camera has been extensively requisitioned for the delineation of the subjects illustrated. These will be found to include, as in that volume, various marine organisms photographed in their native element with possibly an even greater measure of success. The potency of the camera "to hold as 'twere the mirror up to Nature" in almost every conceivable phase and condition of her varying moods and tenses, is indeed established in these pages—at any rate, to an extent that should recommend the more universal employment of this instrument for the portrayal of the protean aspects and metamorphoses of living organisms.

The subjects dealt with in this book include tenants of the sea and shore throughout the entire range of the Australian colonies, from North Queensland, Port Darwin and Thursday Island, to picturesque Tasmania and Bass's Straits. A larger space has undoubtedly, however, been allotted to the products of Western Australia. This colony, although possessing the most extensive land area, has but recently come to the forefront among its compeers, and there has hitherto been but a scant amount of information available concerning its natural history treasures. The marvellous material progress and prosperity that, concurrently with the discovery and exploitation of its auriferous wealth, has so pre-eminently distinguished this colony's career within the past few years will, no doubt, also speedily bring about an equivalent awakening to and development of its latent scientific potentialities.

As an indication of the leading position Western Australia is eligible to occupy with relation to one important biological subject, reference may be made to that Chapter which deals with Houtman's Abrolhos. As there demonstrated, very exceptional facilities prevail at that place for the conduct of reef-boring operations and for the prosecution of all methods of investigation relating to Coral and Coral life. The Islands are situated within a day's journey from the metropolis of the colony and a few hours' sail only from the Port of Geraldton. A permanent biological observatory established there would, consequently, be in near touch with and possess full command of all the conveniences and appliances of up-to-date civilisation.

An element that will highly recommend the Abrolhos Archipelago to the attention of many biologists is the circumstance that notwithstanding the predominating tropical nature of the marine fauna, the Islands are situated so far south that the climate is temperate and consequently exempt from those conditions which are a serious deterrent to investigators of tropical marine life.

The few data chronicled in this volume concerning the remarkable interblending of a tropical and temperate marine fauna that occurs at Houtman's Abrolhos will serve to accentuate the desirability that exists for their further systematic investigation. The question of fully exploring and working out the indigenous faunæ of isolated or otherwise remarkable islet areas is at the present time commanding a large share of attention in scientific circles the world over. Among the multitude of counsellors upon whom will devolve the responsibility of electing which regions shall hereafter receive the advantage of such systematic attention the author would earnestly advocate the undoubted claims upon their suffrages that are possessed by this, biologically speaking, most interesting assemblage of Islands on the western shores of Australasia.

The author has various and sundry obligations to chronicle with relation to the production of this volume. To the Western Australian Government in general and to its enlightened Premier, Sir John Forrest, in particular, he is indebted for opportunities and facilities afforded for acquiring the information herein recorded concerning the marine and other natural history products of that important colony; and to Dr. Ernest Black, Mr. John Brockman, Dr. Williams, Mr. G. S. Streeter, Mr. W. Male, Mr. R. C. Hare, Mr. Broadhurst, and others, for individual assistance towards the acquisition of specimens and for facilities for making studies or taking successful photographs of many of the Western Australian objects illustrated.

To Prof. T. Jeffry Bell, Mr. Boulanger, Mr. Edgar Smith, Mr. R. I. Pocock, Dr. Murray, and other officers of the Zoological and Botanical Departments of the

British Museum, the author is indebted for kindly assistance towards the correct identification of many of the specimens figured. For substantial services in the revision of the proof sheets acknowledgments are especially due to Mr. Robert Bell, M.A.

The author has, finally, to acknowledge his obligations to Messrs. Waterlow and Sons Limited and their staff of expert departmental managers for the painstaking and, to himself, eminently satisfactory manner in which they have carried out his views concerning the production of this volume.

LONDON,
February 1st, 1897.

TABLE OF CONTENTS.

Chapter.		Pages.
I.	General and Introductory ...	1- 38
II	Birds	39- 68
III.	Lizards	69-100
IV.	Termites (White Ants)	101-131
V.	Houtman's Abrolhos	132-152
VI.	Fishes—Phenomenal and Economical ...	153-194
VII.	Pearls and Pearl-Oysters	195-214
VIII.	Marine Miscellanea	215-251
IX.	Insect Oddities	252-265
X.	Vegetable Vagaries	266-289
	Appendix ...	290

LIST OF CHROMO-PLATES.

Plate.		Facing Page
I.	Tidally exposed inshore Reef, Palm Islands, Queensland	1
II.	Gouldian Finches, *Poephila mirabilis*, and *Poephila Gouldæ*	30
III.	The Frilled Lizard, *Chlamydosaurus Kingi*	69
IV.	Madrepora Reef, Pelsart Island Lagoon, Houtman's Abrolhos	144
V.	Abrolhos Nudibranchiate Mollusc, *Doris imperialis*	Between 150 and 151
VI.	Sea Horses and Dragons, Syngnathidæ	153
VII.	A Family Party, Plectognathi...	187
VIII.	A Sea-Star Galaxy	244
IX.	Insect Oddities	254

LIST OF WHOLE PAGE, COLLOTYPE, ILLUSTRATIONS.

Plate.		Facing Page
I.	Natives of King's Sound, Western Australia, with Characteristic Raft. Glass and Quartz Spear Heads...	8
II.	Festive Dance, or "Corroboree," of Natives of Roebuck Bay, Western Australia	15
III.	The Duck-billed Platypus, *Ornithorhynchus paradoxus*, and Spiny Ant-Eater, *Echidna aculeata*	18
IV.	Australian Bears, *Phascolarctos cinereus*, and Flying Phalanger, *Petauroides volans*	26
V.	Western Australian Baobab or Bottle-Tree, *Adansonia rupestris*	37
VI.	Australian "More-porks" or Fern-Owls, *Podargus strigoides*. ♂ and ♀	42
VII.	Australian "More-pork." Male Bird	44
VIII. and IX.	Australian "More-porks." Illustrating Protean Aspects assumed under varying Emotional Influences...	Between 46 and 47
X.	Australian "More-porks." Rain-Bath Manœuvres	50
XI.	"Innocents Abroad"; Young Australian Pelicans	67
XII.	The Frilled Lizard, *Chlamydosaurus Kingi*. Bipedal running attitudes illustrated by Instantaneous Photographs	74
XIII.	Western Australian Spinous Lizards, *Moloch horridus*. ♂ and ♀	83
XIV.	Spine-Tailed Lizards, *Egernia Stokesii*, and Lace Lizard, *Varanus varius*...	93
XV.	Nest Mounds of White Ants (Termites), Meridian variety, Laura Valley, North Queensland	101
XVI.	Nest Mounds of White Ants, Kimberley type, Derby, Western Australia	105
XVII.	Nest Mounds of White Ants, Kimberley type, Derby, Western Australia	108
XVIII.	Nest Mounds of White Ants, Kimberley type, including section	114
XIX.	Nest Mounds of White Ants, Kimberley type, illustrating abnormal shapes and phases of reconstruction	119
XX.	Nest Mounds of White Ants, Meridian variety, Laura Valley, North Queensland	122
XXI.	Nest Mounds of White Ants, Meridian variety, Port Darwin, Northern Territory of South Australia...	124
XXII.	Nest Mounds of White Ants: A. Columnar type, Port Darwin; B. Ovate Termite Mound, with nest burrow of *Tanysiptera sylvia*; C. Small extra-tropical Termite mound, with nest burrow of *Posephotus pulcherrimus*	128

LIST OF WHOLE PAGE, COLLOTYPE, ILLUSTRATIONS.

Plate		Facing Page
XXIII.	Abrolhos Corals, genus Madrepora	134
XXIV.	Abrolhos Corals, *Montipora circinata* and *Madrepora corymbosa* ...	138
XXV.	Abrolhos Corals, *Madrepora prolœriformis*	142
XXVI.	Shell and Coral Beaches, Pelsart Island, Houtman's Abrolhos ...	148
XXVII.	A. Diamond Trevally, *Caranx gallus*; B. Western Australian Snapper, *Pagrus major*	160
XXVIII.	Plaster Casts of Tasmanian Fish	165
XXIX.	A. Bottle-nosed Snapper, *Pagrus major* ♂; B. King Snapper or "Nannegai," *Berys Mulleri*; c. Sea Pike, *Sphyræna obtusata* ...	172
XXX.	A. Sergeant Baker, *Aulopus purpurissatus*; B. Western Australian Jew Fish, *Glaucosoma hebraicum*; c. Cat-Fish Eel, *Plotosus sp.* ...	176
XXXI.	A. Lung Fish, *Ceratodus Forsteri*; B. York Peninsula Barramundi, *Osteoglossum Jardinei*; c. Murray Cod, *Oligorus macquariensis*	180
XXXII.	Western Australian Pearls	195
XXXIII.	The "Southern Cross" Pearl ... ... } Between 198 & 199
XXXIV.	Remarkable Western Australian Pearl ... ...
XXXV.	Experimental Pearl-Shell Cultivation sites: A. Coral Reef, Thursday Island; B. Mangrove Thicket, Roebuck Bay	204
XXXVI.	Natural Clusters of Shark's Bay Mother-of-Pearl Shell ...	209
XXXVII.	Pearl-Shelling Stations, Fresh Water Camp, Shark's Bay, Western Australia	210
XXXVIII.	A. Shark's Bay Golden Pearls; B. Artificially-produced Pearl ...	212
XXXIX.	A. Cygnet Bay Anemones, *Condylactis sp.*; B. Giant Anemone, *Discosoma Haddoni*, with Commensal Fish and Crab	220
XL.	Socially Consorted Anemones and Worm Tubes, *Acrozoanthus australiæ* ...	226
XLI.	Young Turtles, *Chelone mydas*, photographed while swimming ... } Between 236 & 237
XLII.	Jelly Fish, *Discomedusæ*, photographed in water from life ...
XLIII.	A. Coral Rock Oysters, *Ostræa mordax*, Great Barrier Reef; B. Rock Oysters, *Ostræa glomerata*, Keppel Bay, Queensland	247
XLIV.	The Author's Oyster Culture Apparatus	248
XLV.	Queensland Spider, *Nephila fuscipes*, with diminutive males, Commensal Argyrodes and Parasitic Sandfly	264
XLVI.	Western Australian Baobab or Bottle-Tree, *Adansonia rupestris* ...	268
XLVII.	A. Grass-Trees, Cycads and Ferns, Drakesbrook, Western Australia; B. Arborescent Grass-Tree or "Blackboy," *Xanthorrhœa arborea* ...	272
XLVIII.	A. Underground Grass-Tree or "Blackboy," *Xanthorrhœa hastilis*; B. Drumstick Grass-Tree or "Blackboy," *Kingia australis*	274
XLIX.	A. Mangrove-frequenting Fruit Bat, *Pteropus conspicillatus*; B. Orange and White Mangrove, *Rhizophora mucronata*, and *Avicennia officinalis*, Broome Creek, Western Australia	277
L.	A. Trumpet Creeper, *Beaumontia grandiflora*, on a Queensland Verandah; B. Water Hyacinth, *Pontideria crassipes*, Adelaide Botanic Gardens	288

LIST OF PROCESS AND OTHER ILLUSTRATIONS IN THE TEXT.

Fig. No.		Page
	Photogravure—Author's Portrait (facing Title Page).	
1.	Aborigines of King's Sound, Western Australia	1
2.	"Broome Belles," Native Women of Roebuck Bay, Western Australia	1
3.	Carved Baobab Nut and Shell Aprons, with Human Hair Girdles, Kimberley District, Western Australia	10
4.	Natives of Kimberley District, Western Australia, showing Attitude when Holding and Throwing the Boomerang	11
5.	Short-headed Flying Phalanger, showing Position maintained during its Flying Leap	24
6 & 7.	White individual of large Flying Phalanger, showing Tail pendant, and in its characteristically enrolled state	25
8 & 9.	Australian Pouched Mole, *Notoryctes typhlops*, Lateral and Ventral aspects	29
10.	Miner's Tent and Nest Mound of Australian Jungle Fowl, Goode Island, Torres Straits	32
11.	Hollow Wooden Cradle used by the Native Women of the Kimberley District, Western Australia	38
12.	"Dulce Domum," "More-porks," Butcher Bird, and Piping Crow	39
13.	Australian "More-porks," illustrating disparity in Size and Aspect under contrasting influences	44
14.	Australian "More-porks," Nest Building Extraordinary	48
15.	North Queensland Laughing Jackass, *Dacelo Leachii*	53
16 & 17.	Gouldian Finches, Dancing Attitudes	59
18.	Young Australian Osprey, *Pandion leucocephalus*	68
19.	Frilled Lizards at Bay	69
20.	Frilled Lizard at Rest	69
21.	Frilled Lizards Asleep	79
22.	Frilled Lizard as Prototype of Chinese Dragon	79
23—25.	Bearded Lizard, *Amphibolurus barbatus*, Three Attitudes	81
26.	Spinous Lizards, *Moloch horridus*, "A Wayside Greeting"	84
27.	Spinous Lizards, showing Knapsack-like Neck Excrescences	84
28.	Spinous Lizards Feeding at an Ant Track	85
29.	Spinous Lizards, "A Post Prandial Promenade"	85

XIV   LIST OF PROCESS AND OTHER ILLUSTRATIONS IN THE TEXT.

Fig. No.		Page
30.	A Skull, B Tail Sheath of Miolania Oweni ...	88
31.	Spinous Lizards, "A Millinery Novelty"	88
32 & 33.	Stump-tailed Lizards, *Trachysaurus rugosus*, Lateral and Ventral aspects	90
34.	Spine-tailed Lizard, *Egernia depressa*	93
35.	Spine-tailed Lizards, "Tiffin Ahoy!!" ...	100
36.	Termite Mounds, Albany Pass, North Queensland	101
37.	White Ants, Termites, Individual Types	103
38.	Wood-destroying Termites, Kimberley District, Western Australia	103
39.	Termite Nests, Albany Pass, North Queensland	116
40.	Ant-Hill Point, Albany Pass, North Queensland	116
41.	Broadside View of Laura Valley Meridian Ants' Nest	122
42.	Infusorial Parasites, *Trichonympha Leidyi*, of Tasmanian White Ants ...	131
43.	Wreck Point, Pelsart Island, Houtman's Abrolhos	132
44.	Mirage-Elevated Breakers on Outer Barrier, Pelsart Island ...	132
45.	Stag's-horn Coral Growth, Great Barrier Reef, Queensland ...	144
46.	Madrepora Branchlet, showing Re-growth at Broken End	144
47.	Pelsart Island Scroll Coral, *Montipora circinata* ...	146
48.	"Birth of a Coral Island," Houtman's Abrolhos ...	147
49.	Abrolhos Coral, *Madrepora protœiformis* ...	152
50.	Plumed Trevally, *Caranx gallus* ...	153
51.	Ord River Tassel Fish, *Polynemus Verekeri*	168
52.	Tasmanian Toad Fish, *Tetrodon Hamiltoni*	191
53.	Port Jackson Shark, *Cestracion Phillipi*	194
54.	Pearling Luggers, Broome Creek, Western Australia	195
55.	The "Southern Cross" Pearl	200
56.	Fish, *Fierasfer*, embedded in Mother-of-Pearl Shell	202
57.	Pearl Shell Opened, showing Contained Pearl and Commensal Crab	202
58.	Young Cultivated Pearl Shells attached to Parent ...	207
59.	Group of Western Australian Pearls	214
60.	"Rhinoceros Rock," Roebuck Bay, Western Australia	215
61.	Ascidian-covered Rocks, Roebuck Bay, Western Australia	216
62.	Social Holothuriidæ, *Colochirus anceps*	218
63.	Sea Anemones, *Phyonbrachia sp.* ...	222
64.	Sea Anemones, *Condylactis sp.* ...	222
65.	Stinging Anemones, *Actinodendron alcyonidium*	224
66.	Young Peltate Coral, *Dendrophyllia axifuga*, with Expanded Polyps ...	232
67.	*Turbinaria peltata*, Matured Example from Shark's Bay, Western Australia	232
68.	Young Coralla of *Turbinaria conspicua* and *T. peltata* ...	233
69.	Matured Coralla of *Turbinaria conspicua* ...	233

LIST OF PROCESS AND OTHER ILLUSTRATIONS IN THE TEXT.   XV

Fig. No.		Page
70.	Revolute Cup-Coral, *Turbinaria revoluta*	234
71 & 72.	Verticillate and Neptune's Cup Sponges	236
73 & 74.	Soldier Crabs, *Gelasimus coarctata*, Two Figures	240
75.	Army Crabs, *Mycteris longicarpus*	242
76.	Army Crabs, *Mycteris platycheles*, "A Desperate Mêlée"	242
77 & 78.	Coral Rock Oyster, *Ostrea mordax*, var. *cornucopia*, Two Figures	248
79 & 80.	Dwarf Oysters, *Ostrea ordensis*, Ord River, Western Australia, Two Figures	249
81.	Stalactite-like Sand-stone Concretions, Sweer's Island, Gulf of Carpentaria	251
82.	"Butterfly Birthdays"	252
83.	Queensland "X-Ray" Spider, *Argiope regalis*	262
84.	Western Australian Tarantula	265
85.	"Sud Teomine Boabi," Western Australia Bottle-tree	266
86.	Flowers of Baobab Tree, *Adansonia rupestris*	268
87.	Double Stemmed Baobab Tree	269
88.	Prostrate Baobab Tree with Rejuvenescent Trunks	269
89.	A Fallen Monarch, Lightning-Shattered Western Australia Baobab	271
90.	Rosette-Galls of White Mangrove, *Avicennia officinalis*	278
91.	Bird-Pea, *Crotalaria Cunninghami*, Roebuck Bay, Western Australia	279
92.	Salt Bush, *Salicornaria sp.*, Cygnet Bay, Western Australia	280
93.	"Roley-Poley" Grass, Shark's Bay, Western Australia	280
94.	Western Australian Fringed Violet, *Thysanotis dichotoma*	282
95 & 96.	A. Giant Cactus, *Cereus chalybaeus*, Separate Flowers	284
	B. White Cactus, *Cereus nitens*, Adelaide Botanic Gardens	284
97 & 98.	A. Giant Cactus, *Cereus chalybaeus*, Adelaide Botanic Gardens	285
	B. White Cactus, *Cereus nitens*, Separate Flowers	285
99 & 100.	Cereus Grandiflorus, Government Gardens, Perth, Western Australia	286
101.	Tree-ferns and Tropical Plants, Brisbane Botanic Gardens	287
102.	Tree-ferns, Mount Wellington, Tasmania	287
103.	Finis—Breaking Waves, Bunbury Causeway	289
104.	Indian Boomerang (Appendix)	290

CORRIGENDUM.

Page 99, line 3,

For EUMÆUS ALGERIENSIS *read* EUMECES ALGERIENSIS.

ABORIGINES OF KING'S SOUND, WESTERN AUSTRALIA. p. 9.

The Naturalist in Australia.

CHAPTER I.
GENERAL AND INTRODUCTORY.

"BROOME BELLES," NATIVE WOMEN OF ROEBUCK BAY, WESTERN AUSTRALIA. p. 14.

THERE is probably no geographical region of the world that so justly commands the attention of the naturalist as Australasia. In nearly every department of zoology and botany it yields forms and features represented, if at all, elsewhere on the surface of the globe, only by long extinct fossil types.

From another standpoint, as a territorial unit, Australia is of surpassing interest, since, by the light of the most recent biological research and philosophic interpretation of the facts revealed, this insular territory would appear to represent the isolated fragment of a pre-existing, apparently Mesozoic, Antarctic Continent of such vast extent as to have united with its own land area the now remotely separated regions of South Africa, South America, and New Zealand.

The evidence in support of this interpretation, for which science is mainly indebted to the writings and investigations of Huxley, Charles Darwin, A. R. Wallace, W. K. Parker, H. F. Blandford, and H. O. Forbes is accumulating continually, and rendering more logically demonstrable the fact that there once existed two huge northern and southern continents, for which Prof. Huxley proposed the felicitous titles of "Arctogea" and "Notogea" respectively.

As an integral portion of the primitive antarctic continent, or Notogea, Australia, by its apparent earlier isolation from the continental mass, would appear to have preserved alive and in its pristine purity much of that faunal and floral individuality which in other regions of the world was early brought into competition with, and succumbed to, those hardier and more aggressive northern races, that, following the newer land distribution, migrated southward from Arctogea. Had such changes of the apparent previous distribution of the larger land areas of the earth's surface resulted in bringing about the continuity of the Indo-Malay peninsula with Australia, subsequent to the population of the first-named region with a carnivorous mammalian fauna corresponding with, or allied to, that which now inhabits it, it may be safely predicted that but little of the most interesting marsupial and other archaic types of animal life would be now surviving in this Island-Continent. Happily, in the interests of science, and for the intellectual appreciation of all those with whom natural history and the unravelling of nature's mysteries is a continual feast, this catastrophe was averted.

A number of living witnesses may be summoned to give evidence in support of the widely entertained speculations concerning the pre-existence of an extensive antarctic continent wherein the present widely separated lands of Australia, New Zealand, South Africa, and South America were either incorporated or most closely approximated. These witnesses, moreover, are not represented so extensively by that very characteristic Australian group of the marsupial mammalia as by isolated members of various other animal and vegetable types. Up to within the last few months, in point of fact, no marsupial closely related to the several families and many species of this order peculiar to Australia was known to be living in any other part of the world; the insectivorous and carnivorous American opossums, *Didelphyidæ*, possessing no Australian representatives, but having apparently been independently derived from an earlier extensive race of the same order that formerly inhabited both Europe and North America, and that occurs in the fossil state in the Upper Eocene, Lower Miocene and other tertiary deposits of these two continents. It is, at the same time, through these extinct representatives of the *Didelphyidæ* that it has been suggested * that the present marsupial fauna of Australia was originally connected with that of America, though in times probably anterior to that of the division of the earth's surface into the two continents previously referred to.

The discovery of a very interesting marsupial type that opens out a further field for speculation in this direction has been quite recently recorded by Mr. Oldfield Thomas, the mammalian specialist at the British, Natural History, Museum. The species, which has been described by Mr. Thomas under the title of *Cænolestes obscurus* in the Annals and Magazine of Natural History for November, 1895, and the "Proceedings of the Zoological Society" for the past year (1896), has been obtained from the neighbourhood of Bogota, in South America. Although its dimensions do not exceed those of an ordinary rat, it is of peculiar interest with reference to the fact that, while representing an entirely new family among the living marsupials, it may be most naturally assigned to the extinct group of the Epanorthidæ, which has hitherto been known only by fossil remains obtained from the early Miocene, or, according to some authorities, Eocene, deposits of Patagonia. Following up the clue indicated by *Cænolestes obscurus*, it has been found that the form obtained from the neighbouring province of Ecuador, and originally described by Tomes under the title of *Hyracodon fuliginosus*, is a second species of this interesting genus. The derivation of these two types from an originally extreme southern, or notogeal, centre of distribution is thus obviously indicated.

Among the other animal groups that contribute their quota towards demonstrating their apparent common derivation from an original continental centre of development, of which Australia formed a material constituent, that of the freshwater fishes is, perhaps, the most interesting. One remarkable fish, *Ceratodus Forsteri*, now indigenous to but two rivers of Queensland, and known locally as the Mary or Burnett River Salmon, and also as one of several species of so-called Barramundis, is the most familiar with reference to its belonging to an order, the Dipnoi, of which

---

* "*Mammals, Living and Extinct*," by Sir William Flower and Richard Lydekker, p. 135, 1891.

the only other two discovered living representatives, *Lepidosiren paradoxus* and *Protopterus annectens*, inhabit respectively the rivers of Brazil and those of tropical Africa. The true Australian Barramundi, *Osteoglossum Leichardti*, also an inhabitant of certain Queensland rivers, typifies a family group having but a limited number of living representatives, which agree almost precisely with the Dipnoi in the singularity of their geographical distribution. In this relationship Osteoglossum possesses even more significant alliances. For while Ceratodus is represented in Brazil and Africa by allied, but, at the same time, very distinctly differentiated generic types, the rivers of Brazil and Guyana yield a species, *Osteoglossum bicirrhosum*, referable to the same genus as the Queensland fish. A third species, *O. formosum*, inhabits the rivers of Bornea and Sumatra, and a fourth form, *O. Jardinei*, has been discovered and chronicled by the writer within the last few years as inhabiting those rivers of North Queensland which debouch upon the Gulf of Carpentaria. The ally of Osteoglossum in the African continent is *Heterotis niloticus*, common to the Upper Nile and various West African rivers. The most remarkable member of family Osteoglossidæ is, however, the huge *Arapaima gigas*, which shares with *Osteoglossum bicirrhosum* a Brazilian and Guyanan habitat. It is notable as being the largest known of Teleostean, let alone fresh-water fishes. It not unfrequently exceeds a length of 15 feet.

The little native trouts, referable to the genus Galaxias, of the Australian fresh-water streams, and so-called for their somewhat trout-like shape and usually spotted ornamentation, are additional witnesses in evidence of the theory of an extensive Antarctic Continent, wherein, in this instance, New Zealand is distinctly involved. Of the numerous species of Galaxias known to science, an approximately equal number are found in New Zealand and in Australia and Tasmania respectively. Three species of the same genus are peculiar to the fresh-water lakes and rivers of Chili and Patagonia, while one species, *Galaxias attenuatus*— as attested to by Dr. Gunther, "Catalogue Fishes," Vol. VI., p. 211—occurs without any recognisable specific distinctions in the four widely-separated areas of Tasmania, New Zealand, the Falkland Islands, and the southern parts of South America. No stronger evidence could perhaps be adduced in demonstration of the pre-existence of a vast homogeneous Antarctic Continent than that yielded by this little fresh-water generic group.

The fresh-water fishes of the family Haplochitonidæ, including the Australian and New Zealand genus Prototroctes, or so-called cucumber mullets or graylings, and the allied South American genus Haplochiton, yield corresponding though less abundant testimony in the direction recorded of Galaxias. Within recent years a species of

Haplochiton, *H. Sealii*, has been chronicled by Mr. R. M. Johnson[*] as occurring in shoals in the waters of the river Derwent in South Tasmania. It is locally known as the "Derwent Smelt." It is noteworthy in the case of the several Australian fresh-water fish genera—Galaxias, Prototroctes, and Haplochiton—that, while possessing South American allies, they are not, as in the case of Osteoglossum and Ceratodus, previously referred to, represented also upon the African Continent. Both here and in other instances, where a near relationship can be apparently established between Australian and African organic types, the representative forms are essentially tropical or sub-tropical species. Where, on the other hand, the affinities with South American types alone obtain, the specific forms are limited in their distribution to the extreme south or temperate regions of their respective areas. These circumstances would seem to warrant the anticipation that the intercontinental continuity of Australia with the south extremity of South America persisted for a longer interval, and to a much later period, than that between Australia and a greater or less extent of Africa. The very fact, indeed, of the survival of *Galaxias attenuatus* in an unaltered form in the several isolated localities above enumerated affords substantial testimony in this direction. According to the generally accepted biological axiom—"two identical species are never independently developed in remotely separated localities." This axiom, logically applied to the present distribution of *Galaxias attenuatus*, involves the unavoidable inference that its existing widely-isolated colonies must have originated from a common centre, between which and its present habitats there must have been a close land connection down to so comparatively recent a date that even the essential diagnostic characters of the fish have remained unaltered.

The flightless Struthious birds, comprising the ostrich tribe and its allies, are most commonly cited as affording by their geographical distribution the most substantial evidence in demonstration of a pre-existent common centre of origin in the shape of an extensive Antarctic Continent. In this direction, the Australian region is especially rich. It possesses in conjunction with the neighbouring island of New Guinea two representative genera of the order as typified by the Emu, *Dromaius*, and one or more species of Cassowaries, *Casuarius*. While New Zealand is now in possession of but one living generic representative of the order, the well-known "Kiwi" or Apteryx, this now remotely separated island group was formerly the head-quarters of the redoubtable Struthious race, which is exemplified by the giant "Moa"

---

[*] R. M. Johnson, Proc. Zoo. Soc., 1882, p. 128.

Dinornis. Abundant testimony has been adduced to show that the extinction of that remarkable bird has been brought about by the earlier aboriginal inhabitants of the island group within comparatively recent times. The bones of Dromornis, a form very nearly related to the New Zealand Dinornis, if not those of this identical generic type, have been found in the fossil state in the Darling Down deposits of the Australian Continent; as also the remains of a type, *Metapteryx bifrons*, most closely allied to the New Zealand Apteryx.*

In South America various species of the Rhœa, or so-called American Ostrich, now alone survive as representatives of the Struthious order. Several distinct types belonging to the same group have, however, as in the case of the allies of the Marsupial mammal Hyracodon, been recently discovered in the tertiary deposits of Patagonia. In like manner in the African Continent, the typical Ostrich, *Struthio camelus*, is the sole survivor of a number of allied forms that probably populated that and contiguous areas of the earth's surface at an earlier geological epoch.

Among recent references to other bird types which, in addition to the Struthionidæ, have yielded evidence in support of the hypothesis of a widely extending Antarctic Continent, Mr. H. O. Forbes' very interesting paper, "The Chatham Islands and their relation to a former Antarctic Continent" (Proceedings of the Royal Geographical Society, Vol. III., part 4, 1893), is specially worthy of mention. In this communication some very remarkable data are given concerning the large flightless member of the Rail or Wood-hen family, Aphanapteryx, which inhabited the Island of Mauritius, contemporaneously with the Dodo, down to about two hundred years ago. Quite recently, following up the clue afforded by bones transmitted to him from the Chatham Islands, 500 miles east of New Zealand, Mr. Forbes obtained further material that fully established the comparatively recent existence there of a species almost indistinguishable from the Mauritian form. The same deposits in the Chatham Islands also yielded Mr. Forbes remains of several other types identical with existing New Zealand species, including the sheep-destroying Kea, or Mountain Parrot, two hawks, an owl, and the remarkable Tuatara lizard, *Sphenodon (Hatteria) punctatus*. Whilst none of these types possess a direct Australasian association, their record as factors in the composition of a previously continuous land connection between two now widely separated countries in the southern ocean is of high significance.

There remain yet a few familiar Australian birds that are worthy of notice

---

* C. de Vis, Proc. Linn. Soc., New South Wales. Vol. VI., Sec. II., 1891.

in correlation with this Antarctic continental hypothesis. The little group of the Jacanas, greatly resembling and allied to the Rails, but with longer legs and toes of such remarkable length and tenuity that the birds run with ease and great celerity over the surface of the leaves of the water lilies and other aquatic plants, is one of these. The typical genus Parra includes some dozen species which are inhabitants of tropical Australia, Africa, and South America. Their power of flight is essentially feeble, and they could not possibly have arrived at their present widely separated areas of distribution, but for some previous land connection between their respective habitats. A bird of a very different character, but possessing a similar notable distribution, is the gigantic Crane or Jabiru, *Mycteria australis*. Like the Jacana it is a denizen of tropical or sub-tropical Australia, Africa, and South America. As was pointed out many years since by the late Professor Huxley, the essentially south-continental parrot tribe possesses some nearly related family groups in both Australasia and South America.

There are certain other birds indigenous to Australia which, with reference to their distribution further afield, yield, in a less degree, similarly suggestive evidence. The little wood swallows, referable to the genus Artamus, belong to this category. In addition to being found in Australia, where they are most abundantly represented, various species occur in India and in the essentially African region of Madagascar. A genus of true swallows, Atticora, also common in Australia, is yet more significantly distributed, for it has, as with species previously cited, both the South American and African continents recorded among its habitats.

The fauna of Australia is of peculiar interest, viewed altogether apart from those phenomena which appear to justify our regarding that territory as the isolated residuum of a disrupted Antarctic Continent. As an indirect but collateral outcome of that interpretation, it is found that Australia can lay claim to the possession within its boundaries of a fauna that yields the palm to no other one on the earth's surface in the matter of aristocratic and ancient lineage.

The aboriginal population of Australia, such of it as still survives, is of itself a standing monument of the high antiquity of that country's fauna. As is conceded by the common consent of experts in ethnology, the Australian aboriginal represents the most primitive type of humanity. He is, in fact, a surviving relic of the Stone Age, who, in this huge isolated island, has, in company with the marsupial mammals, preserved his primæval simplicity down to the present date. Like all such less civilised, or less effectively equipped, races, he is fast disappearing before the advancing stronger and lethally armed northern arctogeal stocks. He is already extinct through that agency in Tasmania; reduced to his last and most pitiful conditions of existence in Victoria and the Europeanised districts of the adjacent colonies; and it is only in the far north and in areas unsuited for the white man's occupation that he still retains his primæval habits and naturally robust physique.

That the Australian aboriginal represents a race entirely distinct from the inhabitants of New Zealand, New Guinea, or any of the Indo-Malay islands is self-evident to anyone having had the advantage of personal contact with members of these several races. Even in the case of Papua or New Guinea, with whom an alliance might be most reasonably anticipated, it has been pointed out by Dr. A. R. Wallace that the roots of their respective dialects, a most important diagnostic character, are essentially distinct. Among the notable points of divergence exhibited by the Australian aborigines, as compared with the most contiguous races, those relating to their development of the arts of navigation and the manufacture of war or hunting weapons, are particularly prominent.

Navigation as practised by the coastal Australian tribes is of the most primitive description, exhibiting in this association a striking contrast with the Malay and Papuan races, who are expert sailors and boat builders. The aboriginal Tasmanians apparently forswore the sea and possessed no floating craft whatever. Among the Southern Australian tribes, in the wider sense, nothing in advance of a floating log, used in still waters for fishing purposes, has been recorded. On the Eastern, Queensland coast, and in the neighbourhood of the Palm Islands more especially, rough canoes, rarely capable of holding more than two people, are fashioned out of single sheets of bark stripped from the larger Eucalypti, and dexterously fastened together at the two extremities. Higher up the same coast we meet with dug-out canoes having the typical Malay and Papuan outrigger. The form has been most undoubtedly borrowed from the Papuan, and is indeed most frequently derived directly from Papuan sources; a trade in which these canoes form an important factor being carried on between New Guinea and the Torres Straits Islands, whence they filter through to North Queensland.

The North-Western, Kimberley or King's Sound, district of Western Australia, undoubtedly produces the most distinctive type of native craft. This consists of a triangular raft made from poles of the indigenous "Cyprus Pine," apparently identical with *Frenella robusta*, fastened together with wooden pegs, and supplemented at the wider end by a few vertically affixed sticks, upon or between which the successful

PLATE I

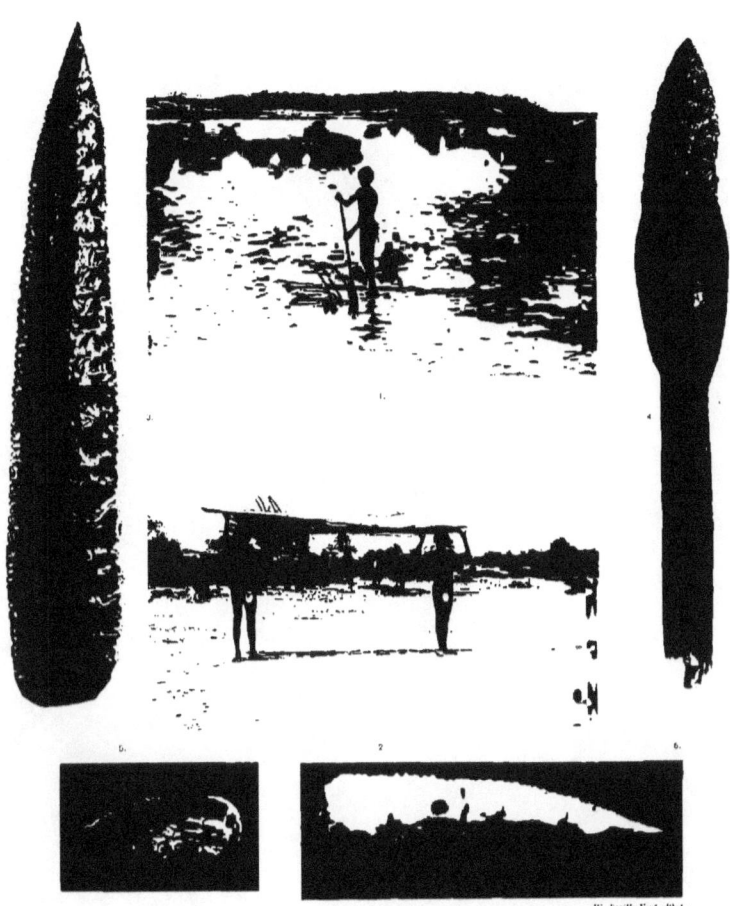

Figs. 1 & 2. RAFT AND NATIVES OF KING'S SOUND, WESTERN AUSTRALIA, p. 9.
Figs. 3-6. GLASS AND QUARTZ SPEAR HEADS OF WESTERN AUSTRALIAN ABORIGINES, p. 12.

fisherman impales his finny trophies. This ingenious raft, in a yet more primitive form, composed of two or three poles only, upon which the native sits and paddles with his hands, is mentioned in Capt. King's "Survey of the Coasts of Australia," Vol. 1, p. 43, 1826. At page 38 of the same work the author refers to the yet simpler boat consisting of a single log, one such craft with its seated navigator being figured on the title page of the work.

Captain King's figure and descriptions relate to observations made by him in the vicinity of Dampier's Archipelago, several degrees south of King's Sound, where the more complex, and what might be designated "fully-rigged" raft is employed. As shown in characteristic photographs of this more advanced type of naval architecture, reproduced in Plate I., Fig. 1, taken by the author at Cygnet Bay, King's Sound, the occupant usually occupies an erect position, paddling the craft with his fishing spear. In this particular instance, a rude paddle, fashioned out of a split rail, is substituted for the customary spear. Seen from a little distance, and more especially when, as often happens, there are no supplementary vertical attachments and the raft is more or less completely submerged, the natives, propelling these frail structures, present to a remarkable extent the appearance that they are walking on the surface of the water, and when thus migrating in social companies from one to another of the King's Sound Island groups, constitute a singular spectacle. The precise form and construction of this King's Sound raft is clearly shown in Fig. 2 of the same plate, which represents the native who propelled the raft in the preceding figure, occupied with a comrade in transporting it to the water.

The foregoing photographic reproductions, together with those selected for a head-piece to this Chapter, constitute characteristic illustrations of the general physique and ordinary apparel and armature of the natives in the immediate neighbourhood of King's Sound. In the first relationship, the figure and build of many of these dusky warriors, excepting for the slenderness of their limbs, is of a shapely, almost classic mould, and in marked contrast to that of the aboriginal tribes further south. This very perceptible fact is, in a large measure, due to the circumstance that the natives here represented have suffered little or no deterioration from contact with the colonizing Europeans and are as yet strangers to those diseases and vices, without the counterbalancing virtues, which to their slow, but certain destruction, are so readily acquired wherever the white man and the primeval savage are brought into intimate intercourse. As a matter of fact, several of the individuals deployed in open line upon the sand ridge, in the heading to this Chapter, had only within the previous day or two, made their first acquaintance with civilisation, having just arrived from an unsettled district to the north of King's Sound to participate with previously engaged kinsmen, in the earnings and perquisites obtainable for their assistance in the Bêche-de-Mer fishery.

The sartorial accessories of the North Australian aboriginal are not such as to demand elaborate description. In addition to his "birthday suit," a polished, variously engraved mother-of-pearl shell, Meleagrina, secured round his waist by a girdle of twisted human hair, probably that of his wife or wives, a stick thrust skewer-wise through the nasal septum, and on festal occasions, a little paint, represents the alpha and the omega of his not very extensive wardrobe. Some of these shell aprons, which may be regarded as an artistic antipodeal adaptation of the classic fig-leaf, are somewhat elaborate in their pattern of ornamentation. Among the examples of these shell aprons, included in the photographic figures reproduced on this page, the one on the top left hand corner might suggest, to the enthusiastic student of the dawn of art, the prototype of the decorative pattern known as the "Grecian Key." It is a trite saying that extremes meet! On carefully examining between the lines in this particular example, it will be observed that small triradiate characters are scattered here and there near the lower edge. These must not be interpreted as the equivalent of the cuneiform symbols of Ancient Nineveh, excepting to the same extent as the broad arrow of the British Government may be traceable to the same source. As a matter of fact, these particular shell ornaments were obtained from natives belonging to the settled district of Roebuck Bay, who have consequently added this, to them, white man's totem mark, to their own repertoire. The most conspicuous character of these shell etchings, reproduced also

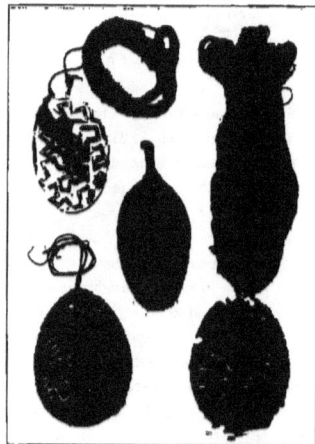

W. Saville-Kent, Photo.
CARVED BAOBAB NUT AND SHELL APRONS, WITH HUMAN HAIR GIRDLES, KIMBERLEY DISTRICT, WESTERN AUSTRALIA. ONE-FIFTH NAT. SIZE.

on the baobab fruit in the centre of the group figured, is the strictly rectilinear plan of all of the several patterns. This character prevails also, so far as the writer has observed, in all the carvings on their wooden shields, woomeras, or throwing sticks, and other articles in ordinary use.

The weapons of the Australian aboriginals are but few, narrow wooden shields, spears, the boomerang or kiley, a truncheon or knob kerri, a stone axe with a wooden haft, and in some parts of Queensland a large roughly fashioned two-handed sword, fairly complete the list. In the boomerang, however, they undoubtedly possess a weapon that is almost (see Appendix A.) unique in the world's armoury. The form and method of use of this instrument will be familiar to most readers. In outward aspect it is merely an elongate smoothly flattened piece of hard dark-coloured wood, most frequently of acacia, bent in the centre and in the same plane at a somewhat varying obtuse angle. Diverse shapes and sizes of the boomerang are used respectively for war, when hunting Wallaby or Kangaroo, or for striking fish. In the hands of its owner this simple weapon, as is well known, can be made

W. Saville-Kent, Photo.
NATIVES OF KIMBERLEY DISTRICT, WESTERN AUSTRALIA, SHOWING ATTITUDE ASSUMED WHEN HOLDING AND THROWING THE BOOMERANG.

to perform a series of most astonishing evolutions. It may be sent skimming along the surface of the ground for hundreds of feet or circling in the air almost out of sight, and, if thrown with skilful hands, returns as though it were a trained, sentient emissary, to the feet of the thrower. The position in which this very characteristic weapon is held on the point of its release from the hand, as also the manner in which reserve boomerangs are stored, like holster pistols, in the hair girdle, previously referred to, is aptly illustrated on this page, where two warriors are ostensibly pitted against each other in an attitude of mortal combat.

The spear of the native Australian is commonly, and more especially as used for striking fish, simply a long pointed stick of hardened wood. For the larger terrestrial game and for conflict with his fellow man, however, its efficiency is usually increased by the addition of a bone barb or trenchant spear-head manufactured out of chert, quartz or other suitably hard stone, which is affixed to the shaft by means of a strong, resinous gum derived from the bruised blades and stalks of the spinifex grass, *Triodia irritans*. Since coming in contact with European civilisation, the natives have shown themselves to be great adepts at turning to account all materials introduced by the settlers, suited to their simple needs. In this manner' the discarded glass bottles of manifold description which bestrew the ground like the leaves in the famed Valley of Vallombrosa, around every North Australian township, and which, in a less marked degree, mark the track of the prospector, have proved a veritable God-send to the native. Out of this most unpromising material he will manufacture spear-heads, pointed like a needle, and of the most exquisite workmanship.* An abnormally elongated example of one of these spear-heads, manufactured by a Kimberley native, with the view rather of demonstrating the workman's skill than for practical use, is reproduced from a photograph of the natural size in Plate I., Fig. 3. Opposite to this, fig. 4, is one of more normal dimensions, blunted by use, and attached to the broken spear-haft by the customary spinifex-gum cement. Figures 5 and 6 represent smaller sized spear-heads manufactured out of white quartz, which are now comparatively rare. The needle-like sharpness of the point in fig. 5 is particularly well defined.

In addition to bottles, the insulating glasses attached to the telegraph posts have unfortunately been found by the Kimberley natives to be equally efficacious for the manufacture of spear-heads, and are not unfrequently appropriated for this purpose in the sparsely settled districts, to the great discomfiture of the telegraph officials. The method by which these glass and quartz spear-heads are manufactured presents points of interest. The flaking off of the superfluous surfaces is not accomplished, as might be imagined, by direct percussion, but by a skilfully applied pressing or gouging action with the aid of another suitably shaped fragment of hard stone, or, if the native can obtain it, a piece of iron. The rough shaping of the

---

* The interesting circumstance has been related to the author by Mr. Henry Balfour, the accomplished Curator of the Oxford University Museum, that the Fuegian Natives of Terra del Fuego and the mainland shores of the Magellan Straits are in the habit of utilising discarded bottles in a closely identical manner for the manufacture of their arrow-heads.

general form and sides is accomplished in a few hours, but the formation and finish of the finer points is a longer process, involving the expenditure of much time and patience, and, as witnessed by the writer, frequent failures before perfection is arrived at.

There is an accessory instrument commonly used in conjunction with the spear by the aborigines of North Queensland and North-Western Australia which is of rare occurrence among spear-armed nations. This is the Throwing Stick, or "Woomera," a piece of flattened wood about two feet six in length and three to six inches wide. A small bone or hard-wood peg is attached at an acute angle to the further, distal, end of the instrument, and this fits into a notch in the proximal end of the spear. This accessory instrument gives as it were double length and leverage power to the arm of the spear wielder, who can thus launch his weapon with irresistible force against all but the most impenetrable objects. Among the Kimberley, Western Australian, Natives, the flat surfaces of this "Woomera," or Throwing-Stick, are, as previously stated, commonly ornamented with carvings presenting various rectilinear patterns, while the handle end, among the North Queensland Tribes, is often decorated with pieces of shell, or the scarlet seeds of that cosmopolitan tropical creeper, *Abrus precatorius*, half embedded in a matrix of spinifex cement. A peculiarity observed by the author as usually distinguishing the Woomeras of Queensland from those of Western Australia is the circumstance that the peg attached to the extremity in the Queensland examples is affixed in the plane corresponding with the broad side of the weapon while in those from Western Australia, it is invariably at right angles to it.

The mechanical means utilised by the Australian aborigines for the production of fire invite brief attention. In Queensland the mechanism usually employed consists of two slender light-wood rods some four or five feet in length. One of these is placed horizontally on the ground and held firmly in this position with the feet, while the second rod is placed vertically upon it, with its tip resting in a slight indentation made in the horizontal one. The vertical rod is now rotated backwards and forwards between the two hands at so high a speed that the lower one produces sparks which are communicated to some dry grass placed close at hand. From this, with a little nursing, a large fire is soon established. For convenience of carriage, the "broader" ends of these two fire-sticks are inserted into a short wooden sheath, which is commonly covered with spinifex gum, decorated, as are the handles of the "Woomeras," with the scarlet seeds of *Abrus precatorius*. This rotatory method of producing fire is, it would appear, practised also in India and the Malay Archipelago, whence doubtless it was originally communicated to North Queensland.

In the Kimberley district of Western Australia, fire is also produced through the friction of two pieces of wood, but on a different principle. The main portion of the apparatus consists of a dry piece of wood about eighteen inches long, such as the butt of a eucalyptus sapling. This is split down for some six inches at its narrower end, a wedge of wood is inserted to keep the split edges apart, and within this cleft a tuft of fine dry grass is fixed. The thin edge of the split half of a second shorter piece of dry wood is now rapidly rubbed across the grass-holding portion of the first one, as in the action of filing or fiddling. Within a few minutes fire is produced, which ignites the accompanying tuft of grass, and is further developed as in the preceding case.

It is, perhaps, worthy of note that these mechanical methods of generating fire are but seldom resorted to. When the fire has been once established, it becomes the duty of the women, after the manner of the classic " Vestal Virgins," to maintain the flame unquenched, and, during migrations, to carry lighted fire-sticks with them. In all except the few remaining absolutely unsettled districts, moreover, the fire instruments produced by Bryant and May, and their compeers, have well-nigh superseded the primitive native methods.

The conventional "place aux dames" has been gallantly conceded to the Australian aboriginal women-folk in the form of a corner illustration in the opening page of this Chapter. Compared with their dusky lords deployed in martial statuesque attitudes immediately above them, they scarcely present a pleasing contrast. Being in point of fact, as is generally the case among the lower races of humanity, the mere slaves and drudges of their lieges, it is but little to be wondered at that at an early date after emerging from childhood they lose, as viewed from a European standpoint, all of such little comeliness and attractiveness as they may have originally possessed. The most notable, though by no means the most handsome, figure in the group occupies the second place on the left. It represents a widow whose recent bereavement is attested to by the special pattern of her coiffeur, her normally matted hair being subdivided into pendant ringlets that are separately stiffened with rolls of moistened clay.

The illustration added as a tail-piece to this Chapter, may be most appropriately referred to on this page. It is the photographic replica of a hollow wooden cradle, such as is customarily used by the native women of the Kimberley district of Western

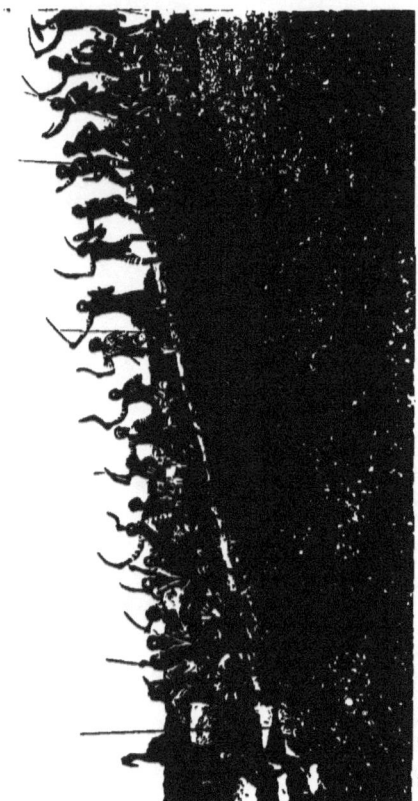

FESTAL DANCE, OR "CORROBOREE," OF KIMBERLEY, WESTERN AUSTRALIAN, NATIVES, p. 15.

Australia, for the reception and porterage of their infants. As not unexceptionally happens with its homotype in modern civilisation—the more or less elaborate perambulator—this infant receptacle, when not in demand for the more legitimate purpose of its construction, is frequently employed for the baser purposes of deporting supplies for the family commissariat. The specimen here figured, is a remarkably fine and well-shaped example, measuring over three feet in length. It was obtained by the author at Derby, in the neighbourhood of King's Sound, and has been contributed by him to the ethnological collections of the Oxford University Museum.

This short notice of a few of the more salient characteristics of the Australian aborigines may be concluded by a reference to the photograph reproduced in Plate II. It depicts about a score of the warriors of the Roebuck Bay district, Western Australia, in one of their festal dances, or so-called corroborees, clad for the occasion in fullest ball-dress costume, and taken, in the majority of instances, literally "on the hop." The shapes and proportionate dimensions of the several weapons previously referred to, as also the physical development of the natives of this particular tribe, are very clearly portrayed in the illustration.

The animal inhabitants of Australasia, other than *homo sapiens*, provide the naturalist with a wide field for speculative interpretation and investigation. The mammalian class alone furnishes evidence of its high antiquity in the fact that, with the exception of the Dingo, or wild dog, which there is strong reason to believe was introduced by human agency, and a few rodents, all its members belong to the primitive marsupialian order, or to the yet lower organised one of the monotremata. This last-named group includes but two Australian specific types, the familiar Spiny Ant-eater, *Echidna aculeata*, or so-called "Porcupine," of the Australian settlers, and the yet more remarkable Duck-billed Platypus or Ornithorhynchus, *Ornithorhynchus paradoxus*. Photographic representations of both of these very singular animals will be found reproduced on Plate III. A third species, *Proechidna Bruynii*, very nearly related to the ordinary Echidna, is, in company with the last-named species, indigenous also to New Guinea. The essential external characteristic of the representatives of this small but most interesting order is the peculiar beak-like modification of the mouth and the accompanying rudimentary nature, or, as in Echidna, complete obliteration of the teeth. The internal structure of the monotremata is yet more remarkable, approximating in certain essential details to that of the Sauropsida or birds and reptiles. This implied affinity has, within recent years, been more conclusively established by the discovery that the young are produced from eggs laid by the parent after the manner of the members of the above-named groups. The antiquity of this monotrematous order is indicated by the circumstance that teeth most closely resembling the temporarily developed ones of Ornithorhynchus have been met with in association with certain obscure mammalian remains that occur as fossils in the mesozoic strata of North America.

The natural food habits of both the Ornithorhynchus and the Echidna are such as to prevent them from becoming forms with which the British public can hope to become very familiar in the living state. The Ornithorhynchus is essentially an aquatic animal. It is for the most part, though not strictly, nocturnal, and dependant upon a pabulum of worms, fish spawn, mollusca and aquatic insects that cannot be easily supplied in a state of captivity. All efforts, so far, even in Australia, have failed to keep it alive for more than a few weeks, and no attempts to bring it to Europe have proved successful. The authority who has cultivated, and published an account of, the most intimate acquaintanceship with the Ornithorhynchus is, undoubtedly, the late Dr. George Bennett, of Sydney. In his well-known work, "The Gatherings of a Naturalist," published in the year 1860, he has given a most interesting and widely-quoted record of his extended experiences in the possession of numerous examples of both young and adult individuals, neither of which, however, he was able to keep alive for a long period. Dr. Bennett was also unsuccessful in solving that knotty question relative to the reproductive phenomena of Ornithorhynchus, which had at this earlier date already attracted the attention of many eminent naturalists, and which was only set at rest in the year 1884 by the investigations of Mr. W. H. Caldwell, who then, for the first time, incontestibly demonstrated that both this type, as well as the Echidna, were oviparous mammalia.

The Ornithorhynchus, or Duck-billed Platypus, has not fallen within the writer's purview to an extent that enables him to place on record any new data concerning its natural habits. At the salmon and trout-hatching establishment on the river Plenty, in Tasmania, this interesting animal had, unfortunately, to be systematically destroyed on account of its too strongly developed proclivities for dieting on the jealously guarded ova of the Salmonidæ. A wounded specimen, obtained from this source, which survived for but a day or so, was the only living one that fell into the author's possession. While investigating and reporting for the Victorian Government upon the fish and fisheries of the Victorian section of the river Murray, in the neighbourhood of Echuca, a well-authenticated instance was reported to the writer of a lad who, incautiously holding a male Platypus that had become entangled in his father's nets, received a severe wound from the animal. The Platypus, it appears, gripped the boy's palm between its opposed spurs, as though with a pair of callipers, and with such force as to pierce the flesh on either side. The result was a festering wound, which refused to heal for some months and deprived the lad for the time of the use of the injured hand. The cicatrice of the scarcely healed wound was shown to the writer, who has no hesitation in accepting this as an authentic demonstration of the capacity of the male Platypus to use its spurs defensively. Much doubt has been expressed upon this point in Natural History works, and with the exception of a somewhat analogous instance recorded by Mr. Spicer, of a Tasmanian example, in the Proceedings of the Royal Society of Tasmania for the year 1876, little or no evidence of an absolutely positive nature has been forthcoming. The spur of the male Platypus is of a somewhat complex structure. In adult individuals it is as much as an inch long, of hard, horny consistence, traversed throughout its length by a minute canal, terminating in a fine longitudinal slit near the point and connected at its base with the duct of a large gland situated at the back part of the thigh. The apparatus, as a whole, in fact, resembles to a most remarkable degree the combined fang and poison-gland of a venomous serpent.

In Flower and Lydekker's "Mammals, Living and Extinct," p. 123, from which the above characters of the spur of Ornithorhynchus have been reproduced, the evidence concerning its nature and functions are accepted as most strongly favouring the interpretation that these structures are employed as aggressive weapons, after the manner of the antlers of deer and other similar organs, in combats between contending males. The peculiar incurved direction, however, in which the spurs are set upon the hind feet and the ease with which, in life, they may be employed to grasp any object of approximate proportions, has led the writer to believe that they are not improbably employed, as are the claspers of the male members of the shark tribe, for the secure retention of the female at the breeding season. The slipperiness of a Platypus and the difficulty experienced in retaining hold of a struggling individual are well known to those personally familiar with the living animal, and the natural advantages of possessing some suitable prehensile structure are self-evident. The analogy of function suggested between the claspers of the shark and incurving spurs of the male Platypus allow of an even further histological comparison, the organs of the shark being ossified appendages of the pubes which have, in like manner, large secreting glands at their bases which communicate externally by tubular canals. The supposed poisonous properties of the glands associated with the spurs of the male

C

Platypus have not been practically demonstrated, and it would seem to be highly probable that, as with many other animals, their functional import and activity are intimately connected with the pairing season. The fact that wounds from these spurs are difficult to heal may be explained by the circumstance that a puncture from any blunt, conically pointed instrument may produce a similar effect.

Dr. Bennett, in his work previously quoted, makes a very brief reference to that near living relation of the Platypus, the Echidna or Spiny Ant-Eater, commonly, but incorrectly, associated by Australian colonists with the title of the "Porcupine." The single example kept by Dr. Bennett does not appear to have ingratiated itself very deeply in his favour, and is finally dismissed with the sentence, "So much trouble was given by its burrowing habits and spinal irritation, that its death was not regarded with much regret." A couple of specimens of the Tasmanian form, *Echidna aculeata var. setosa*, were for some months in the author's possession, and well repaid the care and attention bestowed upon them. While for the first few days excessively shy, presenting an impenetrable *chevaux de frise* of sharp-pointed spines to all friendly advances, and incorrigible burrowers in their endeavours to escape from captivity, they soon showed themselves amenable to kindly influences. After a brief course of domestication, they would follow their owner in the house or adjacent grounds, and were quite accustomed to, and seemingly appreciated, being carried, thrown across the arm, after the manner of a lap-dog. Having satisfied their hunger at an ants' nest or with the artificial food, chiefly bread and milk or oatmeal, provided for them, they especially delighted, when liberated in the garden, in spreading themselves out at full length to bask in the sunniest spot they could find. In the house they displayed an inquisitive turn of mind, peering into every crevice and climbing upon and exploring every accessible article of furniture. This climbing proclivity, in point of fact, occasioned the demise of one of the specimens, which, scaling and accidentally falling from the back of a high chair, injured its spine to such an extent that it shortly afterwards died from the effects.

An adjacent piece of uncultivated bush-land that abounded in ants' nests proved a most happy hunting-ground for the two Echidnæ, which, as recognised members of the family circle, rejoiced in the respective sobriquets of "Prickles" and "Pins." The natural ant-eating propensities of the Echidna do not appear, so far, to have been precisely defined. As clearly demonstrated by observations and experiments made with the examples in the author's possession, adult ants, pure and simple, do not constitute its normal, or even an acceptable, diet. Placed in contiguity to a teeming

Fig. 1. DUCK-BILLED PLATYPUS. *Ornithorhynchus paradoxus*, p. 15.
Figs. 2-5. ECHIDNA OR SPINY ANTEATER, *Echidna setosa*, p. 18.
Figs. 2-4. AUSTRALIAN MAINLAND. Fig. 5. TASMANIAN VARIETY.

ant track, they would, unlike the ant-eating lizard *Moloch horridus* hereafter described, take no notice of it, appreciating the insects only under the conditions obtaining in the nests or hillocks. These edifices they would soon tear open with their powerful claws, exposing to view the white succulent nymphs, larvæ, and pupæ, or so-called eggs, upon which alone they concentrated their attention. A considerable quantity of adult ants and also of earth is no doubt inadvertently consumed with the specially sought pabulum, and it is this circumstance, added to the fact that this adventitious material remains longest undigested, and is, indeed, frequently the only thing found within the alimentary canal on dissection, that has given support to the commonly-received opinion that ants, in the ordinary adult state, constitute the Echidna's food.

The domestic habits of the Echidnæ in the author's possession were surprisingly cleanly, and, notwithstanding the low status of the creature in the mammalian scale, identical with those exhibited by a well-trained cat, rejectamenta being in a similar manner buried beneath the earth by the deliberate rake-like action of the animal's fore-paws. In common with Ornithorhynchus, the male Echidna develops a large horny spur, accompanied by a special gland, upon each of its hinder limbs. As, however, in no instance yet recorded has the animal been known to attempt to use these spurs aggressively, it may be inferred that, as surmised in the case of the last-named species, they are subservient to some as yet imperfectly comprehended sexual function. Being, as compared with Ornithorhynchus, easy to keep in captivity, and adapting itself more or less readily to an artificial diet of bread and milk and minutely chopped boiled eggs, the Echidna is usually on view in the menageries of the Australian Colonies, and has on one occasion been exhibited alive in the London Zoological Gardens.*

The order of the Marsupialia, while representing a very decided advance in both structure and affinities as compared with the Monotremata, are, next to this last-named group, the most primitive of existing mammalia. The familiar and highly characteristic physiological distinction of this Marsupialian order is the general possession by the female animal of a pouch or "marsupium," or otherwise a sphinctered

---

* While penning these lines, July, 1896, a fine living Echidna, *E. aculeata*, has been imported to England and secured for the Hon. Walter Rothschild's admirably appointed Zoological Museum at Tring. Through the facilities extended to the author by its fortunate possessor and the Museum Curator, Mr. E. Hartert, the smaller photographs from life included in Plate III. have been added to this volume.

folding of the abdominal integument for the safe lodgment of the newly-born young. These young are, furthermore, brought forth at a very early and rudimentary stage of their development, and there is not that pre-natal union with the mother through the medium of an allantoic placenta that is the rule among all higher mammals. According, however, to the most recent investigations, a rudimentary development of this embryological structure occurs in one of the Bandicoots, *Perameles*.

Within the limits of the Marsupialian order the modifications in form and habits of its component members are almost as marked as those which are found throughout the entire range of the higher mammalian groups of the Eutheria. True carnivora are thus traced in such predatory types as the so-called Tasmanian tiger or wolf, *Thylacinus cynocephalus*, and the Tasmanian Devil, *Sarcophilus ursinus*. These two forms, while now living only in the southern island of Tasmania, were, as shown by their fossil remains, formerly represented by identical or closely allied species on the Australian mainland. A circumstance, however, of even higher interest and significance is the discovery within the past two years of the fossil remains of a species, *Prothylacinus patagonicus*, in the tertiary deposits of Patagonia, which apparently closely resembled the Tasmanian Thylacine in habits and structure. These remains, together with those of many other Marsupial types having distinct Thylacine and Dasyurine affinities, have been figured and described at length by Florentino Ameghino in the Brazilian "Bulletin of the Academy of Cordova" for the year 1894.

Passing on to the so-called Australian native cats or Dasyures, usually distinguished by their profusely spotted plan of ornamentation, there may be said to be a considerable approximation to the Viverrinæ or Civet and Ichneumon group of the carnivora; while in Myrmicobius, a small Western and Southern Australian type, the habits and correlated conformation of the attenuate snout and abnormally elongate protrusible tongue closely correspond with those of the higher ant-eaters. The only known existing species, *Myrmicobius fasciatus*—frequently misnamed a "squirrel" in those districts of Western Australia where it most abounds—is of special interest, since it possesses a larger number of teeth than any existing Marsupial. In this respect and also in their character, the teeth of Myrmicobius coincide closely with those of certain primitive mammalian types, such as Amphilestes, Amblotherium and allied forms, which occur in the upper jurassic formations of both Europe and the United States.

The Bandicoots, *Perameles* and its allies, include numerous small forms, the largest not exceeding a rabbit in size. Their habits are partly vegetarian and partly insectivorous. In the latter respect, as also in their general form and usually much elongated contour of the snouts, they bear a not inconsiderable resemblance to the shrews and certain other members of the higher mammalian order of the Insectivora. The Australian badgers or wombats, genus Phascolomys, with their thick-set bodies, short legs, and plantigrade gait, are commonly likened to members of the bear family. Associated with this purely superficial resemblance they possess chisel-like incisor teeth and other internal characteristics which more nearly approximate them to the beaver and other of the larger rodents. It is of interest to note that the remains of an extinct species of wombat, which must have equalled the tapir in size, were discovered some time since in the pleistocene deposits of Queensland, and were correlated by Professor Sir Richard Owen with the distinctive title of *Phascolonus gigas*. Various species of wombats still exist both in Tasmania and throughout the Australian mainland.

The general features of the extensive family of the kangaroos, Macropodidæ, will be too familiar to need descriptive detail. In addition to the kangaroos, it embraces the wallabies, wallaroos, kangaroo rats, and other allied types, varying in dimensions from the big "boomer" or "old man" kangaroo, which will over-top a man when standing erect on its hind feet, to diminutive forms which, as their name implies, are not larger than ordinary rats. All the Macropodidæ are distinguished by the preponderating length of their hinder limbs, upon which alone they progress under any stimulus to rapid movement by a characteristic series of leaps and bounds. Being essentially vegetarian in their habits, the larger species of the kangaroo family, where abundant, so seriously tax the resources of the Australian pasture lands as to necessitate the adoption of stringent measures to keep them in check. This untoward necessity, combined with the high value set upon kangaroo skins, has contributed towards the complete extirpation of the "Boomer" throughout a large extent of the prairie-like tracts of Australian pastoral land on which it abounded previous to the advent of the settler.

The most remarkable members of this family group are undoubtedly the Tree Kangaroos, belonging to the genus Dendrolagus, which to the form of the ordinary terrestrial species unite the tree-frequenting habits of the opossums or phalangers. To the three species of this genus originally reported from New Guinea, two additional ones—*Dendrolagus Lumholtzi* and *D. Bennetianus*—their native name "Boongarry," have been recently discovered in North Queensland.

The transition from the Tree Kangaroos to the essentially arborescent Australian Opossums or Phalangers, Phalangeridæ, would seem, at first sight, to be most natural. As pointed out, however, by Flower and Lydekker, "Mammals Living and Extinct," p. 166, the resemblance between the limbs and habits of the Tree Kangaroos and the Phalangers must have been independently acquired. The family of the Phalangeridæ embraces many very dissimilar structural forms, including some of the most beautiful representatives of their order. Almost all the species are remarkable for the extreme softness and richness of their fur, the black Opossum, *Phalangista fuliginosa*, from the colder climate of Tasmania, outrivalling the others in this respect. For symmetry of form and grace and agility of movement, however, the palm must undoubtedly, be given to their near allies the "Flying Opossums," or so-called "Flying Squirrels," pertaining to the genera Petaurus, Petauroides and Acrobates. Here we have a structural modification in all ways identical with what occurs in the true Indian Flying Squirrels, Pteromys and Sciopterus, of the order Rodentia, which consists of a parachute-like membrane that extends between the fore and hind limbs. With this accessory locomotive apparatus the flying Phalangers have little or no occasion to descend, as do the ordinary opossums, to the ground between the component trees of the vast Eucalyptus forests in which they take up their abode, intervening chasms of one hundred feet or more being readily surmounted with the aid of the extended parachute. The largest species of Flying Opossum, *Petauroides volans*, has a body equalling in dimensions that of a large cat, and with its thick fur and long bushy tail it bears a by no means inconsiderable resemblance to the French or Persian variety of our domestic Grimalkin. The portraits of a remarkably beautiful albino example of this species that was in the author's possession in Queensland, are given on page 25. It was obtained near Brisbane. Although the fur throughout was a rich creamy white, the eyes were not pink but retained the rich brown lustre of the normal individuals. Being almost exclusively nocturnal or crepuscular in its habits, the many attempts to secure a successful photograph from life of this Opossum by daylight proved abortive and the one here reproduced, with pendant tail, in the act of feeding, was taken by the author at night, with the aid of a magnesium flash-lamp.

The abnormally long furry tail of this Flying Phalanger is not prehensile, as with the majority of the ordinary opossums, and when the animal is leisurely browsing, as in the portrait referred to, usually hangs laxly at full length. At other times, when walking along or resting on a branch, this animal manifested the singular habit of coiling its tail in a tight revolute coil like that of a watch spring or a butterfly's proboscis, as shown in the drawing from life, reproduced in the lower figure on the same page. The author is not aware that this peculiar comportment of the caudal appendage is exhibited by any other known mammal, though it would appear to be to some extent approximated by certain of the Lemurs and in the Saki monkeys.

The white example here figured and described, was a female, and, when captured, had a half-grown male cub in its pouch. This young one, while having white ears and a white breast, was otherwise of a dark Chinchilla grey tint, like the more ordinary members of his species. A back view of this young individual, well illustrating the great length and thickness of its beautiful fur, is reproduced in Plate IV., fig. 4. Although this species has been frequently imported to Europe, neither of these two Queensland individuals took kindly to any other diet than the foliage of their native gum trees, and more especially the Queensland Peppermint variety, *Eucalyptus microcorys*. Efforts were made to accustom them to a regimen of fruits and farinaceous substances, with the view of bringing them to England. The young animal unfortunately succumbed to the ordeal, and the white one was left in charge of an ardent admirer, who guaranteed it a constant and unlimited enjoyment of its native pabulum.

Among the several species of Australian Flying Phalangers, or so-called Flying Squirrels, the little form known in many districts as the "Sugar Squirrel," *Petaurus breviceps*, is in many respects the one best adapted for making a domestic pet, and is most justly alluded to in Flower and Lydekker's volume, previously referred to, as "the most beautiful of all mammals." Its size is somewhat less than that of the British Squirrel, its thick downy fur most comparable in colour and texture to that of the Chinchilla, and its habits in captivity are most attractive and endearing. Its range in the Australian Colonies is practically cosmopolitan, it occurring in Queensland, the Southern Colonies, and as far north as the Kimberley district of Western Australia. Although apparently not originally indigenous to Tasmania, it has been transported to and liberated there within recent years, and is now tolerably abundant in the midland districts of that island-Colony. In association with the several individuals kept at various times by the writer, it was observed that the leaping, or so-called "flying" properties they so prominently manifest, are somewhat erroneously represented in the majority of works on Natural History. In the figures given of this and other species of Phalangers taking their characteristic leaps, the body is almost invariably depicted as assuming a horizontal position, or, if inclined at any angle, in such manner that the head is the lowermost.

As a matter of fact, the position exhibited at such times is almost identical with that assumed by a trained acrobat performing on the high trapeze, or that of the Langur monkeys, as they have been observed by the writer, leaping across the intervening spaces between the forest tree tops on the outskirts of the Botanic Gardens, at Singapore. The head and shoulders, as represented in the photographic illustration reproduced on this page, are always maintained at the highest level, with the fore limbs outstretched and ready to grasp the first branch or other object reached.

That this is the true position maintained by the Phalangers during their flying leaps was very practically demonstrated by one of the examples of the Sugar Squirrel, *Petaurus breviceps*, obtained from Roebuck Bay, Western Australia, and presented to the author by Mr. G. S. Streeter. This little animal

W. Saville-Kent, Photo.

SHORT-HEADED FLYING PHALANGER.
Petaurus breviceps. SHOWING POSITION MAINTAINED DURING ITS CHARACTERISTIC FLYING-LEAP. ONE-THIRD NATURAL SIZE.

accompanied him for some time on his travels, usually sleeping in its cage throughout the day, waking up to its characteristic activity as soon as the sun had set, and sharing the full liberty of whatever apartment was allotted to its owner. At one resting stage, a room some thirty feet long and over fifteen feet high was placed at the writer's disposal. The flying squirrel speedily signalised its appreciation of the abundant space at command, by climbing up the curtains and cornices and thence gaining access to the projecting frieze close to the ceiling. Arrived at this coign of vantage, the little fellow delighted in launching himself through the air, with limbs and patagium outspread, to objects at the most remote end of the room, the author's person, if standing up to receive him, being a favourite goal to aim at. In all instances, as on other occasions noted, the oblique attitude previously attested to, was in no way departed from. This same individual distinguished himself one day in this apartment by so utterly disappearing, instead of returning as usual to his nest to sleep, that he was given up for lost. Towards dusk, however, Master Tiny, as he was familiarly known, quietly emerged—like Minerva springing from the head of Jupiter—from the summit of one of the stereotyped hollow China dogs that adorned the hotel-room mantelpiece. A small hole had been accidentally broken in the back of the head of this ornament, and into this the little Phalanger had managed to squeeze himself. In fact, so long as the opportunity lasted, he now regularly repaired to this singular dormitory for his siesta, doubtless regarding it as a

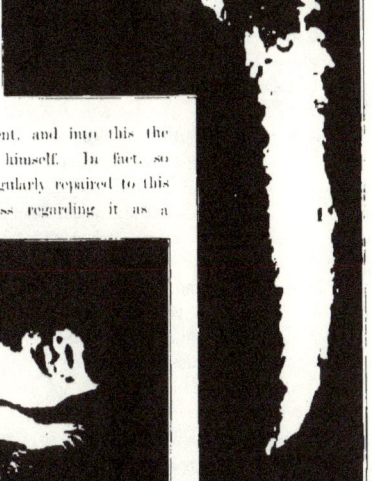

W. Saville-Kent, Phot.,
WHITE INDIVIDUAL OF LARGE FLYING PHALANGER, *Petauroides volans*, WITH TAIL PENDULOUS AND IN ITS CHARACTERISTIC SPIRALLY ENROLLED STATE.   p. 22.
ONE-FIFTH NATURAL SIZE.

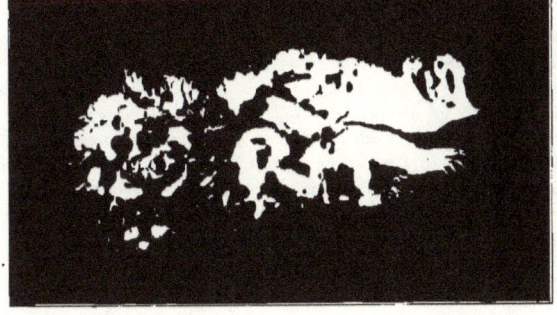

W. Saville-Kent, del.

more appropriate substitute for his native hollow-tree branch than his elaborately constructed cage.

Another member of the superlatively-beautiful family of the Phalangistidæ is the exquisite little Pigmy Flying Mouse or Phalanger, *Acrobates pygmæus*, of less dimensions than the common mouse, furnished with a parachute, and having the hair on its tail pinnately disposed like the barbs of a bird's quill feather. According to Gould, this species also makes a charming pet, and the author regrets not having had the opportunity of making its personal acquaintance. The Cuscuses or typical representatives of the genus Phalanger, while belonging to the Australian region, are now limited in their distribution to New Guinea, and other islands of the East Indian Archipelago as far west as Celebes. Though not hitherto recorded from the locality, the author on one occasion observed an example of these animals in the woody scrub of Thursday Island, Torres Straits. This island is but little over seventy miles from the nearest land of New Guinea, but it may at the same time have been an escaped pet, the species being not unfrequently brought in the boats trading between the two islands. In general form and the remarkable slowness of their movements, the Cuscuses have much in common with the Lemurine Lorises of the genus Nycticebus. The possession of a long prehensile tail, like that of the ring-tailed Opossum, Pseudochirus, is necessarily a prominent external feature in the Cuscus that is conspicuous for its absence in all the Lorises.

The typical little Australian Bear or Koala, technically known as *Phascolarctos cinereus*, may be appropriately styled by way of contrast to the Flying Phalangers, the most droll and bizarre of living mammals. In Natural History works, it is generally represented with drooping head and a most sad and woebegone facial expression. This, however, is a gross injustice to the little fellow, who, as seen under the natural conditions represented in the photographs from life in Plate IV., is a most contented and happy-looking little mortal. The habits of the Koala, like those of the more typical Phalangers, are essentially arboreal. It is, however, a very slow and leisurely-moving animal, contenting itself with abiding in and browsing upon the leaves of one Eucalyptus tree for days and weeks together, and rarely descending to the ground. In general form, in the complete absence of an external tail, and in its slow movements and arboreal habits, the Australian Koala is somewhat suggestive of the South American Sloths, Bradypodidæ. This suggested analogy, is, of course, entirely superficial, for the sloths, while belonging to the true, Edentate, order of the Eutheria, possess that essential anatomical organisation that characterises all members of the

PLATE IV

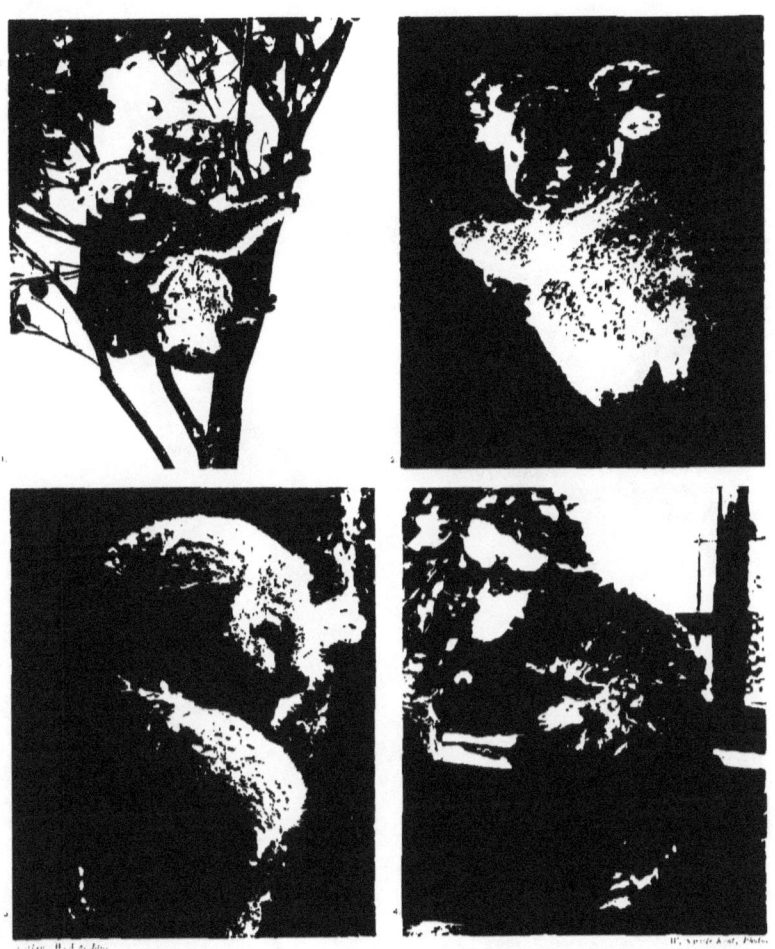

Figs. 1-3. AUSTRALIAN BEAR OR KOALA, *Phascolarctos cinereus*, p. 26.

higher mammalian groups. As, moreover, has been already mentioned on a previous page, a nearer structural approximation to the Eutheria in the matter of placentation has been recently found to obtain in one of the Bandicoots, *Perameles*, than occurs in any other known marsupial type. Although feeding naturally, like certain of the Phalangers, almost exclusively upon the foliage of various species of Eucalyptus, Phascolarctos, while young, adapts itself fairly readily to a milk or farinaceous diet, and has, on one or two occasions, been successfully brought to Europe. At a more advanced age, however, it appears to pine for its native forests and accustomed food, and cannot, as a consequence, be induced to become a permanent resident in this country.

Mr. A. D. Bartlett, the veteran superintendent of our Zoological Gardens, imparted to the author the intelligence of a tragic fate that befel the last Koala exhibited in the Regent's Park Menagerie. At night it was usually brought into the house and had the run of a room which, among other furniture, contained a large swing dressing glass. One morning the little animal was found crushed to death beneath this mirror, upon which it had apparently climbed and overbalanced with its weight. The further information that this specimen was a female, evoked the suspicion that, after the manner of its sex, personal vanity contributed a by no means inconsiderable share towards the compassing of its untimely end. The female Koala produces but one cub at a birth, and this, so soon as it is old enough to leave the pouch, is transferred to and carried about on its mother's back, to which it clings with great tenacity. The very characteristic portrait of a mother and cub represented in Plate IV. fig. 1 originally appeared in the pages of the "Australasian," and to the editor of that excellent journal the author is indebted for the privilege of reproducing the picture in these pages.

Notwithstanding its conspicuously tranquil facial expression, the little Koala or Australian Bear is credited with the reputation of giving way, under provocation, to fits of ungovernable fury. As is a familiar fact to all Australians who are acquainted with the species in its native haunts, two individuals quarrelling, as they are somewhat prone to, can easily give points to our domestic tabby in making night or even day hideous with their denunciatory language. The singular attitude assumed by Phascolarctos when sleeping, is characteristically illustrated by Plate IV., fig. 3. It corresponds to a remarkable degree with that exhibited under the same conditions, by the "Potto" of the Lemurine genus *Perodicticus*, as portrayed in Lydekker's "Royal Natural History." When thus contracted into a homogeneous furry ball and clinging high in the air to a perpendicular branch, the animal may be readily mistaken for a big bunch of moss or for one of those large gall-like excrescences of frequent occurrence on many species of Eucalypti. It may be suitably recorded here that the fossil remains of several extinct forms apparently allied to the Koala, but of huge comparative proportions, have been unearthed from the Australian tertiary deposits. Koalemus is one of these, as also Thylacaleo, formerly interpreted by Owen to be a large, carnivorous mammal, which in accordance with its presumptive habits must have represented "one of the fellest and most destructive of predatory beasts." The further light of more recent investigation has, however, conclusively demonstrated that this Bogie Carnivor was a peaceable vegetarian marsupial, uniting in its ponderous ungainly carcase the combined structural characteristics of both the Phalangers and Kangaroos.

A still larger extinct form, which also appears to have possessed the structural characters of both the Phalangistidæ and the Macropodidæ, is the huge *Diprotodon australis* of Owen, the remains of which have been found in abundance in the tertiary deposits of both Queensland and South Australia, and are most richly represented in the fossil collections of the Adelaide Museum. As shown by its skeletal elements the body of this, the largest of recorded marsupials, must have equalled that of a Rhinoceros, while its habits were probably closely allied to those of Megatherium and others of the extinct giant sloths of South America.

The exploration of the arid tracts of Central Australia has within the past few years been rewarded by the discovery of an entirely new and highly interesting modification of Marsupial morphology. This mammalian novelty is represented by a singular little creature possessing the burrowing habits and much of the co-ordinated structure of the European mole. Upon it the scientific name of *Notoryctes typhlops* has been conferred by its original describer, Dr. E. C. Stirling, F.R.S., the accomplished lecturer on Physiology at the Adelaide University and the Hon. Director of the South Australian Museum. Excellent illustrations and copious descriptions of the general aspect, habits and structural features of this mole-like Marsupial, or Pouched Mole as it is popularly designated, are contributed by Dr. Stirling to the Transactions of the Royal Society of South Australia for the year 1891, whence the accompanying figures and descriptive details are, with cordial acknowledgments, appropriately reproduced. The total length of the little animal scarcely exceeds five inches, and it is covered by a long soft lustrous fur of a generally light fawn colour, but which inclines in some parts to a glistening golden hue, and in others to a considerably lighter silvery tint. As shown in the illustrations overleaf there

GENERAL AND INTRODUCTORY.

is little or no trace of a separate neck, the front of the snout is protected by a hard horny shield, and there are no visible eyes, the somewhat eye-like spot occupying a remotely posterior position in Fig. 2, representing the external ear-opening as purposely exposed to view by the brushing aside of the surrounding fur. The peculiar modifications of the tail, feet, and the extraordinary development of the third and fourth claws of the fore limbs for fossorial purposes, are distinctly shown in both of the accompanying figures.

Regarding the observed habits of the Pouched Mole, it would appear that it is by no means as permanent a subterranean dweller as the typical Moles, Talpa and

Fig. 1. Fig. 2.
AUSTRALIAN POUCHED MOLE, *Notoryctes typhlops*. FIG. 1. VENTRAL. FIG. II. LATERAL ASPECTS, AFTER DR. J. F. STIRLING. TWO-THIRDS NATURAL SIZE, p. 28.

its allies. It frequents the sandy spinifex lands in the neighbourhood of the Finke River water course and the Alice Springs Stations in the southern part of the Northern Territory of South Australia, distant from Adelaide about one thousand miles. The capture of the first example obtained was due to the observation on the surface of the sand of peculiar trails which, on being followed up, resulted in the discovery of the animal reposing under a tuft of the Spinifex or Porcupine grass, *Triodia irritans*. Further observation elicited the fact that its usual habit was to travel underground for from a few feet to many yards at a depth of not more than two or three inches below the surface, its presence and position at such times being revealed by the raising and cracking of the surface of the sand. All efforts to keep the animal alive for a longer period than three or four days have so far failed, the chief difficulty being apparently the food question. The remains of ants were found in the intestines of the first example dissected, and another specimen, while in captivity, is reported to have devoured one of the coleopterous grubs that burrow into and feed upon the roots of the acacia trees, thus demonstrating its normal insectivorous predilections. Not improbably, as is the case with the Echidna, as observed by the author, the favourite food is the tender larval and nymph forms of the ants that would have to be sought for underground, either by burrowing or by the scratching away of the superimposed earth's surface. The experiment of placing adult ants with Notoryctes as a tentative food supply was by no means successful, the animal itself, according to Dr. Stirling, apparently running the greater risk of being eaten. It is much to be feared, under the circumstances so far recorded, that but little chance exists of this interesting little marsupial being established as a permanent tenant of the London Zoological Gardens.

Of the remaining terrestrial vertebrate fauna of Australia, it will hardly be anticipated that any group would possess the remarkable individuality that is exhibited by the Class mammalia. Numbers of birds systematically migrate, while reptiles, amphibia, and their ova can be transported on floating driftwood. As a consequence, we find a very considerable infiltration of Indo-Malay representatives of each of these classes, in the northern, or tropical, Australian districts more especially.

The highly characteristic Monitors or Varani, the largest of Australian Lacertidæ, popularly called "Goohannas," are thus found to be generically identical with forms inhabiting India and North Africa; while a water-frequenting species, *Varanus salvator*, growing to a length of six or seven feet, is specifically the same as the Indian and Malay type. There are a number of smaller forms, such as Geckos, Skinks, and other lizards, which possess a similar tropical Indo-Australian distribution. Australia, at the same time, produces several very remarkable Lacertilian types that are found nowhere else outside its limits, as is made evident in a succeeding chapter that is specially devoted to representative members of this animal group. Concerning prehistoric types, it is of interest to record that Australia formerly produced species of Monitors or Varani, and their allies, that are estimated to have been three or four times larger than any existing species. One of these, *Megalania prisca*, was at least over 20 feet in length. As originally described by the late Sir Richard Owen, this huge monitor was accredited with having been armed with horns and spikes of the same character, and bearing the same proportion to the creature's body, as those of the little spiny lizards, *Moloch horridus*, illustrated and described at length in Chapter III. Had that interpretation been correct, this reptile in the flesh would have undoubtedly represented one of the most formidable monsters that ever trod the earth. The acquisition of further material, however, has demonstrated that the fossil remains of two very distinct reptilian types had been obtained from the same source, and that, while a huge monitor, for which the name of Megalania has been retained, undoubtedly existed, the horned head and tail cuirass, previously supposed to have belonged to the lizard, were the property of a large species of turtle. Upon this Chelonian the name of *Miolania Oweni* has been since conferred by Mr. Smith Woodward, of the Geological Department of the British Museum.

Reference has been previously made in this Chapter to those bird species to whom special importance is attached, with allusion to the peculiar distribution of either themselves or their allies. Apart from these considerations, there are many forms which are of high interest, regarded from the point of view that they are strictly and exclusively Australasian. The Emu, as one of those forms, has already received notice with reference to the suggestive testimony afforded by the surviving allies of the same Struthious order.

A little group of birds that is most essentially Australasian, and probably represents a very primitive stock, is that of the Mound-builders or Megapodidæ. There are three distinct generic types belonging to this family group, all of which agree with one another in their very remarkable habit of constructing huge mounds of earth, leaves and other vegetable substances, within which they deposit their eggs and then leave them to be hatched out by the natural heat generated by the decaying matrix. Of the three known species, the Megapodium or Australian Jungle Fowl, *Megapodium tumulus*, is abundant in the coastal districts of Northern Queensland and South Australia, and also in New Guinea and the intervening islands of Torres Straits. This Megapodium is a plain-looking bird, of mixed grey and brown hues, about the size of a small fowl, and remarkable for the apparently disproportionate dimensions of its thick legs and feet. It is with these abnormally large but most serviceable feet that the birds collect together the materials of their mound-like nests, which may be as much as fifteen feet high and sixty feet in circumference. A characteristic photograph of one of the nest mounds of this species, taken by the author at Goode

Island, in Torres Straits, in close proximity to a miner's tent, which it greatly exceeds in dimensions, is given on this page.

The second species of mound-builder is the Talegalla, or so-called Brush or Scrub Turkey, *Talegalla Lathami*. This bird is larger than the preceding, the male being the size of a turkey, which it much resembles in shape, while the head and neck are bare of feathers and ornamented with a fleshy wattle much after the manner of typical members of that group. While met with most abundantly in the thick scrubs of the extreme north, the area of distribution of the Talegalla extends much further south than that of the preceding type. The habits of this species are essentially gregarious, many birds usually combining towards the construction of the huge mound-like nest; and as many as a bushel of eggs, which are most excellent eating, are not unfrequently abstracted from a single mound. As with the preceding species, the Scrub Turkeys use their powerful feet only in the construction of their nest, grasping bunches of leaves, grass, and all other available substances with one foot, and throwing it backwards towards the selected spot. In this manner they

W. Saville-Kent, Photo.
MINER'S TENT AND NEST-MOUND OF AUSTRALIAN JUNGLE FOWL, *Megapodium tumulus*, GOODE ISLAND, TORRES STRAITS, p. 31.

commence from the outside area of a very considerable space, and gradually work in towards the centre. Examples of this species, were some years since successfully kept at the London Zoological Gardens, and exhibited there their characteristic nest-building habits, and deposited eggs, which were duly hatched. It was observed, under these artificial conditions, that the male bird paid considerable attention to the eggs when deposited, always maintaining a perpendicular, cylindrical opening in the centre of the nest-heap for the purposes of ventilation. The young birds, as soon as hatched, were for the first twelve hours kept covered up in the material of the mound, but on the following day emerged with their wing-feathers well developed, but encased in membranous sheaths, which soon burst, leaving the limb completely free. On the third day the young birds were capable of strong flight. The relatively large size of the eggs, and the egress of the birds from them in a more highly advanced state of development than obtains with any other known species, is common to all of the members of this remarkable family, and is held to be a fact indicative of their remote ancestry. In the case of Megapodium, it has been observed that the mound-constructing instinct is so strongly ingrained by heredity, that young birds taken fresh from the nest, and confined under favourable conditions, have at once commenced to construct mounds after the characteristic manner of their tribe.

The third member of the Megapodidæ or Mound-builders is the handsome bird *Leipoa ocellata*, known in South Australia as the Leipoa or Native Pheasant, and in Western Australia as the Gnow. This species has much the form of a pheasant, but in its shorter tail and ocellated markings more nearly resembles the Indian Tragopan, *Ceriornis Lathami*. The mound constructed by the Leipon is relatively small, compared with that of the Talegalla and Megapodium, rarely exceeding eight or nine feet in diameter, and two or three feet in height. A larger quantity of sand and soil being, moreover, mixed with the vegetable substances, it acquires so much more solid a consistence that it may be readily mistaken for an ant-heap. For the table this species is esteemed more highly than the two preceding forms, the eggs also, of which about a dozen are deposited in a single nest, being greatly prized.

One of the most essentially Australian bird groups is that of the Bower Birds, usually relegated by ornithologists to a position near the Starlings, and remarkable, as in the case of the Megapodidæ, for their architectural propensities. In this instance, however, the edifice raised is a supplementary structure in no way associated with the nidamental functions that characterise the mound of the Megapodidæ, for which purpose an ordinary nest is constructed. The Bower Birds, in point of fact, possess highly advanced æsthetic tastes, and erect their so-called bowers as combined playing halls and veritable museums of arts and natural history wherein they collect together every transportable object that takes their fancy. The location of these bowers is on the ground, usually in the dense scrub or within the sheltering shade of an appropriate bush. The basis of the bower consists of a rough platform of sticks. Upon this is raised on either side a series of vertically disposed twigs, which, meeting at their apices, form a sort of arched corridor that may be two or three feet in length. The furniture and decorations of the building have now to be added. To accomplish this object the whole of the ground inside and around the bower is bestrewn with the variety collection previously referred to. . Shells, bones (often including small skulls), pieces of glass, pottery, and fragments of human wearing apparel, are indiscriminately pressed into service, and mixed in ever varying proportions. Gaudy parrots' feathers, pieces of coloured cloth, or other brightly tinted substances are regarded with especial favour, and when obtained are usually inserted among the interstices of the interlacing branches of the bower's superstructure. The bower, when completed, is regularly resorted to by its architects as a recreation ground, more especially in the early mornings, when, if cautiously approached, they may be seen chasing one another in wanton play, to and fro, through the arched corridor and around the decorated grounds. The same bower is maintained in a state of repair and frequented by the same pair of birds for several successive seasons, while such continual additions are made to the "museum" collections that they not unfrequently accumulate to the extent of several barrow loads. Among the many known species of Australian Bower Birds, the Satin Bower Bird, *Ptilonorhynchus holosericeus*, and the Spotted Bower Bird, *Chlamydera maculata* are the most familiar. Each of these birds is about the size of an English thrush. In the first-named species the male is a rich satiny-black with a purple gloss, and the female bird a deep olive green. The Spotted Bower Bird, as its name implies, is distinguished by its mottled plumage, which is a mixture of soft greys and browns. These quiet tints are, however, diversified by the presence, on the back of the head, of two small patches of longer, silky feathers of brilliant rose-pink, which are particularly conspicuous in the male bird.

The Birds of Paradise, Paradisidæ, which are usually allocated, in systematic works, to a position adjacent to the Bower Birds, while most abundantly represented in New Guinea, and the adjacent islands of the Malay Archipelago, possess one Australian species that is commonly associated with this group in the popular mind, though strictly belonging to the Hoopoes, Upupidæ. This is the so-called Rifle Bird of

Queensland and New South Wales, *Ptiloris paradiseus*. Its size averages that of an ordinary pigeon; the body is a deep velvet black, with purple reflections; the breast and abdomen are of the same colour, with olive green edges to the feathers; the top of the head and the throat are covered with smaller, scale-like feathers that glitter with that brilliant metallic green sheen that is so characteristic of the breasts of humming birds, and many of the typical Birds of Paradise, the two central feathers of the tail being of the same resplendent hue. The two closely allied New Guinea species, *Epimachus magnus* and *E. albus* have been justly described as among the most lovely bird forms that inhabit the face of the earth. The metallic tints of the head and throat in Ptiloris is, in these instances, more extensively distributed, being associated in the former type with a long, resplendent tail, and erectile ruffles developed in the neck and shoulders. In *E. albus*, which is known in the trade as the Twelve-wired Bird of Paradise, there is also a crested metallic tinted collar. In addition to this, the tail is reduced to twelve wire-like elements which represent the shafts only of the ordinary feathers, while the whole posterior half of the body is enveloped in a mass of long, curled and silky plumes of a pure white hue.

The list of notable Australian birds would be incomplete without brief mention of the well-known Lyre Bird, *Menura superba*. The extraordinary lyre-form development of the tail feathers of this remarkable bird is too well known to need elaborate description. Such, in fact, is the popular demand at the antipodes for this tail-plume for decorative purposes, that the extermination of the bird has been accomplished in many districts where it was once plentiful, and its ultimate extinction is threatened if measures are not taken to restrain its present wholesale persecution. The contour of the Lyre Bird, with its long neck and stout, gallinaceous feet, is by no means unlike that of a peacock, and the wonderful tail, which is possessed only by the male bird, fulfils a corresponding rôle of vain display. Like the Bower Birds, the Lyre Bird is an architect, but it is content with a raised earthen mound or platform only, upon which it is accustomed to execute innumerable antics, spreading its wings and erecting its tail, for the fascination of a train of female admirers, or for its personal delectation, after the manner of the true peacock tribe. One bird not unfrequently possesses several of these dancing, or "corroboree" mounds, as they are styled by the colonists, situated at some little distance from one another, to which, when disturbed, it successively repairs. As with the Peafowl, it is only the adult male Lyre Bird that develops the characteristic tail, the females and young males being clad in a plain brown plumage, without any special points of attraction. For a short period of the year, moreover, commencing about January or February, the adult male loses the characteristic plumes of which he is so demonstratively proud, and is not then to be distinguished from his more homely mate. A second, somewhat smaller, species of Lyre Bird, in which the tail is shorter and less handsomely marked, is also found in the mountain districts of New South Wales, and has been associated by Mr. Gould with the title of *Menura Alberti*.

Anterior to the discovery of Australia, that *rara avis in terris*, of Juvenal, a "Black Swan," was regarded much after the manner of the Phœnix, as an ornithological paradox. The Phœnix has not yet been accommodated with a local habitation and a scientific name, but the Black Swan, *Cygnus atratus*, is, as is well known, one of the most characteristic bird species of the southern districts of Australia. It figures, as did the Lyre Bird formerly for New South Wales, as the emblematic national animal type on the postage stamps, and also on the banner of Western Australia, formerly known as the Swan River Colony. The species is now almost as familiar in English ornamental waters as the white varieties. Though perhaps not quite so graceful in its form and movements as the typical Mute Swan, *Cygnus olor*, it is, with relation to the pleasing contrast afforded by the bright scarlet bill and, excepting the white wing primaries, exquisitely shaded dark satiny crenulated plumage, regarded by many as the more handsome bird. In contradistinction to the White or Mute Swan, so-called with reference to its silent habits, the black species produces very pleasing flute-like notes. This is especially effective heard during the silent watches of the night, and produced by large flocks assembled in a neighbouring lagoon or passing overhead. The Gippsland Lakes, in Victoria, afford very favourable opportunities for observing the Black Swan in vast numbers under its natural conditions and in all phases of growth. The passenger and trading steamboats traversing these lakes pass close to the floating flocks, and during the breeding season almost run down stray broods of the little Cygnets. This observation applies with equal force to the long navigable reaches of the Murray River. To the lover of bird life, a steamer trip through this last-named magnificent water-way yields a unique and almost inexhaustible delight. At certain seasons of the year—spring and summer months—when the waters are out and the forest-lands on either side for hundreds of miles are one vast network of lakes and shallows, birds, chiefly of the natatorial and wading orders, are present in countless thousands. Black Swans, Ducks of many varieties, Teal, Cranes of various descriptions, including, in drier spots, the familiar "Native Companion," *Grus australasianus*, Spoonbill,

BAOBAB OR BOTTLE-TREE, *Adansonia rupestris*. NORTH-WEST AUSTRALIA.

Curlew, and many others, abide upon the banks or disport in the open water, while an advance guard of screaming Cockatoos, Ibises, Hawks, and other of the wilder species, herald the approach of the intruding steamboat to their feathered kinsfolk higher up or lower down the stream.

So considerable a space has been devoted to certain more notable Australian birds, such as the quaint More-porks, Piping Crows or Colonial Magpies, Giant Kingfishers, and other less prominent types, in a succeeding chapter, as to obviate the necessity of further comment on them here. This observation will apply also to the fish tribe, which has already received a share of notice with reference to the several peculiar Australian types that, by virtue of the geographical distribution of their nearest allies, point towards the pre-existence of a large central Antarctic Continent whence many of the inhabitants of the more widely separated regions of Australia, Africa and South America would appear to have primarily migrated.

The flora or plant life of Australia is as strikingly distinct in its character as the animal races. The vast forests of Eucalypti, embracing some 150 known species, which form so characteristic a feature of the greater portion of the tree-producing areas of Australia, represent in themselves a most ancient lineage, non-existent at the present day outside the Australasian region, but whose members, as fossil deposits teach us, formerly constituted a dominent feature in European forestry. The Bankseas, Hakeas, and numerous other of the essentially Australian Proteaceæ, tell the same tale, and further evidence in a similar direction might be adduced from the characteristic Grass-trees, or "Blackboys," Xanthorraceæ and Cycadaceæ. The Heath tribe, Epacridæ, spice-perfumed Boronias and numerous other Diosmeæ, which clothe the more open moorlands of Temperate Australia, are also to a preponderating extent unique. In the tropics again the very characteristic Baobab or Bottle-tree, *Adansonia rupestris*, peculiar to the northern territory of Western Australia, is of special interest with relation to the fact that, in common with certain animal types previously referred to, its nearest ally, *Adansonia digitata*, is indigenous to tropical Africa. A characteristic representation of the Australian Baobab in full foliage is given in Plate V., while a fuller reference to and additional illustrations of both this and other types of Australian vegetation are relegated to a succeeding Chapter.

Material evidence yielded by the vegetable kingdom in support of the notogeal continental interpretation is contained in Mr. H. O. Forbes' Paper on the Chatham Islands previously quoted. Taking plant groups that are confined, or nearly so, to the Southern Hemisphere, Mr. Forbes remarks:—" Among the Saxifrageæ, a genus

Donatia, is distributed only in New Zealand, Chili, and Fuegia. Of the 950 species of Proteaceæ, only 25 cross to the north side of the Equator, otherwise they are distributed to all the southern continents with Madagascar, Tasmania, New Zealand, and New Caledonia. The Monimiaceæ, with 150 species, have the same distribution; one genus, *Laurelia*, being common to Chili and New Zealand. The genus *Cryptocarya*, of the Perseaceæ, is common to New Zealand, South Africa, and South America. Of the Cypress sub-family of the Coniferæ the genus *Callitris* is found in Africa, Madagascar, and Australia, and *Fitzroya* in Chili and Tasmania. *Todea barbara* occurs at the Cape of Good Hope and in Australia; *Lomaria alpina* at the Cape, Australia, and South America; *Fuchsia* and *Passiflora* in New Zealand and South America." Respecting the last-named genus, *Passiflora*, Mr. Forbes might also have included Australia, no less than five indigenous species being recorded in Baron von Mueller's systematic " List of Australian Plants."

HOLLOW WOODEN CRADLE USED BY THE NATIVE WOMEN OF THE KIMBERLEY DISTRICT, WESTERN AUSTRALIA. p. 14.

AUSTRALIAN GOULDIAN FINCHES. Poephila mirabilis et P.Gouldi.

"DULCE DOMUM."   *W. Saville-Kent, Photo.*

CHAPTER II.
BIRDS.

THE Avifauna of Australia is too extensive a theme to deal with comprehensively in a volume of the present pretensions. The late Mr. John Gould's magnificent Monograph and his smaller Manual upon this subject may, moreover, be appropriately recommended to all those who desire to possess a fuller knowledge of the wealth of bird-life that is so eminently distinctive of the Australasian Continent. In the present Chapter the writer proposes to draw attention to a few special types only with which he made an intimate friendship during his residence at the Antipodes. The recorded account of these foregatherings will, he trusts, result in enlisting an increased share of public interest in their favour.

By concentrating attention upon some selected species and making its varied aspects and habits a special study, one is astonished to find in the long run what a different conception of the animal is arrived at from that where the acquaintanceship has been limited, possibly, only to the contemplation of the inanimate body, the mangled trophy of the sportsman's gun. Having this manifest truism distinctly in view, the writer made the investment in the streets of Brisbane, Queensland, a few years back, of a pair of young Australian Fern-owls, or Goat-suckers, *Podargus strigoides*, popularly known throughout the length and breadth of the Australian Colonies by the respective titles of the "Mope-hawk" or "More-pork."* The former of these two appellations has been probably conferred upon the bird with reference to its somewhat hawk-like aspect and at the same time retiring habits, it being, like other members of its tribe, strictly crepuscular. The second title, "More-pork," by which this bird is the more commonly known, has a very different and, as it so happens, an entirely mis-applied derivation. To travellers through the vast solitudes of the Australian bush, and to all settlers residing within a reasonable distance of its virgin forests, the weird notes at night of an owl-like bird, which repeats at intervals in melancholy and mournful measures the words "*More-pork*," "*More-pork*," is a familiar phenomenon. In the search for the author of the doleful strains, it commonly happens that a representative of the species, *Podargus strigoides*, now under notice, is discovered in the neighbourhood. It has even been asserted that the bird has been shot in the very act of uttering the fateful words that sealed its death warrant.

In a conversation held some years since with Dr. E. P. Ramsay, the late Curator of the Australian Museum, Sydney, and a most enthusiastic ornithologist, that authority assured the author that he had fully satisfied himself that the night-bird that emitted the "More-pork" note was a true owl, the technical name of which is *Ninox boobook*. The author can as unhesitatingly affirm that Podargus is absolutely innocent of giving voice to the note imputed to it. As hereafter shown, it possesses a vocabulary of many tones, but none of these can be interpreted by the wildest stretch of imagination into the invocation for pork generally attributed to it. A name, however, once affixed by popular fiat to a bird or beast is practically as unalterable as the laws of the Medes and Persians, and so *Podargus strigoides* and its allied variety, *P. Cuvieri*, will doubtless be associated in Australia to the end of time with its mis-

---

* With reference to the wide gape of the bill, or in other words distinctly "open countenances," which the Podargi share in common with their near allies the true Goat-suckers, *Caprimulgidæ*, Mr. Gould has conferred upon the members of the genus, in his previously mentioned Monograph, the highly expressive sobriquet of Australian "Frogmouths."

applied title of the "More-pork." Unfortunately, as an outcome of the popular error rife in regard to the vocal talents of this bird, another gross injustice is rendered it from a social standpoint. Attributing to Podargus the eerie, melancholy call-note of the Boobook Owl, it is, as attested to in Mr. Gould's original description, with reference to this note, commonly regarded by the uneducated settlers as a bird of ill-omen and, if not persecuted on that account, held in high disfavour. It is trusted that the testimony here recorded concerning the author's specimens will assist somewhat towards the dissipation of this most unmerited prejudice.

To return to our own particular, not "Moutons," but "More-porks," the birds, a pair of them, were purchased in the first instance as big balls of fluff, wherein the gleaming of their glorious golden eyes yielded the only sure indication of their correct topography. The company with which they were first consorted was of the most heterogeneous description, consisting of young Parrots, Cockatoos, Magpies (Piping Crows), Butcher-birds and many others. The accommodation provided in the hawker's van, in which the birds paraded the streets of Brisbane, was far too limited to permit of such a luxury as a separate compartment, and hence our More-porks were imprisoned with a mixed assemblage of the various species above enumerated. To this ill-assorted company, and more especially that of the screeching Parrots, they manifested the most distinct antipathy, and, shrinking into the darkest corner of their noisy cage, sought temporary respite from the madding crowd.

The rescue of the poor Podargi and their translation to an independent home, where they were altogether freed from the discordant voices and more unwelcome hustlings of their former comrades, soon wrought a marvellous change in their aspect and comportment. From this time forth, for no less a period than five years, these two birds occupied the position of familiar household pets, and rewarded all the care and attention bestowed upon them by the concession of an undreamt-of insight into their most marvellously Protean moods and tenses. Fortunately, about this time, the chance gift to the writer of a very modest form of camera opened up his mind to the great possibilities afforded by the photographic art for accurately recording and delineating the remarkably divergent aspects and attitudes which these Podargi were capable of assuming. As a means to this definite end, photography was accordingly taken up, and with such a fair measure of success in the accomplishment of the object as is testified to by the illustrations of these pages. While the thirty or more replicas included in Plates VI. to X. may be said to embody presentments of the most conspicuously distinct variations of contour

F

and attitude the birds exhibited, these represent the outcome of but a fractional portion of the numerous "sittings" with which they consciously and unconsciously favoured the writer. A glance at these Plates without a reference to the accompanying context might very pardonably convey to the reader the impression that these likenesses represented the portrait gallery of an extensive aviary, in place, as is actually the fact, of their depicting the multifarious aspects of but a single pair of birds.

Before proceeding to a systematic analysis of the special attitudes and emotions of the Podargi that are here recorded by the camera, a few explanatory words appear to be desirable. The photographs reproduced in Plates VIII. and IX. owe, it may be observed, their in many instances legendary and artistic embellishments to the circumstance that, on account of their quaintness and suitability for the purpose, they have, with the able assistance of the London Stereoscopic Company, been adapted by the writer as ornamental designs for note paper and menu cards. In all instances these photographic replicas have also been much reduced in size, having been originally taken as whole or half plate negatives as illustrated by the examples reproduced in Plates VI. and VII.

The most typical presentments of the normal form and aspect of *Podargus strigoides*, when in a state of complete repose, are probably afforded by the photographic likenesses reproduced in the two Plates just quoted. In these portraits an essentially hawk-like aspect is predominant. At the same time the delicate mottlings of the plumage, which in their pencillings and gradations of mingled greys and browns wonderfully resemble the pattern and tints of the tree branches on which the birds are accustomed to perch, are very distinctly shown. This notable correspondence of the bird's plumage with its environment is habitually utilised by Podargus in a remarkable manner for the purpose of concealing itself from the observation of recognised enemies or possible assailants. Should, for instance, a hawk appear in sight or any other object of an apparently unfriendly aspect, this bird will at once straighten itself up stiffly and, with its mottled feathers closely pressed to its body, assume so perfect a resemblance to a portion of the branch upon which it is seated that, even at a short distance, it is almost impossible to recognise it. Under these conditions, in fact, it so readily escapes detection that several instances have been related to the writer in which people have actually placed their hand on the bird, when seated on a rail or log fence, before being conscious of its presence. Trusting in its wonderful mimicry of nature, it will thus remain stiff and motionless, and not attempt to fly away until forcibly removed.

AUSTRALIAN "MORE-PORKS" OR FERN OWLS. *Podargus strigoides.*

The two examples in captivity furnished the writer with abundant opportunities of witnessing and portraying this singular mimetic phenomenon. Both of them, after the manner of their species, were excessively timid, and, while they displayed with delightful abandon all their natural habits and idiosyncrasies before members of the household, the near approach of strangers, and more especially of children, was an almost invariable signal for their assuming the above-described erect, stick-like attitude. When this posture was resorted to in association with more natural influences, such as a passing hawk or a cat stealing along the hedgerow, it was always preceded by the utterance of a very characteristic alarm note, entirely different in character from the several vocal variations referred to later on, to which they give expression under normal conditions. This warning note of Podargus may, indeed, be so accurately reproduced by the human voice pronouncing the words "Chup', chup'," shortly and sharply, with distinct intervals, that the writer was able at will to compel the birds to "stand at attention." When the alarm note was emitted by the birds naturally it was often a difficult task to detect the cause of their anxiety. After eager scannings, a small speck in the empyrean would probably reveal the well-nigh invisible presence of a wedge-tail eagle or may be a flight of wild duck crossing the sky at the same high altitude. A characteristic portrait of one of these birds in its erect, mimetically defensive pose is given in Plate IX., fig. 15. The correspondence of the plumage pattern in this figure with the rough corrugations of the branch it rests on; the customary closing of the eyes, with the exception only of the narrowest slit-like aperture; and the analogy of the stiffened plumose feathers that spring from the base of the upper mandible, to a ragged tuft of bark fibre, are all represented here with photographic fidelity. Another example in which both of the birds have assumed the attenuated attitude associated with alarm is supplied by Fig. 17 of Plate IX. In this instance the female bird, to the left, is exhibiting its not uncommonly manifested tendency to shrink into itself, as it were, in such a manner as to occupy the smallest possible amount of space. Notwithstanding the very closely corresponding dimensions of this pair as instanced by Figs. 2, 7, 8, and 10, occasions would sometimes happen in which the one shrinking into itself from alarm and the other standing up and distending its plumage defiantly, might be supposed from their photographs to represent two quite distinctly dimensioned birds. This artificially induced disparity in their respective calibres is very strikingly illustrated in the portrait of them reproduced on page 44.

Many additional illustrations may be cited by way of demonstrating the very remarkable diversity of aspect and contour that may obtain in one and the same bird under the influence of conflicting emotions. Taking as the extreme in one direction the attenuate, rigid aspect of the bird already described and illustrated by Plate IX., fig. 15, and placing against this, as its antithesis, such forms as those presented by Plate VIII., fig. 6, and Plate IX., fig. 13 (which are all photographic presentments of the same individual), the impression would certainly be conveyed to anyone not cognisant of the fact, that the two portraits represented absolutely distinct species. In a less marked degree a corresponding diversity of aspect is found also among many other figures in this pictorial series.

Special external influences were productive of the somewhat remarkable posturings illustrated by the two figures last quoted. Both of them are indicative of strong emotional excitement. There is thus represented in Fig. 6 an attitude which was commonly assumed by the male bird by way of a greeting to the writer at first sight of him on his returning home

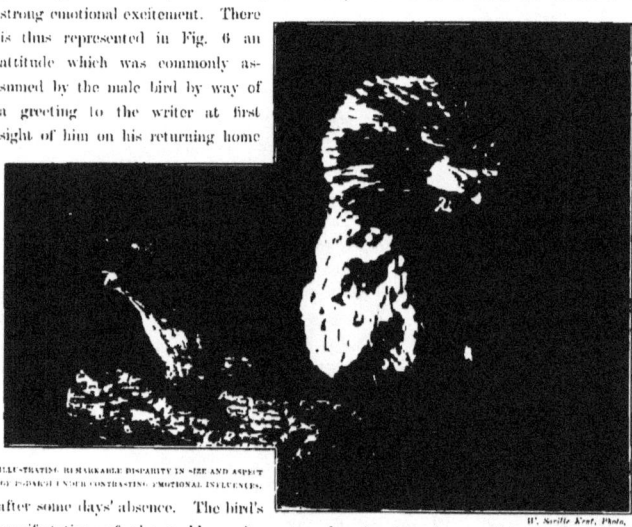

ILLUSTRATING REMARKABLE DISPARITY IN SIZE AND ASPECT OF P-BARGE UNDER CONTRASTING EMOTIONAL INFLUENCES.

W. Saville Kent, Photo.

after some days' absence. The bird's manifestations of pleasurable excitement under the foregoing conditions were on all occasions most unmistakeably demonstrated.

In Plate IX., fig. 13, the erected condition of the feathers is somewhat analogous, but the facial expression, if it may be so designated, typifies a very

PLATE VII.

AUSTRALIAN "MORE PORK." *Podargus strigoides.* (MALE BIRD).

distinct emotion. In this instance defiance, together with a certain amount of terror, enters into the composition of the excitement manifested. The assumption by the bird of this remarkable pose was first observed by the writer in connection with the chance opening of an umbrella in its presence. The necessary domestic parapluie would appear to exercise a very awe-inspiring influence upon *feræ naturæ* generally. We have a dim boyhood's recollection of a story in which a royal Bengal tiger was put to ignominious flight, when on the point of making its fatal spring, by a lady who had the presence of mind to suddenly unfurl her umbrella in the animal's face. In the case of the Podargus it possibly mistook the offending article for some huge form of bat, somewhat resembling, though far surpassing in size, the Flying Foxes, *Pteropi*, which share with it its natural haunts, but whose too familiar approach it would undoubtedly resent. However this may be, though familiarity in the long run bred contempt, for a long while the production and sudden opening of an umbrella elicited a like emotional manifestation, one of which occasions was turned to good account in securing the photograph reproduced.

Some of the most remarkable of the many metamorphic phases exhibited by these two Podargi, in which the erection of the feathers was a conspicuous feature, was manifested in the presence of a rain shower. On such occasions, humouring the birds most plainly suggested desires, they were usually allowed to enjoy a shower bath. Then, whether clinging to the perch or to their owner's wrist, they would pass through the most extraordinary evolutions. Every feather would stand on end, imparting to the birds the largest possible dimensions. The wings, separately or collectively, would be elevated or depressed, the bird meanwhile balancing its body first on one side and then on the other in order to expose all parts to the welcome rain, and even hanging head downwards to accomplish its purpose. Satisfactory photographs during falling rain, and of such excitedly restless subjects, in order to record some of the more bizarre attitudes assumed by these birds while enjoying their shower bath, proved somewhat difficult of achievement. A few of the most successful shots, out of many attempts, reproduced in Plate X., will communicate some idea, however, of their quaint appearance. In the last but one on the list of these figures, No. 9, head, wings, and tail appear to be mixed up in the most inextricable confusion.

A jocose friend having suggested to the author the peculiar fitness with which these "Rain bath" photographs would lend themselves to a humorous interpretation, we have indulged his whim to the extent of supplementing the formal title by a second, with which, in harmony with textual renderings of the respective figures, the collective one of "Autumn Manœuvres" might be associated, and the male bird—in pursuance of the foregoing suggestion—having authoritatively persuaded his companion to yield him the monopoly of the perch, a drill performance follows, in which the routine exercises of "shoulder arms," "ground arms," "trail arms," "reverse arms," &c., are faultlessly executed, and culminate finally in the assumption of an attitude of the most rigid "attention."

Glancing briefly at the remaining figures illustrative of the special habits or attitudes of these Podargi, reference may be made, among others, to Fig. 18, of Plate IX. In this instance the birds are sound asleep, literally "caught napping" in broad daylight. It is worthy of remark, in this connection, that to sleep throughout the day and be awake all night, after the manner of ordinary owls, by no means represents the customary habit of this species. As in the case of the British Fern-owl or Goat-sucker, *Caprimulgus*, Podargus is essentially dusk-loving, or crepuscular. On several occasions when the birds were visited in the middle of the night they were found to be as sound asleep as any ordinary diurnal species. While occasionally sleeping for a short interval, as in the instance portrayed, they are for the most part wide-awake, though not actively disposed, during the day. It is on the approach of dusk and throughout the twilight hours, however, that they display most energy. Indulging their very distinctly manifested inclinations, it was customary at these times to give them a free run in the garden or to carry them about hawkwise, but untethered, on the hand. For safety's sake, and to guard against the possible chance of their straying into neighbouring premises where cats or other enemies might assail them, it was considered desirable to keep one of each of their wings cut. On such occasions, however, as their feathers were allowed to grow sufficiently long to be serviceable, the only use they made of them was to fly to or after their owners.

An explanation of the remarkable tameness and domesticity exhibited by this pair of Podargi is probably to be found in the circumstances relating to their commissariat; cupboard-love, the world over, is the most persuasive of moral levers. The difficulty of keeping these birds in captivity has been recognised at most of the various Zoological Gardens where the attempt has been made. The chief obstacle encountered in all instances has been the food question. The natural pabulum of Podargus consists almost exclusively of insects, including moths, beetles, and more especially the large Heteropterous Cicadæ which abound in the Australian Eucalyptus forests, and which in their sleeping positions on the tree trunks and branches fall an easy prey to these crepuscular birds. Living-insect-food being

PLATE VIII.

AUSTRALIAN "MOREPORKS," *Podargus strigoides*. ILLUSTRATING PROTEAN ASPECTS—

PLATE IX.

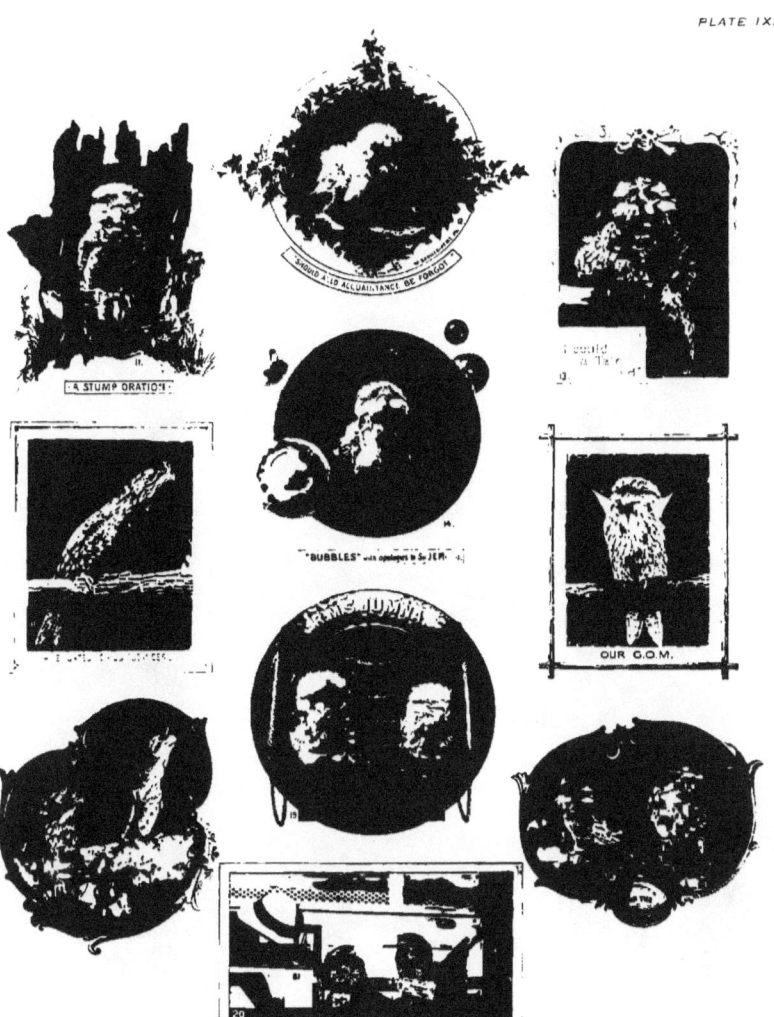

—ASSUMED UNDER VARYING EMOTIONAL INFLUENCES.

almost impossible of attainment in sufficient quantities for these birds in captivity, it has been found possible to reconcile them to a diet of raw meat. It is a noteworthy fact, however, that they will very rarely help themselves voluntarily to this food and are consequently dependent upon hand-feeding. The specimens in the writer's possession were no exception to this general rule, and throughout the five years of their companionship were accustomed to be hand-fed at least three times a day. With such continual and implicit dependence as they thus placed upon the care and attention of their owners, it is scarcely to be wondered at that they should have acquired at the same time the utmost amount of confidence.

For the advantage of those who may be tempted to make similar household pets of Podargi, it may be mentioned that the raw beef which formed their staple food was always steeped in water before being given to them, and that except for such supplementary moisture they required and would take no other liquid nutriment. As an addition to the above-mentioned meat diet, insects such as beetles, grasshoppers or moths were offered whenever procurable and devoured with the greatest appreciation. Common houseflies also they would, while carried round on the hand, pick off the walls or furniture with great dexterity. During their sojourn in London mealworms as a substitute for other insect food were much relished. The most anomalous taste manifested by these birds, however, was the predilection they exhibited, more especially the male bird, for English strawberries and in a less degree for raspberries and other soft fruits. The female bird, on the other hand, developed a fancy for small garden slugs and occasional worms. The London cockroach—of which many households enjoy a supply which may be said to exceed the demand—was greedily devoured by both birds. Like caviare, however, this piquant diet was found to be unwholesome if too liberally indulged in. The circumstances attending participation in this orthopterous delicacy were somewhat droll. Home supplies being scarce, a self-denying neighbour more blessed with abundance would at all times place a trapful at the birds' disposal. A few insects at a time were on these occasions liberated in a large white enamel basin, beyond whose slippery sides escape was difficult. When it was placed before the birds, an exciting game of "snapdragon" speedily ensued, and within a few minutes' interval the "plums" had entirely disappeared.

A highly important element in the successful custodianship of the author's Podargi was undoubtedly the provision for them of congenial perch accommodation. In concession as far as possible to their natural tree-dwelling proclivities, a portion of a tree-branch with the rough bark intact was lashed to their customary resting perch, which was of the ordinary metal kind made for parrots. Previous to this addition, which was most appreciated, the metal perch was, as shown in the lower figures of Plate X., adapted for the better prehension of the birds' feet by a close splicing of whip-cord. Other portable perches were extemporised by a rough log fastened transversely across a shallow wooden box, or raised to a convenient height by a small block only at each end. In no instances were the birds fastened to their perch. During the day time they rarely manifested any inclination to leave it, and when taking exercise in the evening systematically returned to it after making brief excursions in the immediate vicinity.

It was only for their ocean travels that a cage had to be provided, and even under these conditions they occupied their cage only at night. The two illustrations of this series, Plate IX., figs. 19 and 20, in juxtaposition depict incidents on the birds' voyage from Brisbane to London. In the latter one, seated on their long box perch, they are comfortably ensconced in the top berth of their owner's cabin that was placed at their disposal. This tedious voyage was safely and happily accomplished, though not with entire exemption from that common lot which befalls most sea-farers, whether mortals or "more-porks," at some one or other of the more tempestuous periods of their earlier voyages. Fig. 19 of the same Plate, with its attached legend, obviates all necessity of lifting the veil upon later harrowing details.

Among the remaining photo-reproductions of these Podargi inviting notice, that of Plate VIII., fig. 2, is of special interest. It represents an episode in the birds' domesticated career wherein they evinced a most energetic determination to construct a nest and enter upon the cares

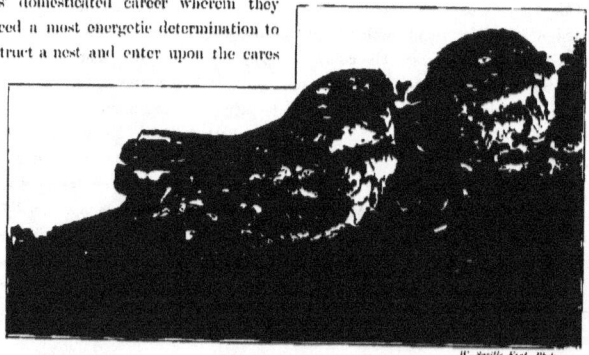

NEST BUILDING EXTRAORDINARY. p. 49.

of housekeeping. Unfortunately, being birds of the same clutch and the female somewhat a cripple in consequence of a fall in her early days, no definite results were arrived at. With each returning spring, nevertheless, the nest-building instinct was strongly manifested and afforded both themselves and onlookers intense amusement. The male bird, more particularly, would take straws or any suitable material offered him from our hands and weave them assidiously into the substance of the nest-foundation provided them in the shape of a few sticks secured around the rim of a shallow box lined with a little hay. Taking turn and turn about, the two birds would spend several hours each day in the modelling or reconstruction of their nest and subsequent occupation of it in quiet content. Their happy and philosophical contemplation of the bantam eggs experimentally introduced for artistic effects is faithfully portrayed in the figure last quoted.

The vocal notes of Podargus, briefly referred to on page 43, with reference only to the one expressed as a signal of alarm, invite some further notice. The cry commonly attributed to this bird and with which its popular title of the "More-pork" is associated has, as already explained, no foundation in fact. In addition to the characteristic alarm note the two examples in the possession of the writer gave expression to several very distinct vocal utterances, each of which possessed a special application. In their early youth or babyhood, their vocal powers were limited to crowing and gurgling noises, much akin to those common to young owls of the ordinary type. These, like the infantile expletives of homo sapiens, vanished with their advancing growth. For the benefit, nevertheless, of those specially interested in infantile language, avian or otherwise, we tentatively reproduce that variation of their baby-phraseology which presented itself to our mind as being the most articulate. Rendered as phonetically as possible in written characters the words, if such they can be called, "m-mow, wow, wow, wow-wow," repeated in unvarying cadence and with but rare cessation, represented that infant cry. To the Australian maternal mind the interpretation of this hieroglyph will probably take the form of a continued appeal for "tucker," and that interpretation will probably be correct.

In its adult state, the most ordinary vocal sound produced by Podargus is a soft soothing cooing note not unlike that of a dove, but with a different accent and capable of easy phonetic reproduction by the repetition of the words, "hoo-doo, hoo-doo," but with somewhat varying timbre and rapidity. These notes were always produced under conditions of placid contentment and were most energetically expressed by the birds when waking to more vigorous life on the approach of twilight and also during their enjoyment of a rain-bath, as described on a previous page. Imitating this note, the author found it possible to establish quite a friendly interchange of greetings with examples of this bird which occupied cages in various of the Australian Zoological Gardens, and to whom the advent of an individual sympathetically versed in Podargian language was distinctly welcome. The transformation from listless melancholy apathy to an attitude of pleasurable excitement and eager expectation was on many such occasions most conspicuous; so much so that one left them deploring the absence of power to open their prison doors and let them free—it seemed so like deserting companions in adversity.

Another very distinct vocal note to which the Podargi gave utterance was manifested with relation only to the near approach of other birds towards whom they entertained no hostile but apparently friendly sentiments. If, for instance, their comrade the Butcher-bird came near them he was always greeted with this note, which was a combined quacking and chattering sound, difficult to place phonetically on paper, but which may be approximately rendered by the, to the human mind, inane words, "quackaty-quack, quackaty-quack." Sparrows hopping on the lawn or perched on neighbouring bushes were vociferously hailed in the same language, under the impression, possibly, that they were poor relations of the Butcher-bird. The most remarkable incident, however, associated with this vocal note was the circumstance that it constituted the greeting of welcome which was commonly accorded to my wife on her first appearance in the morning, and less frequently on other occasions. This form of salutation was not extended to myself or to any other personal acquaintance, and the only plausible interpretation that can be attached to it is that, borrowing a scriptural metaphor, these birds appraised the value of my better half's company as equal to or beyond that of many sparrows, and greeted her accordingly.

The fourth and last vocal sound, yet unreferred to, uttered by the Podargi, bore relation to their mating instincts, and was of a very singular character. During the nest-building season this amatory song, as it may be designated, was frequently indulged in by both birds and may easily be described. It resembled simply a repetition of the words "toot, toot, toot, toot," repeated with comma, or staccato, intervals for a space of two or three minutes, or even more. Then suddenly, as though the accumulated words had all been wound up on a spring which now over-reached its utmost tension, they all ran down again with a rush, but with gradually diminuendo and finally piano and pianissimo expressive force.

PLATE X.

AUSTRALIAN "MORE-PORKS," *Podargus strigoides*. RAIN-BATH MANŒUVRES. p. 45.

While few, if any, other of the many Australian bird pets kept by the writer rivalled the Podargi in the interest evoked by their Protean aspects and engaging habits, certain of them exhibited, under domesticated influences, traits and peculiarities that invite attention. Among the more conspicuous of these, an example of the Queensland Shrike or Butcher-bird, *Cracticus torquatus*, lays claim to brief notice, with reference, more especially to the pathetic circumstances that attended its decease. Various species of this Shrike genus inhabit all of the Australian Colonies, but are chiefly confined to their southern, temperate, limits. The loud garrulous song of certain of these species—which consists of a rapid alternately descending and ascending scale on a gamut of a few notes only, with harmonious intervals—has won for them in those colonies, such as Tasmania, to which *Dacelo gigas* is not indigenous, the local appellation of the "Laughing Jackass." All of the species are very readily tamed, and, being in captivity inveterate mimics, with marked whistling and talking tendencies, are in considerable favour as domestic pets.

A young Queensland Shrike was obtained by the writer at the same time and from the same travelling caravan which supplied the Podargi. It early developed its special whistling properties, and these were so trained that within a few weeks he rehearsed the first stanzas of the Cambrian air "The Rising of the Lark" with a vigour and correctness that would have won for him honourable mention, if not a prize, at a Welsh Eisteddfod. This became the customary merry reveillé with which from his cage in the verandah he was accustomed to rouse the household in the early morning. The courage and spirit of these birds is notable; they will readily attack and drive away Hawks, Piping Crows, *Gymnorhina*, and almost any larger birds that happen to poach upon their preserves. In captivity they evince the strongest likes and dislikes to individual people. In the example under notice children became a special subject of hostility, insomuch that it was found desirable to confine him to his cage when young people were about, upon whom otherwise he was disposed to unpleasantly exercise the aggressive prowess of his sharp beak. Allowed to run loose in the garden during the greater portion of the day, he developed a marked penchant for English sparrows, which to the Australian settler's sorrow have been introduced and have multiplied to a ruinous extent around the greater number of the leading industrial centres. The captured sparrows he was accustomed, after the manner of his kind, to store up for a future banquet in some conveniently accessible spot, such as the forked branch of an adjacent bush, or may be betwixt the bars of his roomy cage. To approach his extemporised larder on such occasions was to incur his most vehement indignation. He would fly at the intruder, attacking him fiercely with beak and claws, all the time giving utterance to a torrent of "chatteration," which, to polite ears, would probably not bear free translation. With a captured sparrow or any other trophy of his bow and spear, these pugilistic manifestations were all in the grimmest earnest. The bird, however, was brimful of fun, and nothing pleased him more on other occasions than a mock battle on the same lines, with as much noise, but as it were buttoned foils, over say a piece of rag or an empty banana skin, which, pushing out or snatching away from between his cage bars, he would dare the writer to steal from his clutches. In other ways this bird was most gentle and tractable with his owner. Among the little tricks he indulged in, he would lie on his back in his cage stiff and rigid as though he were dead, allowing himself to be picked up and swung to and fro by his legs, as though on a pendulum, without moving a muscle. This little trick was acquired by him so easily and in so natural a manner that it favours the suspicion that it represented a hereditary, instinctive habit occasionally resorted to by these birds as a stratagem wherewith to lie in wait for and capture the birds and other small animals upon which they naturally prey. The author is indeed inclined to believe that the sparrows this bird captured in the garden were taken by some such stratagem, for, his wing being cut, he would scarcely have caught one of these alert birds on even terms. He has, however, never seen in the act.

As with many birds and other creatures capable of forming strong attachments, jealousy was with this Shrike a ruling passion, and proved his ruin. With the Podargi, through uninterrupted early association, he was on the best of terms. In an unlucky hour, however, the writer consented to take temporary charge of a neighbour's Piping Crow, the so-called Australian Magpie, *Gymnorhina tibicens*. With the arrival of that Magpie the Butcher-bird's peace of mind received a shock, though unrecognised at the time, from which it never recovered. No sooner did he commence a bar of his favourite air than the Magpie would at once overpower it with a louder and, for the occasion, harsher note, and this unfriendly stratagem, continually repeated, had the effect, in the long run, of entirely silencing his song. Further than this, the Butcher-bird now began to mope, and lost his appetite and more winning ways. The recognised cause of his discomfiture was removed, but unfortunately too late. All attempts to resuscitate his health proved fruitless, and, affectionately responding to our caresses to the last, he peacefully passed away. Surely, as the elder Agassiz has previously suggested, there will be a resurrection

BIRDS. 53

for our departed dumb friends, for heaven without them would lack to many the highest potentiality of happiness.

The true "Laughing Jackass," *Dacelo gigas*, of the Australian Colonies will be a familiar object, both with regard to its form and its extraordinary vocal powers, to

W. Saville-Kent, Photo.

NORTH QUEENSLAND LAUGHING JACKASS, *Dacelo Leachii*.

all visitors to the Zoological Society's Gardens in Regent's Park. To other than lay readers it is scarcely necessary to explain that this bird is a giant member of the Kingfisher family, so modified in its habits that it will thrive in an entirely waterless country, feeding there upon snakes, lizards, insects, and small birds in lieu of fish. The above-named largest and best known species of the genus is plentifully met with throughout the southern colonies of Australia, excepting Tasmania, and farther north is represented by several allied species of scarcely less dimensions and frequently more brilliant plumage. One of these species, *Dacelo Leachii*, obtained by the writer from the vicinity of Normanton, in the Gulf of Carpentaria, North Queensland, yielded during its some months' membership of the family menagerie several very characteristic camera studies. One of the most interesting of these, reproduced on this page, represents the bird in the act of having killed and snapped up a mouse, whose delicacy of flavour he is anticipating by meditative contemplation preparatory to swallowing it whole. This species of Dacelo is conspicuous for the brilliant turquoise blue feathers developed by the adult male bird in patches on the wing coverts, and also above the base of the tail. In this direction it more nearly approaches the yet more magnificent *Sauromarpus gaudichaudi* of New Guinea on the opposite side of Torres Straits.

The vocal accomplishments of the species here figured differ essentially from those of the familiar *D. gigas*, and so much resemble the most characteristic note of the Australian Crane, or so-called "Native Companion," *Grus australasianus*, that the writer on first hearing it in York Peninsula attributed it to that species. The sobriquet of "Laughing Jackass" would appear under these circumstances to be somewhat inappropriately applied to this type. It can scarcely be said to laugh, and its "smile," at close quarters, is so loud and ear-piercing that unsympathetic neighbours most uncharitably defined it as a compromise between the shriek of a locomotive engine and a policeman's rattle.

The Kingfisher family, Alcedidæ, is most richly represented throughout Australia, from the relatively huge Dacelos to the pigmy tropical Queensland species, *Alcyone pusilla*, scarcely two inches in length, including as members in the same district the remarkable Raquet-tailed Kingfisher, *Tanysiptera sylvia*, in which the two central tail feathers are so prolonged that they may equal or exceed twice the total length of the bird's body. Many of these Australian Kingfishers are further interesting from the fact that in lieu of river banks they burrow holes and construct their nests in the hills of the White Ants or Termites that form the subject of a subsequent Chapter. In Plate XXII., fig. B., an illustration is given of a termite nest, showing the burrow entrance of the Raquet-tailed species above-mentioned.

The very commonly expressed assertion that Australian birds are devoid of song is by no means supported by facts, as anyone who is extensively acquainted with the various districts of the Island-Continent will testify. Few, if any, European birds produce a melody as rich and varied as the several species of so-called Magpies or Piping Crows, *Gymnorhinæ*, already alluded to. Among the smaller birds there are several species somewhat resembling the English reed warbler, and one of these, *Acrocephalus australis*, is, indeed, specially mentioned by Gould as possessing a stronger and more melodious song than its European congener. Allied to these again are the many varieties of so-called White or Silver Eyes, genus *Zosterops*, which are veritable garden warblers, in habits, note, and aspect. As with the European birds, to whom they are here compared, they take a considerable toll from soft fruits such as gooseberries or grapes, but in compensation destroy a vast number of insect pests. Some half-a-dozen adult birds of the common Tasmanian species, *Zosterops dorsalis*, came into the writer's possession and were experimentally liberated and supplied with food in a large glass-enclosed verandah, where they had abundant room to enjoy full flight. It was noted then how diligently they sought out and devoured the aphides with which many of the plants in the flower-stands were affected. By the time this source of supply was exhausted the birds had become so tame and accustomed to the writer's presence that, on his bringing a green branch in from the garden, they would pitch upon it while held in his hand, and search for and appropriate its insect treasures. An observation recorded of these experimentally domesticated birds was the circumstances that they always burst into the fullest song when the rain was pattering on the verandah roof. Wild individuals roving freely in their native haunts were observed to sing most vigorously in a similar manner during falling rain.

Apart from the generally expressed, but by no means correct, statement that no Australian birds are songsters, the assertion that the brightest members of the feathered tribe are mute or productive only of discordant sounds is accepted as a universal truism. As obtains, however, in the case of most general rules, several more or less conspicuous exceptions occur. In this special association a remarkably brilliant one is yielded by the avifauna of North Australia. Writing of the beautiful Grass, or Gouldian, Finch, *Poephila Gouldæ*, in his monumental work, the " Birds of Australia," the late Mr. John Gould remarks : "It is beyond the power of my pen to describe, or my pencil to portray, anything like the splendour of the changeable hue of the lilac band which crosses the breast of this little gem, or the scarcely less beautiful green of the neck and the golden yellow of the breast; the latter colour is only equalled, certainly not surpassed, by the crest feathers of the Golden Pheasant. Whenever this bird becomes so far common as to form a part of our preserved collections, or to add a living lustre to our aviaries, it cannot fail to become a general favourite."

Since the publication of Mr. Gould's work in 1848, this finch and the yet more brilliant Scarlet-headed species or variety, the *Poephila mirabilis* of the same authority, have come to be imported extensively into the English market, and, though somewhat high in price, are at most times procurable at the first-class dealers.

Much of the otherwise necessarily elaborate description of the marvellous tints of these most exquisite little finches may be saved by a reference to the coloured Plate, Chromo II., facing page 39, which has been specially executed for this work by the talented bird artist, J. G. Keulemans, from living specimens in the author's possession. In this illustration a pair of the black-headed variety, the original *Poephila Gouldæ*, occupy a position side by side of one another in the lower portion of the picture, while two males and a female of the red or scarlet-headed race, the typical *Poephila mirabilis*, are depicted immediately above them. The females of both these species or varieties are readily distinguished, in their adult garb, by their relatively sober tints which to many tastes are more pleasing than the brilliant contrasts presented by their partner's plumage. The rich violet and cadmium yellow, common to the breasts of the males of both types, is replaced in the females by a delicate satiny lilac-pink and pale primrose. In *P. mirabilis*, the brilliant scarlet-carmine head adornment of the male bird is represented in the females by a patch of feathers of a darker ruby red, which would appear to rarely occupy as extensive an area as obtains in the male. In both varieties it is further notable that the bright grass green plumage of the back in the male individuals is replaced by a darker grey or olive in the females, and there is also in the latter sex a corresponding relative softening of the tone of the narrow band of colour which encircles the head immediately behind the black or scarlet, and which, in the male birds, placed in different lights, reflects the most brilliant tints of emerald green or turquoise blue. In a second larger picture of these Finches, executed for the author by Mr. Keulemans, the birds, some dozen in number, are posed in a semi-circle, and constitute so perfect a prismatic arch of colour that the painting has been appropriately entitled "An Avian Rainbow."

Mr. Gould's personal acquaintanceship with these tropical Australian finches was limited to the possession of preserved skins supplied to him by his agent and co-adjutator, Mr. Gilbert, from whom he also received a few meagre data relating to their habits. Had Mr. Gould lived but a few years longer and availed himself, as he probably would have done, of the opportunities that would have been at his disposal of observing the living birds, he would doubtless have placed on record an account of some of the phenomena chronicled in this Chapter.

Anticipating that full particulars concerning these birds would be probably included in Dr. A. G. Butler's Monograph of "Foreign Finches in Captivity" (Reeve and Co.), of which six parts have been published, the author consulted that work. Although, however, containing much information relating to the many attempts, few of them successful, to breed these finches in this country, and the testimony recorded by various authorities concerning the specific distinction or otherwise of the Black and Scarlet-headed forms, the work makes no mention whatever of their pronounced vocal talents and remarkable terpsichorean accomplishments. Dr. Butler's omission of reference to these phenomena is doubtless to be accounted for by the fact that his personal experience with these finches was apparently limited to their maintenance in an aviary, in company with a variety of other birds. In accordance with the writer's experiences, these finches will only manifest their natural habits and propensities to their fullest bent when, under conditions of untrammelled liberty, and a temperature coincident with that of their native habitat, they are consorted exclusively with individuals of the same species. In addition to making the acquaintance of these Poephilæ in their native haunts, in both Queensland and Western Australia, the author is at the time of writing, and has for about a year past been, the happy possessor of half-a-dozen living specimens. One pair of these represents the scarlet, and the other two the black-headed, varieties.

As attested to in a previous paragraph, considerable doubt still exists among ornithologists as to the possession by these red and black-headed finches of a sound claim for separate specific recognition. When first discovered, the scarlet-headed form was supposed to represent the male, and the black-headed one the female, of the same specific type, and there are those who still hold to the opinion, and who assert that any gradation between the two may be produced. One fact—that the two forms are, as has been personally observed, commonly, if not usually, found in the same flock—lends much strength to this hypothesis. On the other hand, while the immature individuals of both forms do present what appear to be intermediate links, the author, in common with other observers, has experienced no difficulty in relegating the adult birds to their respective category. This is more especially easy of accomplishment in the case of living examples, and where their respective vocal talents furnishes a ready clue to their sexual identity. Whichever of the two interpretations, however, may be the correct one, these birds constitute a most puzzling evolutionary conundrum. For if they be two separate species or even merely sub-species, how is it that, consorting in the same flocks, and living under precisely parallel conditions and environments, they have come to acquire their very definite colour distinctions? If gradations between the two were of more or less common occurrence, the circumstance would not be so remarkable, there being other birds—notably the Ruff, *Philomachus pugnax*—of which it is asserted that two male birds are never precisely alike in colour, though, on the other hand, every phase of variation occurs between the most extreme types.

The evidence concerning the specific distinctness of these Poephilæ adduced by the latest authorities, and recorded in Dr. Butler's book, most strongly favours the interpretation that the scarlet and black-headed individuals are variations only of the same bird, for which the title of *Poephila mirabilis*, as first applied to the scarlet-headed type, is specifically retained. The black-headed bird is accordingly distinguished by the name of *P. mirabilis var. Gouldæ*, while a third accredited variety, in which the scarlet or black area of the head is replaced by yellow, has received the title of *P. mirabilis var. armitiana*. Since, however, this yellow-headed type has been met with only among examples bred in confinement, there is every reason for anticipating that this form is only a weak and imperfectly developed phase of the red-headed species. According to some of the writers quoted by Dr. Butler, scarlet-headed birds may result as the offspring of black-headed parents or *vice versâ*, while by others it is maintained that they breed perfectly true to their race. Mr. Abrahams, one of the oldest and most extensive importers of these Australian finches, quoted by Dr. Butler, attests to experiencing no difficulty whatever in relegating the adult birds to their respective race. Previous to their first moult the young birds of both races and sexes are a dull grey-green throughout, but after this period the distinctive red or pure black feathers begin to make their appearance, and become more conspicuously developed with each successive moult.

Leaving the settlement of their specific distinction or identity to ornithologists, an account of the very interesting traits, or, it might be written, accomplishments, that both the red and black-capped Poephilæ share in common, may be proceeded with. In Gould's standard work and other books on birds, copied from that authority, the only sound attributed to them is a mournful piping added to a double twit. This, as reported to Mr. Gould by his agent, Mr. Gilbert, is all that a traveller or collector would probably hear of them in their wild state, under which conditions, at first alarm, they usually utter these warning notes—the male piping and the female twitting—while they fly for refuge to the tree tops. Under domesticated influences, however, all fear is soon cast aside, and though the females never distinguish themselves as vocalists beyond the aforesaid twitter and a few supplementary notes, the males unburden themselves incessantly in song. This song is not a loud one, but remarkable for its sustained volume and most peculiar timbre. To the writer and others enjoying with him the privilege of the first rehearsals, the most appropriate simile that this Poephila's song suggested was that of a whole choir of birds, such as a company of swallows, singing in a distant grove. The notes, though small and feeble, are very clear and sweet, and come pouring out in so rapid a stream that it is difficult to realize a single one only is emitted at a time. So deceptive, in fact, is the effect produced, that an uninitiated listener, soon after they were first brought home, announced to the author that three or four of the birds had been singing simultaneously, and that all of them commenced together after a preliminary signal note from one individual. As a matter of fact, there was only one singing bird at that time, and a wee piping note, as though from a Liliputian boatswain's whistle, is the customary prelude to the male bird's song.

It will seem to many readers, probably, very much like romancing when to the substance of the foregoing paragraph the information is added that these birds *dance* as well as sing,—and not only this, but one will pipe and the other dance, or both will dance and sing together, or vary their most amusing performances in a variety of manners. The discovery of the terpsichorean accomplishments of these Poephilæ was made very soon after giving them their liberty in a small congenially heated room supplied with convenient perches. Accustomed at first to sleep in a cage at night, the primary act that usually followed their liberation in the morning was the repairing of the little flock to their favourite top perch. After a brief interval, two of the males, a scarlet and a black-capped individual found themselves next door neighbours, and the ball, or more strictly speaking, the *pas de deux* began. As is *de règle* at such functions associated with human participants, there was a ceremonious preliminary interchange of courtesies, which was in itself a remarkable performance. Both birds, turning towards each other, would bring their beaks down nearly to the level of the perch, and while retaining them in this deflected position, vibrate their heads with great rapidity, at the same time uttering their somewhat plaintive

INTRODUCTORY SALUTATIONS.

W. Saville-Kent, del.   THE PAS DE DEUX.

piping note. This head-quivering would continue for several seconds, or even minutes, and then, stretching themselves up to their full heights, they would commence hopping up and down on their perch, exchanging meanwhile their former plaintive note for the hilarious warble previously described. This hopping-step, with both feet at the same time raised clear of the perch, was performed with the most even rhythm, while the grotesque aspect of the performers was frequently enhanced by their flexing their tails at a stiff angle towards the right or left. Some slight idea of the attitudes assumed by the finches in the course of the above-described performance is given in the diagrammatic illustrations overleaf. In the first of these the preliminary ceremonial of bowing and head-shivering is depicted, while in the second one the dance is in full swing. To add to the grotesqueness of the spectacle the favourite object usually selected by the birds for their performance was, as delineated in the illustration, the long-pointed beak of a stuffed Australian Gannet, which adorned the top of a bookcase in the author's study.

Considerable diversity is manifested in the dancing manœuvres of these Poephilæ. While two male birds more frequently dance and sing together in the manner just described, it sometimes happens that one of the two will maintain the bowing and quivering action and piping note while the other dances and sings, or one will continue piping to the other's dance, maintaining at the same time an attitude simply of apparent fixed attention. Sometimes, but not so frequently, the writer has observed a male and female bird take part in one of these performances. In such instances, however, the female usually only participates to the extent of bowing, vibrating her head, and piping. She does not dance and neither does she sing.

It is only quite recently (May, 1896), that an instance has fallen under observation in which a female bird has participated more extensively in the terpsichorean exhibition. This female is of the black-headed race, and has paired with a male of the same colour. She customarily indulges several times a day in this phenomenal pastime, and is not unfrequently the initiator of the sport. On these occasions she will commence bowing and shivering her head to her mate, but only for a brief period. So soon as he responds in like manner, she discontinues this action and pipes only a single plaintive note at rhythmical intervals to his head shaking. This may continue for several minutes, but as soon as he elevates his head and begins to dance, she introduces a new element in the form of expanding and vibrating her tail with extreme rapidity, and at the same time exchanging her previous piping note for a diminutive clucking or croaking sound. The male bird usually breaks off abruptly and flies away after sustaining his dance for one or two minutes. The performance as shared in by both the male and female birds, as above described, no doubt represents the fullest development of the singular phenomenon.

The other two male birds, scarlet and black-headed varieties, in the author's collection, are at the present time, although provided with mates, utterly indifferent to feminine blandishments, but are most devotedly attached to each other, dancing together or warbling to one another alternately, while the other extends his head and neck towards his companion in an attitude of the most rapt attention. These two inseparable friends were, it must be confessed, a few months since at daggers drawn, the disturbing element being, as might be anticipated, a lady bird. Since, however, the removal by death of that or those particular bone or bones of contention, peace has been restored to the camp, and, like burnt children who dread the fire, they have apparently forsworn, for the love of each other's company, all further dealings with the fairer sex.

One of these male Poephilæ, when first received by the author in August, 1895, while the finest individual and the most persistent songster, was, it may be mentioned, remarkably cross-grained. If set at liberty with the others, he maltreated and tyrannised over them unmercifully, and would not respond to any friendly overtures. Consigned, however, to durance vile, while they were at liberty, he would, if they alighted on or near his cage, both dance and sing to them, and he would also do the same to a preserved skin of his own species if held near him. It is this bird which has recently become very sociable, and the most devoted companion of the scarlet-headed male.

While no mention whatever is made of either the singing or dancing accomplishments of these Poephilæ in Dr. Butler's Monograph, a reference is made to certain other types in which a so-called love dance is indulged in by the male bird for the delectation of his mate. The well-known White-headed Finch or Diamond Sparrow of New South Wales, *Steganophora guttata*, is among the forms notable for this peculiarity. The male, when courting, stretches his neck up to an extravagant height, draws in the breast, and expands the chest and abdomen in such manner as to somewhat resemble an oil flask, and bobs up and down on his perch with, may be, a grass-stalk in his beak, and giving voice to his song, which is described by Dr. Russ as consisting of a flute-like call-note and a monotonous bass one. The author has also observed a very similar exhibition by the Firetailed Finch, *Estralda bella*, further referred to in a succeeding page, which in some respects bears a considerable resemblance to *Stegnnophora*. No members of the finch tribe, however, as so far observed and described, would seem to possess any approach to so varied a répertoire of song and dance as do the Gouldian species, which, added to their marvellously brilliant colouring and merry sociable habits, qualifies them for occupying a foremost rank in the estimation of all bird-fanciers.

Some highly interesting examples of the dancing of birds other than finches are recorded in W. H. Hudson's fascinating book "The Naturalist in La Plata." The performances there chronicled are, however, chiefly associated with larger birds, such as Rails, Pheasants, and Lapwings, and a number of individuals mostly acting in concert. Quoting, however, another author's treatise, Mr. Bigg-Wither's "Pioneering in South Brazil," Mr. Hudson's volume makes reference to a small bird of the size of a tom-tit, with blue plumage and a red top-knot, known to the natives as the "dancing bird." It assembles in flocks on the ground and adjacent shrubs, and, while one warbles, all the others keep time with wings and feet in a kind of dance, at the same time twittering an accompaniment. There can be but little doubt that the dancing performances of the Poephilæ here described are, to a large extent, purely sportive, and indulged in for their individual amusement, quite independently of the mating season or any inherent desire to commend themselves, from an æsthetic point of view, to the notice of the female members of the flock. As stated in a previous paragraph, these finches can now be almost always obtained in London at the leading dealers, so that any interested in this subject can readily obtain specimens and observe for themselves. To keep these little tropical finches in health, it is, of course, desirable to make provision against their being exposed to too low a temperature, and more especially against draughts; while, in order to witness the interesting phenomena here recorded of them to full advantage, it is essential that they should be allowed abundant liberty. The accommodation provided for the author's specimens, and under which they have thriven remarkably well, is the free run of his study, a sunny, well-lit room some eighteen feet by twelve, and eleven high, in which abundant provision for fresh air and ventilation is insured by the fitting of movable wire-netting screens to the window-sashes. Perches and other congenial playing and resting objects are placed for the birds on the top of two book cases, including, among other items, several hollow cork-bark cylinders, within which one pair has already commenced to build a nest. A couple of small cages being kept on the table with open doors with supplies of food and water and shell sand, they repair to these systematically for their meals, and can, under such conditions, be easily shut in for transport or temporary confinement when any special necessity may arise. Notably among these fortunately unfrequent occasions must be reckoned that veritable *dies irœ* to the scientist or practical naturalist known as "cleaning-down" day, when both he and his avian or other living pets must, like Br'er Fox, lie low, or altogether abandon their camp to the hands of the enemy.

Although many attempts have been made to breed these Gouldian finches in captivity in this country, the successes have been very few and far between. Coming from the Antipodes, their summer, or breeding season, is usually our winter, and the female birds, being particularly susceptible to ovarian disorders, resulting from cold, commonly succumb at this critical period. As an example of their fertility, it may, however, be mentioned that one female of the black-headed race in the writer's possession laid no less than sixteen eggs. The first of these were deposited at intervals of one or two days only, but the latter ones with intervening periods of as much as a whole week. This little bird apparently overtaxed her strength, being found dead one morning soon after the deposit of her sixteenth egg. The most successful instance of the Gouldian finches breeding in this country is that given by Mr. Reginald Phillips in Dr. Butler's Monograph, and is hereunder summarized. A pair of the scarlet-headed species built a large domed nest of fine grass, with an aperture nearly at the top, in a dead tree affixed in a large aviary-cage. The first egg was laid on the 5th May, 1891, the last on the 9th, when the hen commenced to sit in earnest, the cock taking her place when she came off to feed. The first young voice was heard on the 24th May, and on the 16th June two young birds in full feather were enticed out of the nest by their mother. An examination of the nest on the following day revealed the presence of a third young bird and three clear eggs. The birds while nesting fed on spray and white millet and a little canary seed, and fed their young by regurgitating this food from their crops. These young birds, on their first appearance, were a dull olive green, with horn-coloured beaks, and were fed by the mother until able to take care of themselves. The first moult of these English-born birds took place when they were a year old, May, 1892; the hen's face became black with a brick-red sort of tinge, and that of the cock bird remained a dull greyish green, except for the presence of a few red feathers dotted here and there. In the second moult, 1893, both the cock and hen developed typical scarlet faces, that of the hen, however, being still much mixed with black. It was also only with this second moult that the male bird developed his characteristic long central tail feathers.

Taking Mr. Phillips' successful experience as here recorded as their guide, bird fanciers should encounter no serious obstacles in breeding these beautiful finches for the market. The essential elements for success are no doubt the maintenance of a congenial temperature, more especially at the moulting and breeding seasons, combined with a large margin of liberty. On the question of temperature, Dr. Butler, and other authorities quoted in his Monograph, are, in the author's opinion, inclined to lay too little importance. It must be borne in mind that the natural distribution of the Gouldian finches is restricted to tropical Australia, and that they are not obtained south of these limits in common with the many other Australian species that are extensively imported. Dr. Butler attests to having kept one male bird in an outdoor aviary throughout the winter, but, against this isolated instance, numberless individuals perish indoors at the first approach of the winter season. Were they, in fact, naturally adapted for a colder climate, their geographical distribution, instead of being limited, as now, to the north, would be co-extensive with the Australian continent. It is, at the same time, an undoubted fact that Gouldian finches can withstand considerable extremes of temperature. At Normanton, in the Gulf of Carpentaria, close to their native haunts, and where the author saw a number that were breeding in an outdoor aviary, making their nests in the hollows of some rockwork, the thermometer at night was down to, and below, 50° Fahr. In the daytime, however, by way of compensation, it rose in the shade to from 70° to 80° or more, a range of temperature under which, in captivity, these finches are most supremely happy. The contrast, in fact, exhibited by the demeanour of the examples in the author's possession under the varying conditions of our fickle English climate is convincing. But a few weeks since, May, 1896, a brief premature spell of sunny summer weather maintained the temperature of their room to a height of from 65° to 70° Fahr. with windows open and without the aid of fires; all the birds were full of life, the males unburdening their pent-up hilarity in a continual stream of song. To this succeeded a return of bitter north-east wind, driving the highest day readings of the thermometer to below 60°. The birds at once began to mope, the males ceasing or indulging little in melody, while the females crouched forlornly with their heads tucked under their feathers, and only roused themselves to take their food. But, in fact, for the prompt supply of artificial heat, more serious complications would have undoubtedly ensued.

A brief space may be appropriately devoted to one other noteworthy Australian finch. This is the so-called "Firetail" or Firetailed finch, *Estrelda bella*, of the southern provinces of the Australian continent, which is especially abundant in the picturesque colony of Tasmania. This little bird can lay claim neither to the array of brilliant tints nor to the vocal accomplishments of the Poephilæ. As befits its citizenship of a more temperate climate, its plumage is distinguished for the most part by its delicate, alternating, transverse pencillings of dark and light ashen greys. The upper tail-coverts, however, stand out conspicuously from the rest of its plumage by reason of their intense scarlet-carmine hue, which, when the bird flits along the hedgerows or across the woodland glades, seem to glow with the incandescence of a burning coal. No member of the finch tribe, not excepting even the justly belauded British Bullfinch, probably, is so amenable to human influences or becomes so tame and engaging a companion as does the little Australian Firetail. Tasmanian friends have attested to individuals in their possession which would accompany them in their walks abroad, and one of several belonging to the author would, after disporting himself in the garden, return to his cage at a signal whistle. Among themselves, these birds are eminently sociable, assembling and building in company. When several are kept together and allowed their liberty in a dwelling room, they are up to all sorts of frolics, and there is hardly anything that they like better than to join a companion at opposite ends of a strand of cotton and to pull for dear life one against the other in a veritable "tug-of-war." One bird would even address himself so vigorously to this game of cotton-pulling as to allow himself to be lifted off the ground by one end of the fabric, while he held on with his beak to the other, and so hung suspended in the air for several seconds. With their owner, whom they speedily grow to know, Firetails place themselves upon the most familiar terms, exploring his pockets, penetrating up his coat sleeves and taking no end of liberties. One special pet, who travelled home to England, extended his friendship to visitors and manifested a marked partiality for a lady acquaintance, who possessed the, to him, irresistible attraction of a tiny gold mine in one of her front teeth. The ambition to attain access to, and to exploit the profundities of that glittering El Dorado, of which his keen eyes would detect the most momentary display, was his one endeavour, and with that end in view his pertinacious attentions were somewhat embarrassing.

Of song, in its true sense, the Firetail is deficient, its vocal powers being limited to a somewhat plaintive piping. So much does this piping resemble that of a remote boatswain's whistle that on the voyage to England an example one day deceived the veteran skipper of the ocean liner, who, while seated at the breakfast table, was astonished to hear the summons, as he thought, without his authority, for the hands to shorten sail. Amidst much mirth, the true culprit was unearthed from an adjacent cabin, and, after due admonition, was released on parole. The author's Firetailed finches did not dance to one another, as do the Poephilæ, but, at the same time, one especially fine little fellow, the hero of the previous episode, would hop up and down opposite to his own image in a mirror, and, to indulge his peculiar predilection, he had a small piece of looking-glass fixed to his cage. This bird would also sometimes dance in a similar manner on the table to his owners, with a piece of grass or cotton in his beak. This particular bird, like one of the Poephilæ, was a most confirmed and crusty bachelor, or possibly was only awaiting the advent of his astral affinity. At any rate, he behaved so tyrannically to all other comrades, both male and female, as to necessitate the provision of a separate establishment, in which he was familiarly distinguished by the somewhat derogatory title of "the Gaol bird."

Among the claimants for recognition as Australian songsters, a prominent place must be allotted to the common House-swallow of Western Australia, *Hirundo neoxena*, whose melody always struck me as being considerably more varied and prolonged than that of its European congener. A company of these birds, seated in a row along the street telegraph wires just outside the hotel window, as commonly happens in an Australian township, will regale you with a most *recherché* serenade. As a combined symphony, however, there are probably no birds that produce so singularly pleasing an effect as the little Bell-bird, *Myzanthus melanophrys*, of the Southern Colonies. The Gippsland district of Victoria is especially favoured with its presence. Riding through the dense eucalyptus forests of that province, the traveller, such being the writer's experience, suddenly comes upon a spot, commonly a glade near a running stream, from whence on all sides he is greeted with, as it were, the tinkling of little silver bells, all harmoniously blending with one another, though differing individually in their precise timbre. A hasty glance around fails to discover the authors of the fairy music, and it is only by a painstaking search that the hidden musicians are revealed in the form of a small olive-yellow bird that is most difficult to detect among the masses of foliage of the same tint.

Some of the exquisite little so-called Australian Wrens, genus Malurus, are by no means indifferent songsters, and are so naturally tame that, by imitating their note as nearly as possible with the mouth, they can be attracted to within a few feet distance only of the observer. In shape, size and deportment these little birds very closely resemble the English and European Long-tailed Titmouse, *Ægithalus vagans*, but are far more resplendently arrayed, their gay liveries including tints in which,

PLATE XI.

according to their respective species, the brightest cobalt blue or scarlet commingled with black predominate. A near and interesting relative of the foregoing species, but of even less dimensions, is the little Australian Emu Wren, *Stipiturus malichurus*, whose more homely tints of brown and lilac-grey are compensated for by the unique character of the elongate tail feathers. The pinnæ of these feathers are individually separated and hair-like, as obtains normally in those of the Emu and other Struthionidæ.

As a pendent to these brief bird notices, reference may be made to the several photographs reproduced in Plate XI. These illustrations, which possess somewhat of the comic element, represent a pair of young pelicans, *Pelicanus conspicillatus*, taken from the neighbourhood of the Lacepede Islands, which were brought on board the ss. "Albany," during the writer's recent passage by that vessel up the Western Australian coast. In their infantile, featherless condition, they presented a remarkable likeness to plucked geese, and were regarded with such dismay by a young fox-terrier puppy experimentally introduced to them, that, as portrayed in Figs. 2 and 3 of our snapshot pictures, he, after one moment of open-mouthed agony and distress, incontinently turned tail and fled. The persevering toilet performances of these "Innocents abroad," with scarcely a feather on which to lavish their exuberant energies, as portrayed in Fig. 4 of the same Plate, proved an unending source of mirth to the numerous company on board the good ship "Albany."

On one other of the writer's voyages up the Western Australian coast an interesting bird passenger was embarked in the shape of a young Osprey or Sea-Eagle, *Pandion leucocephalus*. This bird, which had been taken and kept in captivity for a brief season at Shark's Bay, was so tame that it allowed itself to be lifted with the hands and posed by the writer for its portrait overleaf. This species of Eagle is met with throughout the tropical and sub-tropical Australian Coast-line. At Thursday Island, Torres Straits, a pair habitually build a huge nest of interlacing sticks on one of the cruciform wooden beacons that define the boundary of the navigable channel. On Adolphus Island, Cambridge Gulf, in the extreme north west, there is another such nest that is of almost classic interest. It is built in a large Baobab or Bottle-tree, *Adansonia rupestris*, and would seem to be identical with one first noticed by Captain Phillip King in his "Survey of the Coasts of Australia" (1826), and has apparently been tenanted by successive generations down to the present time. A nest, occupying the position indicated by Captain King, was, at all events, found on Adolphus Island by the officers of H.M.S. "Myrmidon" under command of Captain the Hon. Foley Vereker, when making a detailed survey of Cambridge Gulf in the year 1888. The writer enjoyed the privilege of participating in this cruise as the Captain's guest, and having manifested a special interest in this nest, a couple of blue-jackets were told off to scale the tree and to obtain all available data concerning its dimensions and present or recent tenancy. Not being the breeding season, the nest was then empty, but had evidently been tenanted the preceding year. Its diameter was found to be no less than twelve feet.

W. Saville-Kent, Photo.
YOUNG AUSTRALIAN OSPREY, *Pandion leucocephalus*, p. 67.

Chromo Plate III

AUSTRALIAN FRILLED LIZARD. Chlamydosaurus Kingi
½ Nat size, p 71.

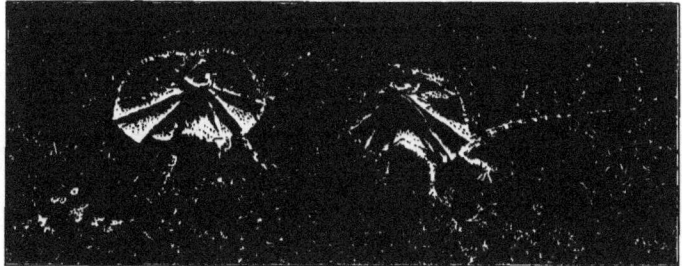

FRILLED LIZARDS AT BAY, p. 71.   W. Saville-Kent, Photo.

CHAPTER III.

LIZARDS.

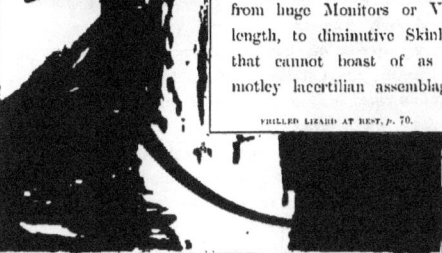

FRILLED LIZARD AT REST, p. 70.

W. Saville-Kent, Photo.

THE Australian Continent, with its vast expanses of virgin forest, rugged rock, and arid sand, is, *par excellence*, the dwelling place of a wonderfully diversified number of representatives of the Lizard tribe. These are found to range in dimensions from huge Monitors or Varani, six or seven feet in length, to diminutive Skinks, Geckos and other forms that cannot boast of as many inches. Among this motley lacertilian assemblage there occur several types which are remarkable for their bizarre form and structural or physiological peculiarities. It has, as a consequence, been determined to devote a short chapter to the illustration and record of the life aspects and habits of a few of the more prominent of these lizards with which, through the possession of living specimens, the author has enjoyed special opportunities of making himself familiar.

The place of honour among this selected series of noteworthy Australian Lacertilia must unquestionably be awarded to the so-called Frilled Lizard, *Chlamydosaurus Kingi*, which is restricted in its distribution to the northern or tropical districts of Australia and is, within these limits, indigenous to the Colonies of Queensland, Western Australia and the Northern Territory of South Australia. The earliest record of this singular species is contained in Captain Phillip P. King's "Narrative of a Survey of the Intertropical Coasts of Australia" (1826), wherein it is named, figured and described in a Natural History Appendix edited by the late Dr. J. E. Gray, F.R.S. This original type specimen was obtained by Mr. Allan Cunningham, the botanist to Captain King's ship, the "Mermaid," at Careening Bay, on the north-west coast of Western Australia. Living examples of this very remarkable Lizard were secured by the author in both North Queensland and Western Australia. He was fortunate, moreover, in bringing a specimen, the first of its kind imported to Europe, safely to England, and in chronicling the greater portion of the data here recorded concerning its very singular aspect and habits.

The natural habitat of the Frilled Lizard is essentially sylvan. It affects the more or less thickly-wooded scrub-lands, and passes the greater portion of its existence on the trunks and lower branches of the trees. At first sight, when seen in repose, as shewn in the photographic illustration which flanks this Chapter heading, there is but little to distinguish this lizard from the ordinary lacertilian type. The hind limbs are relatively long and the front ones short, as obtains, though in a less degree, in many species of Grammatophora (Amphibolurus). The head is somewhat abruptly truncate, the tail long, rough and attenuate, and there are no abnormal spines or protuberances such as occur in *Moloch horridus* and other structurally conspicuous species. On a nearer examination, however, it will be observed that a neatly folded plicated membrane with denticulated edges envelopes, sheath-like, the hinder region of the head and neck and extends backwards to the reflexed angle of the fore-limb. In order to appreciate the remarkable form and development of this membranous structure in the Frilled Lizard it is necessary to come upon the animal unawares or to otherwise submit it to exciting influences. Under such conditions it is suddenly transformed from the most placid-looking of reptilian types into a creature of an essentially forbidding and threatening aspect. The membranous frill, previously applied closely to the neck, is suddenly, and synchronously with the opening of the mouth to its fullest width, erected in such manner that it stands out at right angles around the animal's head. The dimensions of this frill in adult individuals may be as much as from eight to ten inches in diameter and in its erected condition the lizard's body, when facing the observer, is almost, if not entirely, concealed from view. A fairly correct presentment of the Frilled Lizard thus aroused to anger or standing defiantly at bay is embodied in the two photographs from life reproduced as the heading to this Chapter. The ferocious appearance of Chlamydosaurus here depicted only in monochrome is materially enhanced in the living subject by the glowing tints of vermilion red, yellow, and steel blue that suffuse that portion of the frill which covers the front of the neck and chest, combined with the bright yellow distinctive of the tongue and the lining membrane of the mouth and throat. The dashes and pencillings of rosy red which accentuate the outlines of the upper and lower jaws and the margin of the singular ciliated irides, also add sensibly to the creature's angry aspect.

With the assistance of the talented natural history artist, Mr. F. W. Frohawk, the author has succeeded in obtaining a highly characteristic portrait of the example which he recently brought alive to England, with which has been incorporated the colours, copied from life, of one of the brighter male examples immediately after its capture in Roebuck Bay. This portrait appeared originally as a black and white drawing in the pages of the "Field" for August 3rd, 1895, and subsequently with the colours, as added by the author, in conjunction with a paper descriptive of the animal contributed to the "Proceedings of the Zoological Society" for the same year. The author's obligations to the "Field" Newspaper and the Council of the Zoological Society for permission to reproduce the illustration in these pages may be here most appropriately recorded.

With regard to the significance and utility of the erectile frill in Chlamydosaurus, the fact that this organ is of insignificant proportions in young examples, and attains its full development only in adult individuals, would appear to indicate that, as a structure, it has been developed within comparatively recent times and does not represent the residual heritage of a remote ancestry. Respecting its function, there can be but little doubt that it fulfils simply the *rôle* of a "scare-organ," wherewith it terrifies, and diverts the projected attack of, many ordinary enemies. Instances have, in fact, been related to the author of dogs which will readily rush upon and kill other and larger lizards, such as Varani, refusing to come to close quarters with so formidable-looking an object as Chlamydosaurus, when it turns upon them with gaping mouth and suddenly erected frill. The erection of the feathers of an owl or the fur of a cat is correlated with a like "scaring" function, but the inflation of the hood of the Cobra, and in a less degree the neck-membranes of other snakes, furnishes, perhaps, a more appropriate analogy. In one other Australian species, *Amphibolurus barbatus*, commonly known as the Jew Lizard, which is also figured and described in this Chapter, the throat membrane is likewise inflated under the influence of irritation in such a manner as almost to constitute a frill. As hereafter recorded, in fact, this species, in districts south of the habitat of Chlamydosaurus, is commonly distinguished by the popular name of the Frilled Lizard.

It is worthy of note with reference to the elevation and depression of the membranous frill of Chlamydosaurus that the species is not unfrequently delineated in natural history works with this structure more or less fully extended, but with the mouth completely closed. The author has also observed mounted specimens in museums displaying a corresponding relationship of the organ and structure indicated. As a matter of fact, the opening of the mouth and the erection of the frill are synchronous actions which cannot be exercised independently of one another. An explanation of this circumstance is afforded by the presence of slender processes of the hyoid bone which extend on either side through the walls of the membranous frill. The relative elevation of the frill is consequently in direct proportion with the depression of the mandible, and it is only under the condition of the mouth being opened to its widest extent that the frill is so conspicuously displayed as to stand out at a right angle from the animal's neck, as indicated in several of the accompanying illustrations.

Although presenting so weird and formidable an appearance, the Frilled Lizard possesses but feeble powers of aggression. Its teeth are small, the jaws comparatively weak, and it is but rarely that the animal attempts to bite, relying apparently on the discomfiture and retreat of its would-be assailants through the terror-striking appearance of its gaping jaws and erected frill. Individual specimens were, however, found by the author to differ materially in the exercise of their defensive or aggressive proclivities. With the majority of some half-a-dozen examples kept in confinement the non-aggressive form of defence was alone displayed. Two exceptional individuals, however, manifested an essentially hostile and pugnacious disposition, springing at and biting any object placed near them, uttering a hoarse, hissing noise, and also striking savagely whipwise with their rough attenuate tails. The blows thus delivered were, in fact, dealt with such vigour as to smartly sting the hand if exposed to the impact. This flagellating method of attack manifested by Chlamydosaurus must, it may be anticipated, prove very disconcerting to a foe previously unfamiliar with the animal's peculiar aggressive tactics. After a very short interval of confinement and hand-feeding, however, even these two irascible individuals abandoned their previous aggressive tendencies, and became quite domesticated.

An even higher scientific interest than the abnormal development of the frill-like membrane that encircles the creature's neck attaches itself to the very remarkable manner in which Chlamydosaurus progresses along the surface of the ground. In this respect it is apparently unique among the existing members of its tribe. The rumour that the Frilled Lizard was in the habit of running erect on its hind legs only was communicated to the author some years ago, and is also recorded by Dr. Henry Woodward, F.R.S., in the Quarterly Journal of the Geological Society for the year 1874, in an interesting article by that accomplished geologist entitled "Forms intermediate between Birds and Reptiles." The present writer failed, however, to verify this assertion through the single living Queensland example he had in captivity for a short interval; and neither was a friend in the northern district of the Colony more fortunate, who, at special request, experimented with several specimens. It was, on these grounds, anticipated that the rumour, which had been previously received, was the outcome of an optical illusion; more particularly since many lizards, such as certain of the slighter-built Amphiboluri, run so erect on their haunches that it might be imagined their fore-limbs were raised from the ground.

It was consequently with much gratification and delight that, on becoming the owner of several specimens, including the one brought to England, obtained with the assistance of the aborigines of Roebuck Bay, Western Australia, the writer found himself in a position to scientifically demonstrate for the first time the truth of the report concerning the erect gait of Chlamydosaurus that had been received in Queensland. Possibly the specimens previously experimented with had been slightly injured during capture, and lacked the stamina to walk upright. At all events, the Roebuck Bay examples, brought in straight from the bush, were in vigorous health, and at the first trial when left at liberty, save for a light retaining cord, ran along almost perfectly erect, with both their fore-limbs and long tails elevated clear of the ground.

K

The attempt was made on the spot to permanently register, with the aid of the Kodak camera, the absurdly grotesque appearances these lizards presented when progressing in this bipedal fashion. Such, however, was the speed at which the animals ran, that the shutter of that instrument did not work fast enough to secure anything better than a blur at close quarters, and it was only by bringing an Anschutz camera, with its more rapid roller-blind shutter to bear on the specimen after its arrival in London, that the several bizarre figures reproduced in Plate XII. were secured. While even these partake much of the nature of silhouettes, they will serve to indicate the very singular running attitudes which this lizard may assume.

Fig. 1 in this series carries with it so essentially human an aspect that one is sorely tempted, even at the risk of scientific contumely, to place a cricket bat in its right hand. Fig. 2, again, might equally do duty as a parody on the celebrated dance of the "Lord Chancellor" in the play of "Iolanthe"; the exuberant judicial wig of his lordship finding its counterpart in the lizard's semi-erected frill, while the dancing pose is ludicrously perfect. The distance Chlamydosaurus will traverse in this remarkable erect position may average as much as thirty or forty feet at a stretch, when, after resting momentarily on its haunches, as in the attitude illustrated in Fig. 8 of the same Plate, it will resume its running course. When, however, a short space of a few yards only has to be covered, the animal runs on all-fours, sitting somewhat high on its haunches after the manner of many ordinary lizards, such as the Amphiboluri, previously referred to. The figure last quoted, as also the bipedal perambulating one represented by Fig. 7, are reproduced, it should be here mentioned, from drawings from the living animal, originally executed for the "Field" by Mr. F. W. Frohawk.

As is incontrovertibly demonstrated by the unimpeachable accuracy of the accompanying instantaneous photographs, all doubt that has hitherto been entertained respecting the correctness of the reported erect mode of locomotion of the Frilled Lizard may be finally put at rest. Fortunately, moreover, through the medium of the example of this remarkable lizard brought safely to England, the writer has been enabled to afford other scientists an opportunity of witnessing its very abnormal locomotive performances. An anecdote might be even related of how one of our most eminent zoologists, carried away in his enthusiastic delight at the singular spectacle of the creature careering on its hind legs and defiantly erecting its marvellous frill, so far unbent as to follow it excitedly on hands and knees. Verily, here stood revealed, (pardon the hibernianism), the spirit of a genuine naturalist! The specimen in question

PLATE XII.

THE FRILLED LIZARD, *Chlamydosaurus Kingi*.

Figs. 1-6. BIPEDAL RUNNING PHASES, FROM INSTANTANEOUS PHOTOGRAPHS TAKEN BY THE AUTHOR WITH THE ANSCHUTZ HAND CAMERA.
Figs. 7 and 8. RUNNING AND RESTING ATTITUDES, DRAWN FROM LIFE BY F. W. FROHAWK.

was presented by the writer, in August, 1895, to the Zoological Gardens in the Regent's Park, where, for several weeks, its phenomenal aspect and habits proved to be a source of much attraction.

The gait of the Frilled Lizard when running erect is remarkable for the widely-extended and highly-elevated positions which the hind-limbs successively assume. This is shown with marked distinctness in the photographs reproduced in Plate XII. As also indicated in these figures, the pendent fore-limbs during bipedal progression would seem to be subservient to the function of balancing the body, being, as illustrated more particularly by Fig. 5, extended synchronously with the elevation of the opposite hind-limb. Interest of a special kind is attached to the bipedal perambulation of Chlamydosaurus on account of the fact that, while apparently unique among all the existing representatives of its class, such a method of deportment was typical of the long-extinct Reptilian group of the Dinosauria. That such huge representatives of that tribe as the colossal Iguanodons—of which a restored skeleton in the Natural History Museum measures no less than thirty feet in length—skipped along the ground in the jaunty style of the species now under notice is scarcely to be imagined, but it may, at the same time, be inferred that with many of the smaller members of the Dinosaurian class this mode of progression was largely approximated. On the other hand, it might, of course, be argued that the Frilled Lizard's mode of progress has been evolved at a comparatively recent date with reference only to its special environments. As indicated on a previous page, the species is essentially arboreal in its habits, abiding for the most part on the trunks and lower limbs of the trees in the tropical scrubs. It was observed repeatedly by the author of the examples he had in confinement that, when liberated in a room or elsewhere, they invariably made towards and ascended, as far as possible, any vertical object, such as a table leg, or even that of an onlooker. In accomplishing this manœuvre their vertical perambulatory deportment necessarily placed them at the greatest advantage. It is noteworthy, however, in this direction that, while certain of the Australian Lizards belonging to the genus *Varanus* are equally or even more essentially tree climbers, their perambulation is exclusively quadripedal.

This question of the analogy of the gait of Chlamydosaurus to that of certain of the Dinosauria, together with a discussion as to the possible zoological affinities of this remarkable living lizard with members of that extinct group, is entered into at some length in two articles contributed by the author to the "Proceedings of the Zoological Society" for November, 1895, and "Nature," of February 27, 1896, respectively. A few of the more salient points submitted in these papers may be appropriately reproduced in these pages.

Attention was more particularly directed in the foregoing communications to the singular bird-like aspect presented by Chlamydosaurus when running erect, as illustrated by the instantaneous photograph reproduced in Plate XII., fig. 5, wherein the superficial resemblance to a running long-tailed bird, such as a pheasant, is especially indicated. The writer's comments respecting this particular illustration were as follows:

Special interest is attachable to this avian-like ambulatory deportment of Chlamydosaurus by reason of the generally accepted interpretation that the birds are modified descendants of a reptilian archetype. The temptation is naturally also very great to institute comparisons between, and to suggest possible affinities with, this peculiar lizard and the extinct group of the Dinosauria, among whose representatives a bipedal locomotive formula was apparently a characteristic feature. A reference, however, to the skeleton of Chlamydosaurus does not encourage any sanguine anticipations that may have been previously entertained in this direction. It yields no indication of that peculiar avian modification of the pelvic elements, adapted for bipedal locomotion, that are so essentially diagnostic of the more typical Dinosauria, while in all general points it is indistinguishable from that of the ordinary Agamidæ.

Though, as a consequence, no serious attempt would be justified to correlate the erectly progressional Chlamydosaurus with such ponderous specialised Dinosaurs as, say, Iguanodon or Brontosaurus, there are some few species at the lacertilian end of the chain that probably presented, when living, a by no means remote likeness to this existing type in both aspect and gait. The *Compsognathus longipes* of A. Wagner, from the lithographic stones of Solenhofen, is more especially worthy of mention in this connection. In size, some three feet long only, and in the proportions of the limbs and other points, it must have been almost a counterpart of *Chlamydosaurus Kingi*. It is particularly noteworthy of it, moreover, that, as pointed out by the late Prof. Huxley ("Anatomy of Vertebrata," p. 262, Ed. 1871), the pelvic elements of Compsognathus correspond more essentially with those of the ordinary lizards than with those of the aviform Dinosauria, the pubes in particular being apparently directed forwards and downwards, like those of lizards. This type, as likewise Stenopelyx, is also referred to by the same authority (p. 263) as indicating that the more typical modification of the pelvis, in which the pubes are directed backwards parallel with the ischia, as in birds and Iguanodon, "was by no means universal" among the Dinosauria or Ornithoscelida, as Prof. Huxley preferentially named them.

Notwithstanding the distinctly recognised lacertilian character of the pelvis of Compsognathus, Prof. Huxley had no hesitation in assigning to this type an erect bipedal method of locomotion. Writing of it in the "Popular Science Review," 1866, that illustrious biologist remarks: "It is impossible to look at the conformation of this strange reptile, and to doubt that it hopped or walked in an erect or semi-erect position after the manner of a bird, to which its long neck, slight head, and small anterior limbs must have given it an extraordinary resemblance."

The silhouette presentment of Chlamydosaurus, reproduced in these pages, forms a not inapt embodiment of the flesh-clad skeleton that must have suggested itself, ghost-like, to the learned Professor's mind. And it is among the author's keenest personal regrets that, through the recent decease of Prof. Huxley, he should have been deprived by so short an interval of gladdening his former teacher's eyes with the sight of a living organism which, if only in the direction of superficial analogy, so nearly realised one, among the many, of his most sagacious interpretations of the fossil past.

A remaining point in the erect running gait of Chlamydosaurus invites brief attention. Such is the construction of the hind foot and its component digits that, when thus running, the three central digits only rest upon the ground. As a consequence of this structural peculiarity, the track made by this lizard when passing erect over damp sand or other impressible soil, would be tridactyle like that of a bird, and would also correspond with the tracks that are left in Mesozoic strata by various typical Dinosauria. This tridigitigrade formula of the gradation of Chlamydosaurus, induced by the great relative shortness of the first and fifth digits, is distinctly indicated in fig. 1 of the Plate previously referred to.

Whether or not the bipedal locomotive comportment of Chlamydosaurus has been transmitted by heredity from a lizard-like Dinosaurian such as Compsognathus, or has been re-developed independently among its allocated family group of the Agamidæ, is a question concerning which it would be unbecoming temerity on the writer's part to pronounce a verdict. The phenomenon, while dominant among the Reptilia of bygone ages, is, with the exceptional instance afforded by Chlamydosaurus, apparently extinct among living types, and is, on that account alone, of unique interest.

Without overstepping the bounds of prudence, it may be finally suggested that the occurrence within the Australian region, embracing New Zealand, of a wealth of archaic types, such as the Mesozoically related lizard Sphenodon and the fresh-water fish Ceratodus, as also a dominant mammalian fauna that pre-existed, but is now extinct, in other continents, would justify the anticipation that a reptile inheriting the phenomenal habits of a Mesozoic race might be sought for with the greatest prospects of success upon Australasian territory.

The food of the Frilled Lizard in its natural state consists almost exclusively of beetles, the crushed up remains of which were found by the author to constitute the bulk of the excreta of all recently captured specimens. In captivity the regular supply of a sufficient quantity of living coleoptera or other insects being impracticable, their customary diet was experimentally exchanged for one consisting of raw meat. While flourishing upon and apparently liking this pabulum, they could never be prevailed upon to take it of their own accord, it being necessary by slightly exciting the animals to induce them to erect their frills and open their mouths, when, on meat being placed therein, it was immediately masticated and swallowed with unmistakable relish. Feeding the lizards under these conditions was almost as simple a matter as dropping pennies in a slot. On those rarer occasions when living beetles were placed at their disposal the insects were picked up and secured by the rapid protrusion of the fleshy glutinous tongue. The example recently presented to the Zoological Society's Gardens was the survivor of four specimens taken on board ship by the writer at Broome, in Western Australia. One of the finest of these four unfortunately escaped overboard through the apparent omission on one occasion to sufficiently secure the fastening of their cage. A second specimen, through biting off and swallowing the end of the stick by which its daily allowance of raw beef was administered, induced internal inflammation to which it finally succumbed; while the third wasted and died from injuries primarily received from the natives who effected its capture and through whose aid all of the specimens obtained from the Roebuck Bay district were secured. It being the custom of these aborigines to make such captures as snakes and lizards safe by tying them, at the full stretch, tightly to a stick, it is scarcely to be wondered at that one or more of the specimens should have received fatal injuries in the process. The results of the writer's experience concerning the keeping of the Frilled Lizard in captivity would seem to indicate that it is by no means of a hardy nature. A single specimen kept for a short interval in Queensland some years previously, succumbed apparently to sunstroke through its cage having inadvertently been exposed for too long an interval to the sun's direct rays, and, as it happened, before any confirmatory evidence could be obtained concerning the remarkable ambulatory attitude of the species as is here placed on record of the West Australian examples. On the voyage home in the "Glengarry" from Singapore the several specimens were, while within the tropics, allowed a daily constitutional run on deck at the end of a liberally long tether, their liberty being as highly appreciated by themselves as was the sight of their strange gait and attitudes by an interested circle of fellow voyagers.

It is worthy of note that the habits of Chlamydosaurus were found to be strictly diurnal and in that respect contrary to those of the large arboreal Monitors or Varani which, as is well known, take advantage of the dark hours of the night in outlying settlements for visiting and robbing the henroosts of both eggs and young chickens. The Frilled Lizard is guilty of no such nefarious practices and retires to rest with the sun, creeping into a hollow log or clinging perpendicularly to a tree-trunk or other suitable support in the attitudes indicated in the top corner illustration of this page. This picture portrays a fine pair that shared for a while, with the author, the accommodation of a mosquito-proof room in the verandah of Mr. G. S. Streeter's establishment at Broome.

W. Saville-Kent, Photo.

In concluding this notice of the Frilled Lizard attention may be directed to the remarkable manner in which this singular species lends itself to certain of the presentments of the mythical dragon depicted by Chinese artists. The photograph from a roughly stuffed skin, reproduced on page 79, will amply suffice to illustrate this very suggestive resemblance. Although the lizard under discussion is entirely unknown in China, it is a well-known fact that from time immemorial the Malay traders have been accustomed to visit the northern coast line of Australia for the prosecution of the Bêche-de-Mer or Trepang fisheries, the produce of such fishing going to the Chinese market. It is by no means improbable that through such a source skins of the Frilled Lizard found their way to quarters where they were artistically immortalised in the manner suggested. It is quite within the bounds of possibility, indeed, that a careful investigation into the wealth of conventional pictorial art, for which the Chinese are so eminent, would reveal the fact that many of the animals peculiar to the northern districts of Australia were known to them centuries previous to the discovery and opening up of that Island-Continent by Europeans.

The publication by the writer, in "Nature" and elsewhere, of the data here recorded concerning the bipedal perambulation of Chlamydosaurus has elicited the statement in the "American Naturalist" for July, 1896, that a Mexican lizard, *Corythophanes Hernandezii*, was described some years since by M. F. Sumichrast as possessing similar locomotive peculiarities. A reference, however, to the original publication quoted, "Note sur les Mœurs de quelques Reptiles du Mexique": (Archives des Science Physiques, T. XIX, Geneva, 1864), by no means substantiates the correctness of the suggestion made. The passage relating to the locomotion of Corythophanes reads as follows :—"Quant il court il relève le haut au corps presque verticalement, tout en fouettant le sol avec sa queue, ce qui lui donne alors une allure fort singulière." The vertical elevation of the body here described would appear to correspond with the erect locomotive attitude already attested to on page 73 with relation to certain species of Amphibolurus which is also frequently assumed by Chlamydosaurus when traversing short distances. There is, at any rate, no indication of the animal progressing in an absolutely bipedal fashion, while it is distinctly stated that the lizard continually strikes or flogs the ground with its tail, in place of carrying that appendage raised above it as obtains under corresponding conditions in Chlamydosaurus. It is, at the same time, quite possible that Corythophanes raises its fore feet from the ground when running, and the practical demonstration of this fact in such a manner as has now been conclusively substantiated in the case of the Frilled Lizard would be highly interesting.

A characteristic Australian lizard that invites brief notice in proximity to Chlamydosaurus is the so-called "Jew" or "Bearded Lizard" of the Southern Colonies, originally distinguished by the title of *Grammatophora barbatus*, but more recently allocated to the genus Amphibolurus. The popular title of the Bearded Lizard as applied to this reptile is readily explained by a reference to the lowermost of the three photographs illustrative of the species reproduced on this page. As there indicated, there is a voluminous development of the integument around the creature's throat, which, when the mouth is widely opened, is erected much after the manner of the "frill" of Chlamydosaurus, but, being developed on the under surface of the

W. Saville-Kent, Photo.

BEARDED LIZARDS, *Amphibolurus barbatus*. ONE-THIRD NATURAL SIZE.

neck only, presents a not inconsiderable likeness to a stiffened beard. As an outcome of the conspicuous resemblance which to a considerable extent subsists between the abnormally developed neck membranes of the two species between which comparisons have been instituted, it is found that the two forms are apt to be confounded with one another. The name of the Frilled Lizard is not only frequently misapplied to this Amphibolurus, but the species is figured under the name of Chlamydosaurus in one or more standard Natural History works. In addition to the material structural differences between the two forms, it is worthy of remark that the true Frilled Lizard, Chlamydosaurus, is an essentially tropical type, while *Amphibolurus barbatus* is as characteristically a denizen of the southern or temperate areas of the Australian Continent.

The Bearded Lizard is one among the several Australian representatives of its tribe that is not unfrequently brought to England and placed on view in the Reptile House of our well-appointed Zoological Gardens. Although it so happened that this particular species was not included among those which the author kept alive, and had an opportunity of making a special study of, during his residence in Australia, the facilities were afforded him while engaged upon this work of securing the accompanying photographs and chronicling the data here recorded of an English imported specimen, very kindly placed at his disposal by Mr. A. E. Harris, F.Z.S., an enthusiastic admirer of the Lacertilian Zoological group.

Among the points inviting investigation with reference to Amphibolurus, was that of its method of perambulation, its share with Chlamydosaurus of the title of the Frilled Lizard having brought about the attribution to it of a corresponding bipedal plan of locomotion. The contour of the body and the small relative size of the hinder limbs, however, by no means encouraged sanguine anticipations in this direction, and as, in fact, practical experiments proved, the gait of Amphibolurus differs in no way from that of the generality of Lizards. One peculiarity, however, which it was found to share with Chlamydosaurus was the free use it made of its tail as a weapon of defence, striking vigorously with this organ right and left at the writer's hand or other presented objects on occasions of abnormal excitement. With the Varani, as hereafter noted, the caudal appendage constitutes a most effective offensive and defensive weapon. As compared with Chlamydosaurus, the body scales of Amphibolurus are considerably larger, those bordering each side of the spine being most conspicuously so and also notable for their sharp trenchant edges. It was found, in fact, that these larger scales could

PLATE XIII.

W. Saville-Kent, Photo.

W. Saville-Kent, Photo.

cut the skin to the extent of drawing blood if the animal was held in the hands while struggling to escape.

The Australian Lizard that next in order to Chlamydosaurus claims attention on account of its bizarre aspect is the so-called Spinous Lizard, or Mountain Devil, *Moloch horridus*. This species is found very abundantly in the semi-tropical district of the Gascoigne, in Western Australia; it also inhabits parallel latitudes of South Australia, and has been rarely taken in central Queensland. The earliest notice of this singular lizard occurs in the Hon. Sir George Grey's "Journals of Discovery in North-West and Western Australia," published in the year 1841. In this volume there is also given an excellent figure of it, accompanied by its technical description by Dr. J. E. Gray, then Keeper of the Zoological Department of the British Museum, to which the specimen was consigned.

The reason why such formidable appellations as "York Devil" and "Mountain Devil," by which this species is popularly known in Western Australia, have been conferred upon it, is difficult to find, for there is probably no other representative of its class that is possessed of less potentiality for hurt or evil. Doubtless the appearance conveyed by the *chevaux de frise* of spines and tubercles, with which its body and limbs are armed at all points, added to the horn-like development of the two anterior head spines, represent the chief attributes that have linked this harmless lizard with so dire a name. Did this small species, which rarely exceeds six or seven inches in length, attain to the dimensions of a crocodile, or even that of some of the larger Monitors, it would reasonably take rank among the most formidable-looking of living creatures. It is not so many years, indeed, since such a huge fossil representative of *Moloch horridus* was supposed to have been discovered, and was described and dilated upon as such in eloquent terms by the late Professor Sir Richard Owen, under the title of *Megalania prisca*. According to that accomplished physiologist's original description, this terrible reptile was no less than fourteen feet or more long. It possessed a flattened skull somewhat resembling that of an ox in shape, 1 foot 10½ inches in breadth, and that was armed with as many as nine horn-like prominences. This skull and several joints of a long, hollow tail sheath, which bore a singular resemblance to that of the extinct gigantic Armadillo, Glyptodon, represented the basis upon which this lacertilian type was established. Further material, however, derived from the same source, the Australian tertiary deposits, and also those of Lord Howe Island, off the Australian Coast, elicited the fact first demonstrated by Professor Huxley, that these fossil remains

SPINOUS LIZARDS, *Moloch horridus.*
"A WAYSIDE GREETING," p. 87.
W. Saville-Kent, Photo.

belonged to a huge Chelonian, a member of the turtle tribe, and it is now accordingly allocated to that group under the new name of *Miolania Oweni*. A facsimile of Owen's original illustration of this remarkable type is reproduced on page 88, immediately above a group of Molochs photographed from life to approximately the same scale.

Numerous examples of *Moloch horridus* have been kept in captivity by the author, affording by their grotesque aspect and habits many hours of enjoyable relaxation. The chief difficulty attending the maintenance of this species under artificial conditions is the very important one of their food

SPINOUS LIZARDS, SHOWING KNAPSACK-LIKE NECK EXCRESCENCES, p. 87.
W. Saville-Kent, Photo.

supply. In works in which this topic is referred to, including one of the British Museum Palæontological Guides, Ed. 1886, *Moloch horridus* is pronounced to be a vegetable feeder. As, however, obtains in the *Chlamydosaurus*

Kingi, previously described, the natural pabulum of Moloch consists exclusively of insects, but in this particular type is restricted to ants of the minutest size. The small black evil-odoured species, common in both South and Western Australia, was always a prime favourite with the specimens kept by the author, and wherever these ants abounded, in conjunction with a sufficiently warm temperature, no difficulty was experienced in maintaining these lizards in perfect health. *Moloch horridus* is by no means a rapid traveller, its utmost speed being one which is easily overtaken by a man's moderate walking pace. Consequently, no risk of losing them was hazarded by turning them loose by the roadside, or in a garden walk where a banquet of ants was readily accessible. A favourite pasture ground of some half-a-dozen specimens brought by the author from the Gascoigne district to Perth, in Western Australia, was the Government Gardens in the last-named City, many of the paths of which picturesque grounds abounded with ant-tracks. Liberated there, they would soon settle down to feeding in a row, and the number of ants an individual lizard would assimilate was somewhat astonishing. On several occasions experi-

MOLOCH OR SPINOUS LIZARDS FEEDING AT AN ANT TRACK, *p.* 86.
W. Saville-Kent, Photo.

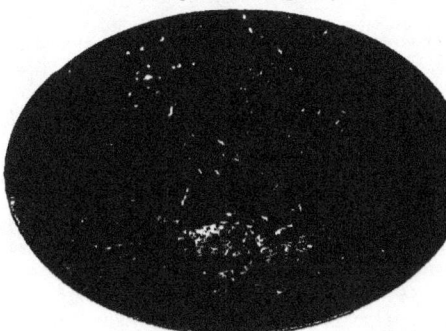

"A POST PRANDIAL PROMENADE," *p.* 86.
W. Saville-Kent, Photo.

mental reckoning elicited the fact that no less than from one thousand to fifteen hundred ants were taken in successive order at a single meal, each ant being separately picked up by a flash-like protrusion of the slender adhesive tongue. An approximate idea of the aspect presented by this little family party when enjoying a mid-day repast may be gained by a reference to the instantaneous photographs taken by the author reproduced on page 85. In the first of these illustrations all the specimens portrayed are crouched closely to the ground and absorbed in the delights of gastronomy. The second photograph serves to illustrate another interesting detail in the life habits of this lizard. Having satisfied their healthy appetites, two of the examples in this picture are indulging in a post prandial promenade, and it may be observed that they walk erect with their tails high in the air, much after the fashion assumed by contented kittens. This bizarre comportment of the caudal appendage is almost invariably exhibited when the little animals are in full marching order. Another conspicuous feature in the ambulatory gait of *Moloch horridus* is the peculiarly uncertain, vacillatory movements they often manifest at the commencement of their course, balancing themselves to and fro with one foot elevated before starting, as though they could not quite make up their minds to risk the *premier pas*, and not unfrequently repeating the action after making a short advance. This desultory mode of progress communicates to these lizards a very grotesque appearance, suggesting to the observer the idea of a mechanically wound-up toy. A somewhat similar vacillating method of perambulation is exhibited, it may be observed, by the common Chameleon.

The defensive capacities of Moloch are of an essentially low order, being limited chiefly to the passive resistance offered by the projecting spines. These thickly distributed thorn-like defensive weapons are not unfrequently of such needle-like acuteness as to readily pierce and draw blood from the hand that grasps the animal incautiously. When so handled, Moloch will also occasionally open its mouth and emit a slight hissing noise, and at the same time a blast of very offensive breath, due probably to the unsavoury odour of the ants it feeds upon. On one or two occasions the author observed these lizards comport themselves in a distinctly hostile manner. In these instances, the aggressor has rushed at and vigorously pushed and butted a comrade with its horns with a dash that would do credit to the reputation of a Liliputian buffalo. Two examples will also occasionally mutually disagree and butt at each other with much outward display of mortal enmity, but with little or no resulting damage to either combatant. As a general rule, the number of individuals kept together were most peaceably disposed. As observed of other members of the lizard tribe, they were much given to saluting one another *en passant*, when allowed their liberty, with the momentary tactile protrusion of their respective tongues. One such "Wayside greeting" is fairly well portrayed in the snapshot heading illustration to page 84.

Considerable difficulty has been experienced by the writer in keeping these singular lizards for any length of time. So long as the weather was warm and the supply of ants plentiful they have thriven well. Exposure to a colder temperature than that of the districts they frequent, however, speedily affects them, and they become lethargic and gradually pine away. In captivity the females frequently deposit, individually, half-a-dozen or more whitish eggs. These are large in proportion to the lizard's size, over half-an-inch in length, and, like those of the majority of lacertilia, invested by a tough leathery integument. Owing probably to their uncongenial surroundings, these eggs never arrived at maturity, but at the end of a few days commenced to shrivel up and decompose. So far as it has been possible to ascertain, the eggs of Moloch are naturally deposited at some little depth in the ground, and it would appear also that the lizards themselves hibernate under similar conditions. No efforts on the part of the writer to obtain very young examples of this species were rewarded with success, the most diminutive specimen received through the assistance of Dr. Williams of Carnarvon, Western Australia, being a little over three inches in length.

A very singular structural element in *Moloch horridus* that would appear to be worthy of the investigation of physiologists is the peculiar knapsack-like protuberance that is developed on the back of the neck. It is armed laterally with two large defensive spines, while many smaller ones are interspersed over its remaining superficies. The most striking aspect of this remarkable structure is afforded by the back and lateral views photographically reproduced at the bottom of page 84. In the latter of these portraits it might be further suggested that the head contour is grotesquely rhinocerotic.

Notwithstanding the difficulties attending the conservation of Moloch for a lengthened period, a specimen has been brought to, and was for a short time on view, at the Zoological Society's Gardens so recently as last year, 1895. The living lizard species that bear the strongest resemblance to *Moloch horridus* are not, as might be anticipated, natives of the Australian Continent, but denizens of California and Mexico, where they are popularly denominated "Horned Toads,"

Phrynosoma cornutum being the commonest form. Although belonging to a separate family, that of the Iguanidæ, their shape and size correspond very nearly with those of Moloch, but defensive spines of conspicuous dimensions are chiefly limited in their distribution to the region of the head. Probably, though the fact would not appear as yet to have been established, Phrynosoma, in common with Moloch, is an anteater.

A SKULL, ONE FOOT TEN INCHES IN DIAMETER, AND B TAIL-SHEATH OF *Miolania Oweni*, ORIGINALLY DESCRIBED BY SIR RICHARD OWEN AS THE FOSSIL REMAINS OF A GIGANTIC LIZARD ALLIED TO *Moloch horridus*. AFTER OWEN, p. 83.

Life-sized photographic representations of male and female individuals of *Moloch horridus*, taken by the author from living specimens, form the subject of Plate XIII. In addition to portraying the relative proportions and spinous developments of the respective sexes, these figures serve also to indicate their more ordinary colour patterns. While varying much in tint in different individuals and even in the same individual at different times, a corresponding configuration of the colour areas is mostly dominant. These, on the dorsal surface, usually consist of bilaterally arranged patches of varying shades of brown, light red, or an almost crimson hue upon a lighter ground of grey, brown, or it may be cream-colour or a brighter yellow, while in any case there is invariably a yet lighter median streak separating the right and left halves of the colour patches. A dark Y-shaped patch, with a lighter linear border, is developed on the head and there are irregular blotches of the same tint on the four limbs.

W. Saville-Kent, Photo.
A MILLINERY NOVELTY, OR "WHERE DID YOU GET THAT HAT?"

In taking leave of this interesting type, we reproduce the accompanying photographic life study for the special benefit of those among our fair readers to whom an original millinery notion may be as the breath of life. The novelty of the design, and the rare nature of the "trimmings" brought from the earth's uttermost end, may be unhesitatingly guaranteed. And, notwithstanding that grave objections might possibly be raised by paterfamilias or in ecclesiastical quarters concerning the scriptural injunction against "sacrificing our daughters to Moloch," the charms of piquant originality added to those already possessed by the fair wearer will be such as to triumphantly overcome all scruples, and to secure for her the highest mede of envy and admiration.

Among the lizard pets that proved specially amenable to domestication in the author's hands during his later Western Australian peregrinations, a prominent position must be given to the species photographically represented by two illustrations on the next page. This type, *Trachysaurus rugosus*, locally known both as the "Stump-tailed" and the "Sleepy" Lizard, is of relatively common occurrence throughout the temperate districts of the Australian Continent, the individual here figured having been obtained from the neighbourhood of Pinjarrah, W.A. The conspicuous features of this lizard are the massiveness of the head, the large size of the scales, the shortness and thickness of the tail, and the diminutive dimensions of the limbs compared with the corpulent body. Its colour varies through innumerable tints of golden, yellow, red or purple browns, relieved on the sides and under surface by marbled patterns in which a creamy or ivory white predominates. From fifteen to eighteen inches represents its normal adult length, the specimen, a comparatively young one, here photographed from life in two characteristic positions, measuring close upon one foot. The photograph forming the lower figure on that page serves particularly well to illustrate the relative small size of the scales on the under surface, the fine reticulated pattern of the surface markings, of his, so . to say, delicately embroidered waistcoat, and the grotesquely small, and at the same time uniform, proportions of the four limbs. To successfully obtain this rather peculiar pose the animal was held struggling in the folds of a rough bath towel while the camera was brought to the right focus and the shutter snapped. This article—a bath towel—it may be mentioned, forms a most admirable background when photographically immortalising animals of this description, and has, as will be easily recognised, been made to do duty on a variety of occasions for the pictorial illustrations of this volume.

When first captured this individual, in common with other examples of the same species, exhibited a highly pugnacious disposition, hissing vehemently and snapping savagely with its widely distending jaws. A bite from this lizard is, as a

M

STUMP-TAILED LIZARDS, *Trachysaurus rugosus*,
ONE-FOURTH NATURAL SIZE.

matter of fact, a by no means pleasant experience, the crushing power of the massive jaws being very considerable. In addition to this, it will retain its hold upon any subject seized with bull-dog tenacity. In their natural state, Stump-tailed Lizards are accredited with being the inveterate enemies of snakes, which they will fearlessly attack and destroy, and they are therefore naturally regarded with a considerable degree of favour by the more intelligent settlers.

The specimen here photographically immortalised, who, to distinguish him from his companions in captivity, was familiarly known as Moses, very speedily abandoned his earlier aggressive attitude and accommodated himself to civilised surroundings. Within a few days after capture he permitted himself to be handled without attempting to bite, and a few weeks later was accustomed, by way of salutation, to lick the hand presented to him with his broad blue-black flannel-like tongue. Kept customarily in a box with a variety of other lizards, including Molochs and two species of Egernia, he never manifested the slightest hostility towards them, even though the latter more especially would snap up the food he was partaking of from under his very nose. Moses' temper, in fact, as might be anticipated by a glance at his large soft-expressioned eyes, was of the most even and amiable description.

As in several allied types, the gastronomic proclivities of Trachysaurus are essentially omnivorous. The particular individual that forms the subject of this notice fed heartily and indifferently on animal or vegetable substances. Raw meat was eagerly devoured, but as substantial and welcome a repast was made off lettuce leaves, green peas, raw or cooked, and fruit of any kind. Moses' special treat, however, was garden snails, a diet to which he was a stranger on his native heath, but was introduced to experimentally at the Adelaide Botanic Gardens. This was veritable turtle in his estimation. He would crush up and swallow these gasteropods, shells and all, with the greatest gusto, and would even chase one on the floor if rolled before him, like a kitten following a cotton bobbin. It would seem probable that the presence of the calcareous shell enters extensively into this lizard's predilection for snails. Some unexpected evidence of his craving for lime was evinced on a previous occasion, when he happened, while at liberty in the garden, to stumble across a clump of roughly-dried oyster shells; these he immediately attacked, breaking off with his powerful jaws fragment after fragment which were forthwith crushed up and swallowed.

A somewhat grotesque account of the Stump-tailed Lizard is contained in the record of its discovery at Shark's Bay, in Western Australia, by Captain William Dampier, "A Voyage to New Holland," in the year 1699. It reads as follows:— "The land animals that we saw here included a sort of guanos, of the same shape and size with other guanos described, but differing from them in three remarkable particulars, for these had a larger and uglier head and had no tail, and at the rump, instead of a tail there, they had a stump of a tail which appeared like another head, but not really such, being without mouth or eyes; yet this creature seemed by this means to have a head at each end, and, which may be reckoned a fourth difference, the legs also seemed all four of them to be fore-legs, being all alike in shape and length, and seeming by the joints and bending to be made as if they were to go indifferently either head or tail foremost. They were speckled black and yellow, like toads, and had scales or knobs on their backs like those of crocodiles, plated on to the skin, or stuck into it, as part of the skin. They are very slow in motion, and when a man comes nigh them they will stand still and hiss, not endeavouring to get away. Their livers are also spotted black and yellow, and the body, when opened, hath a very unsavoury smell. I did never see such ugly creatures anywhere but here."

Trachysaurus rugosus is included among the many species of Australian Lizards that are not unfrequently imported to this country. While going to press a fine example is on view in the Reptile House of the Zoological Gardens that arrived there nearly two years ago, and between which and the foregoing description given by Captain Dampier comparisons may be instituted. While it will be at once recognised that all of the structural details respecting this singular form given by that early explorer are essentially correct, it will, it is anticipated, be conceded that a most gross injustice has been rendered the animal from an æsthetic standpoint. The photographic portraits here reproduced should suffice, indeed, to clear our client from the cruel aspersions upon its good looks implied by Captain Dampier's closing sentence.

An interesting circumstance connected with the life history of Trachysaurus is the fact that it is not only ovo-viparous, bringing forth living young in place of laying eggs, as do the majority of the lizard tribe, but the young, when born, a single one only at a birth, are so fully developed as to be nearly half as large as the parent.

A reference was made in the foregoing account of Trachysaurus to two species of Egernia which were the comrades of a captive specimen. These were the Spine-tailed lizards, *Egernia Stokesii* and *E. depressa*. The former species, Plate XIV., fig. B., which is the larger of the two, was obtained by the author abundantly on that member of Houtman's Abrolhos, off Geraldton, Western Australia, which is distinguished by the name of Gun Island, where it was caught basking in the sun or taking shelter underneath the low scrubby bushes with which the greater portion of the island is covered. The smaller species, *Egernia depressa*, which appears to be considerably rarer, was forwarded to him from Carnarvon, in the Gascoigne district by Dr. Williams. Apart from their size, which in immature specimens may be equal, the two forms may be readily distinguished by the diverse character of the pattern of their colour-markings. In the larger species, *Egernia Stokesii*, the pattern invariably consists of cream or tawny yellow spots, and

PLATE XIV.

SPINE-TAILED LIZARDS. A. *Egernia depressa*. B. *E. stokesii*. Two-thirds natural size. p. 92.

reticulations upon a dark, almost black ground, while in *E. depressa*, the spots or reticulations are black, or very dark, upon a pale pearl-grey ground. As a rule, colour characters are held by systematic biologists to be of but little value; when, however, such colour distinction is found to obtain without exception throughout an extensive series of individuals, colour marking, and more especially the pattern of its distribution, undoubtedly becomes a reliable element in specific diagnosis. These sculpturing of superficially very their scales. The closely allied contour of the species are found body of the to differ materi- smaller variety is ally in their more also, as its tech- minute structural nical name im- organisation, plies, very much notably in the more depressed.

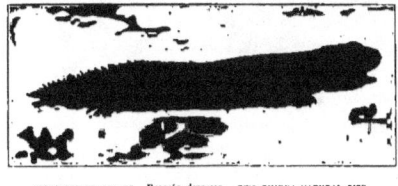

SPINE-TAILED LIZARD, *Egernia depressa*. TWO-THIRDS NATURAL SIZE.

In captivity these two species of Egernia manifested a much more shy and retiring disposition than the several lizards previously described. At all times, unless feeding, they displayed the most active desire to take advantage of the nearest covert that presented itself, and, under such conditions, much patience had to be exercised in securing the several photographic representations of the species that are reproduced in these pages. The food question with these lizards was, on their first acquisition, a troublesome one. They utterly refused to take of their own accord, animal food of any kind which, judging from the carnivorous habits of an allied species, apparently *E. Kingii*, also abundant in the Abrolhos Islands, it was presumed they would appreciate. In order to keep them alive, small pieces of meat were daily administered them by gentle persuasion, and these, on being placed well within their mouths, were contentedly swallowed. Later on, the chance placing of a fragment of vegetable material within reach of their perambulations elicited the fact that they were essentially vegetable feeders, and from that date they regaled regularly and voraciously on fruit and vegetable substances of the most varied description, lettuce leaves, however, usually proving the prime favourite.

A little episode in their pre-salad "beef-eating" days will be found somewhat quaintly represented in the photographic reproduction appended as a tail-piece to this Chapter. It portrays three specimens of *Egernia Stokesii* sitting up patiently awaiting their mid-day meal. In order to repress their otherwise too ardent tendencies to run to cover, they are carefully swaddled for the nonce in cambric handkerchiefs. The "Bob-tailed Lizard" being the local appellation by which this species of Egernia is most familiarly known in the colonies, these three worthies were, as a matter of convenience, respectively distinguished by the euphemistic titles of "Robert," "Bobbie," and "Bob."

Although there are no less than twelve other known species of the Australian genus Egernia, none of these have the peculiarly depressed, abbreviated spinous tails that characterise the two varieties here figured, and consequently possess but little to distinguish them from the more familiar lizard types. One member of this smooth, elongate, cylindrical-tailed group was observed by the author to be particularly plentiful on Pelsart Island, in the Abrolhos Archipelago, replacing *E. Stokesii*, which appeared to be exclusively represented in Gun Island, a few miles further north. This species was apparently identical with *Egernia Kingii*, previously reported from these islands, and was, as mentioned in a preceding page, essentially carnivorous in its habits. This circumstance was practically illustrated by the familiarity of an individual which had taken up its quarters in the mess-room of the camp, and which habitually came out after meals to appropriate any small bones or fragments of meat which had fallen from the table. Probably the carnivorous propensities of this species, in addition to its larger size, will account for the observed absence of the smaller vegetable-feeding *Egernia Stokesii* on this Island, in which it, *E. Kingii*, was so abundantly represented. One of the larger long-tailed species of Egernia, *E. Cunninghami*, from the Australian mainland, is on view, while going to press, in the Regent's Park Menagerie.

Among the numerous varieties of Lizards kept in temporary confinement by the writer, no others proved themselves so amenable to humanizing influences as the several types previously described and illustrated. Several examples of the Australian Monitors or Varani, locally dubbed "Gooannas," and of species pertaining to the genus Grammatophora, fell thus within the author's purview. Of one of these Monitors or Varani, *Varanus varius*, popularly known also as the "Lace Lizard," with reference to its skin markings, that came into the author's possession at Brisbane, Queensland, a little anecdote may be appropriately related. The specimen was a handsome one, adorned throughout its body and limbs after its kind with a complex reticulated pattern, and having its throat resplendent with interblending tints of sky-blue and lemon yellow. He was at the best of times a sulky animal and, though he fed well, repulsed all friendly overtures and continually strove to make good his escape from the extemporised cage provided in the garden for his occupation. One night his efforts proved successful, and after vain though patient searchings he was reluctantly given up for lost. The astonishment that was experienced ten days later may be better imagined than described, when the returned prodigal was seen in a very emaciated and dilapidated condition, struggling vehemently to regain access to his former prison-house. During his voluntary absence he had evidently fallen upon evil times, possibly been surprised, after the manner of his tribe, in robbing a henroost and so narrowly escaped the wrath of an avenging Nemesis as to receive on his hinder quarters a blow that was doubtless intended for his head, and which, had it attained its mark, would have brought its career to a speedy and ignominious termination. The fact that he returned minus his long, handsome tail, which had apparently been chopped off at the very stump, lends substantial support to the foregoing tentative interpretation. The most interesting point in the episode is the circumstance of the creature's voluntary return to captivity, associated with which he probably carried in his brain the reminiscence of a more liberal dietary than fell to his lot in the outer world.

On relating this anecdote of the truant Varanus recently to Dr. G. D. Haviland, that naturalist informed the writer that he had had a very similar experience with a member of the same genus in North Borneo. A captive specimen in his possession similarly made good its escape, and after an absence of four or five days returned to the scene of its internment, having evidently a preference for the "flesh-pots of Egypt," albeit accompanied by durance vile, to freedom and a precarious commissariat in his native jungle.

Certain of the Australian Monitors attain to very considerable dimensions, and are by no means desirable subjects to encounter at close quarters with unprotected hands. The writer possesses a skin of *Varanus varius*, previously referred to, from the Eucalyptus forests of Gippsland, Victoria, which measures over seven feet in its total length, and has claws attached that are as formidable as those of a large tiger cat. Even smaller examples, such as the one signalized in the foregoing anecdote, which, previous to curtailment, measured a little over three feet, could use its hind talons to such effect that an incautious attempt to pick it up on the occasion of the "prodigal's return," resulted in the most gruesome scarifying of the hands of the experimenter. The species of Varanus here referred to is an essentially arboreal type, preying to a large extent, in its adult state, on the opossums and their young, to whose holes, high up in the hollow gum-trees, they will lay patient siege until the poor victims, if not otherwise accessible, are starved into surrender. Birds and their eggs, insects, and even lizards of smaller size are all equally acceptable as food to these giants of their race.

Examples of the Lace Lizard are usually on view in the Reptile House at the Regent's Park, including, at the present time, specimens that measure as much as four or five feet in total length. In none of these imported individuals, however, do the bright blue and yellow tints that decorate the throat exhibit the vividness characteristic of specimens fresh from their native "bush." The individual possessed by the author being of too erratic a disposition to be trusted to sit for his portrait in the open, the best had to be made of the opportunities afforded of photographing him through the wire netting of his cage. The most satisfactory of these attempts, reproduced in Plate XIV., suffices, notwithstanding the intervening wirework, to impart a very fair idea of its most characteristic attitude, the natural involute coiling of the tail exhibited by the creature when in complete repose being especially noteworthy. Their tail constitutes with the Monitors a very formidable weapon. It is frequently longer than the body compressed from side to side, and as tough as leather. Independently of its armature of teeth and claws, the animal can by virtue of this appendage transform itself into a sort of animated stockwhip and severely punish incautious aggressors at close quarters. An attendant at the Zoological Gardens was assailed in this manner and had his neck severely lacerated by the tail of one of these reptiles while cleaning out its den, a short while since. Varani skins are, it may be mentioned, extensively utilised nowadays for the manufacture of purses and other "fancy leather" or "Lizard Skin" articles.

There is a species of Monitor, *Varanus giganteus*, allied to the arboreal variety, *V. varius*, that is accredited with attaining to even larger proportions, a length of as much as eight feet having been recorded of specimens from the northern territory of South Australia. The habits of this species are, however, very distinctly amphibious, on which account it was originally referred by Gray to the genus Hydrosaurus. *Varanus*, or *Hydrosaurus salcator*, is another huge Monitor possessing similar amphibious habits and attaining to like dimensions, which, in addition to inhabiting Queensland, is met with in India and throughout the Malay Archipelago. A remarkably fine example of this Monitor, shot by Lieutenant Stanley Flower at Singapore, has been recently presented by him to, and has been admirably set up in, the Zoological Galleries of the British, Natural History, Museum. Excepting to those who may take delight in the society of full-grown crocodiles and alligators, the recommendation of either of the foregoing Australian Varani as household pets would be a matter of supererogation. Large as are the dimensions of these living Monitors, the Australian tertiary deposits have produced conclusive evidence of pre-existing species which very much exceeded them in size. Certain of these, which have been referred to the genera Notiosaurus and Megalania, would appear to have attained to a length of no less than from fourteen to eighteen feet. A relatively small and handsome species of Monitor, technically known as *Varanus acanthurus*, or the Spiny-tailed Monitor, occurs in tolerable abundance in the north-west district of Western Australia, under conditions corresponding with those which yield the Frilled Lizard, Chlamydosaurus. An example of this Monitor was obtained for the writer at Roebuck Bay, and was also brought to England and presented by him to the Zoological Gardens. As with the species kept in Queensland, it always manifested a hostile attitude, attacking and worrying the Chlamydosauri, with which it was at first associated, and repelling all friendly overtures from its keeper. Its arrival in England and participation in the polite society of an innumerable company of other lizards, does not appear to have exercised any ameliorating influence upon this creature's temper, which is as short and uncertain as when it was first captured, now over twelve months ago.

A noteworthy feature in the Varani is their possession in a more readily recognisable degree than in any other lizards of that vestigeal structure known as the "pineal eye," a minute functionless median optic organ situated on the top of the head, and intimately connected with the so-called pineal gland. In bygone ages this median eye possibly represented the chief or only optic organ of a doughty race of reptiles to whom the title of the Cyclops would have been both figuratively and literally appropriate. Some admirable dissections of this structure, made by Professor Charles Stewart, F.R.S., may be seen in the Museum of the Royal College of Surgeons.

One of the rarest and handsomest lizards at the present time on view in the Zoological Society's Menagerie is undoubtedly the Australian type *Physignathus Leseuri*, known as Leseur's Water Lizard, from New South Wales. This example is over two feet long, with a fine massive head and particularly large intelligent eyes, having their lids delicately ciliated like those of Chlamydosaurus, which, in many respects, allowing for the entire absence of a frill-like membrane, it considerably resembles. As its name implies, its habits are semi-aquatic, and it displays a marked appreciation of the ample water supply provided for it in its allotted cage. Among the more commonly imported lizard types are the Australian Cyclodi, or Blue-Tongued Lizards, *Tiliqua (Cyclodus) gigas* and *T. scincoides*. Several examples of these species are almost always to be seen at the Zoological Gardens, and are frequently procurable from the animal dealers. Except for its relatively slender head, small scales, and tapering, but not elongate, tail, *Tiliqua scincoides* bears a considerable resemblance to Trachysaurus, the body being thick and the four limbs small in proportion, and closely resembling one another as in that species.

The data, chronicled in this Chapter, concerning a few of the more remarkable of the Australian Lizards, will, the author hopes, secure for this hitherto much neglected and to a large extent unjustly maligned animal group a more equitable share of that interest and sympathy which is so freely bestowed upon other races. To many, even among the highly educated classes, the very name of reptile is held to be synonymous with a creature of repulsive aspect and accredited poisonous properties. As a matter of fact, there are but two known poisonous lizard species. These are the Helodermæ, *H. horridum* and *H. suspectum*, which inhabit Mexico. The Monitors or Varani are, as already testified, for the most part unamiable, and, from their size and redundant equipment of teeth and claws, capable of serious aggression. The great majority of the lizard tribe are, however, eminently susceptible of domestication, speedily losing their initial antagonism induced through fear of man, and becoming attached to their owners. For singularity of form, if not for brilliancy of colouration, this animal group possesses few, if any, rivals, while even in this last-named relationship it yields many remarkable exceptions. Birds or fish, by way of example, are rarely more beautifully coloured than the South European Ocellated Lizard, *Lacerta ocellata*, a species some fifteen or sixteen inches long, that is more or less frequently imported by the animal dealers. The ground colour of the body in this lizard is usually a most brilliant green, and has been likened to an armour of emeralds, while the sides are ornamented with spots of the richest turquoise or azure blue. The smaller Green or Jersey Lizard, *Lacerta viridis*, which is so abundantly imported from the Channel Islands, affords another familiar instance of notable lacertilian colouration which might be indefinitely multiplied.

That the Lizard tribe is rapidly coming to the front as one well worthy of the attention of amateur zoologists, is attested to by the increasing numbers in which they are now imported into England and purchased from the dealers by private individuals. Geckos, Skinks, Chameleons and many other lizard types are indeed commonly advertised with other imported pets for sale in the columns of the " Field " and kindred sporting papers. Several varieties of lizard pets at the present time in the writer's possession were obtained through such media. Included in this collection is a handsome grey, red-striped Skink, *Eumœus algeriensis*, from Mogador; a couple of specimens of the Ocellated Lizard, *Lacerta ocellata*; two or three Egyptian and other Geckos; and an example of the specially interesting New Zealand or Tuatara Lizard, *Sphenodon punctatus*. All of these types are essentially carnivorous and thrive on a diet of beetles, worms, snails and every other description of "small game" commonly accessible within a garden boundary. Apart, in fact, from their intrinsic interest, lizards can be most highly recommended as a companion hobby to the horticulturist, to whom probably no better additional stimulus towards keeping his grounds clear of insect and kindred pests could be provided than the pampering of the somewhat fastidious appetites of his reptilian pets. Earwigs, caterpillars, cockchafers, woodlice, *et hoc genus omne*, will at all times prove a most welcome *bonne bouche* to some one or other of his interesting pensioners. The maintenance of their larder in an efficiently replenished state will be found at the same time to exert a most beneficial influence towards the permanent reduction of the ranks of these garden enemies.

The New Zealand Lizard, *Sphenodon punctatus*, may indeed be recommended as a most useful auxiliary in the garden, independently of any natural history claims. The specimen in the writer's possession has shown itself to be partial not only to worms, beetles, caterpillars, and all manner of insects, but it has displayed a most distinct penchant for slugs, and will attack and devour with the greatest gusto the huge field slugs, *Limax ater* and *Arion rufus*. Sphenodon being to a large extent nocturnal in its habits, it is abroad and on the *qui vive* at the season commonly elected by these gasteropods for their depredations. It has been proved, moreover, at the Zoological Gardens that this lizard is quite capable of withstanding our English climate without any protection beyond its own earth burrow, within which it retires by day and hibernates during the winter months.

Much of the increasing popularity of the Lizard class has no doubt been brought about through the recent addition of an admirably organised Reptile House to the Zoological Society's Menagerie, in the Regent's Park. To promote the study and dissemination of knowledge of and interest in all varieties of reptiles, including snakes as well as lizards,—at which many will doubtless elect, as is done in this Chapter, to draw the line,—an association, styling itself the "Reptilian Society," has been recently established under the able Presidentship of Dr. Arthur Stradling.

There is also lying on the author's table at the present moment the first number of a new magazine, the "Vivarium," which is being started as the recognised organ of the Reptilian Society. This serial, as is notified in its introductory chapter, will seek to supply information of a more practical nature than has been hitherto available for those who make reptiles their particular hobby and desire fuller knowledge concerning the habits and treatment of their adopted favourites. While this earliest number of the "Vivarium" has been produced, through the talented energies of the Society's Secretary, the Rev. J. E. Spain, of the Rand Rectory, Wragby, in a most creditable and artistic manner by a lithographic process only, it is hopefully anticipated that a sufficient number of supporters of the Society and those subjects which the periodical specially advocates, will be forthcoming to justify the early advancement of this magazine to the dignity of print. The author experiences an especial pleasure in directing the attention of those who may elect to adopt for their hobby the very interesting subject of reptile life, to the facilities and advantages provided for their particular benefit by the Society and Journal named.

"TIFFIN AHOY ! !" p. 93.   W. Saville-Kent, Photo.

PLATE XV.

CHAPTER IV.

TERMITES (WHITE ANTS).

IN comparison to their individual size, the only terrestrial animals, probably, that by their concerted efforts produce anything approaching the colossal fabrications of the Coral Polyps, are the so-called White Ants or Termites. In this hypothetic parallelism it is of course not suggested that the organisms and their correlated products possess any homologous relationship, since, while the Corallum or Coral Mass is literally the internally secreted skeleton of the associated Polyps, in the case of the White Ants the huge fabrication is purely external or architectural.

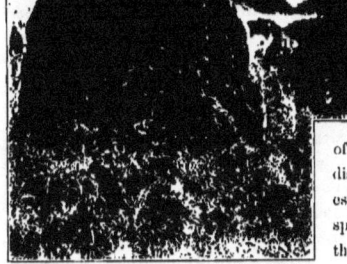

W. Saville-Kent, Photo.
TERMITE MOUNDS, ALBANY PASS, NORTH QUEENSLAND.

The very considerable diversity of both size and shape that distinguishes the mounds or termitaries constructed by different species of White Ants is familiar to most travellers in tropical climates. In the northern territories of the Island-Continent of Australia some highly remarkable distinctive types are found, and it is with especial relation to a few of the more conspicuous of these, which have fallen within the writer's observation, that the present Chapter has been written.

For the information of those non-scientific readers, who are not conversant with the classificatory systems adopted by entomologists for the distinction of the leading groups and orders of the insect world, it may be stated that the so-called White Ants, or Termites, are not true ants in the scientific sense of the term. For, while the typical ants belong to that order, the Hymenoptera, which notably includes the wasps, bees and ichneumon-flies, the White Ants have, up to within a recent date, been regarded as modified representatives of the Neuroptera, an order typically illustrated by the Dragon-flies, Ant-lions and Lace-wing-flies. According to the latest researches many entomologists regard them as representing an independent group, occupying a structural position midway between the Neuroptera and the Orthoptera or Cockroach and Cricket tribe.

As with the Hymenopterous groups of the Bees and Ants, the Termites form large social communities. These, however, are frequently made up of a greater number of specially modified individuals than obtains in either of the two above-named tribes. As demonstrated by the investigations of Dr. Fritz Müller with reference to the European species, *Termes lucifugus*, no less than eleven diverse individual types may be found inhabiting the same nest. These include : 1, the youngest larvæ, which in their earliest condition present no recognisable distinctions; 2, the semi-matured larvæ of the soldiers; 3, adult soldiers; 4, semi-matured larvæ of workers; 5, adult workers; 6, nymphs (with imperfect wings) of the first order developing to kings and queens; 7, king; 8, queen; 9, nymphs of the second order developing to supplementary males and females; 10, adult supplementary males; 11, adult supplementary females. It would appear, however, that all of these individual types are not developed in every Termite community. In some of those examined by Dr. Müller a single presiding king and queen were wanting and their place supplied by a number of supplementary males and egg-producing females.

According to the more recent researches of Grassi it is possible to recognise as many even as fifteen distinct individual modifications; the four added to Fritz Müller's list including extra substitution forms of kings and queens or so-called "neoteinic" individuals in both their immature and matured conditions.

The woodcut illustrations reproduced on the next page, chiefly from M. Grassi's Memoir, portray all of the more important of the individual types recognisable in a Termite community. As a rule, on breaking into a large termitarium or laying open the Termites' excavations in wood or other substances, the only individuals conspicuously represented belong to the two categories of the ordinary workers and

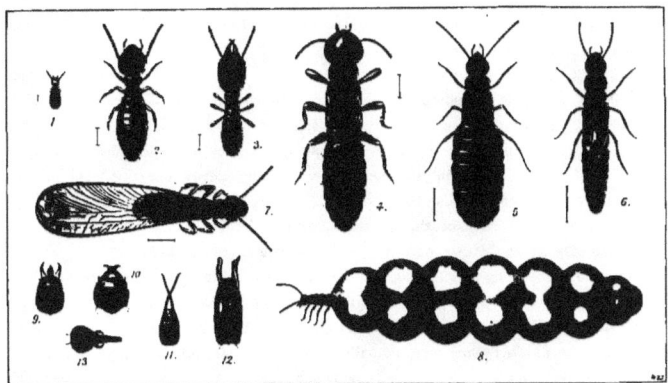

FIGS. 1 TO 7, WORKERS, SOLDIER, NYMPHS, WINGED MALE AND MATURE FEMALE OF THE EUROPEAN WHITE ANT, *Termes lucifugus*, MUCH MAGNIFIED, AFTER M. GRASSI; FIG 8, GRAVID QUEEN OF *Termes bellicosus*, AFTER SMYATHMAN, NATURAL SIZE; FIGS 9 TO 13 MODIFICATIONS OF HEADS AND JAWS OF SOLDIER INDIVIDUALS OF VARIOUS SPECIES, MAGNIFIED, AFTER HAGEN.

WOOD-DEVOURING WHITE ANTS, *Termes sp.*, OF THE KIMBERLEY DISTRICT, WESTERN AUSTRALIA. TWO-THIRDS NATURAL SIZE. FROM AN INSTANTANEOUS PHOTOGRAPH. AT + + TWO SOLDIER INDIVIDUALS, *p.* 104.

soldiers, and of these orders the workers are usually present in preponderating numbers. An instantaneous photograph taken by the author at Derby, in Western Australia, of such a disturbed community, including in the field of view over one hundred individuals, is reproduced on the lower half of the page above quoted. The species here portrayed two-thirds of life-size represents one of the most destructive species in tropical Australia. It erects no mound, but lives in subterranean passages, or in chambers excavated in the wood that chiefly constitutes the field of its depredations. Among the number delineated, two soldier individuals may be easily recognised, in the vicinity of the + marks, by the darker tint of their bodies, and more especially by the larger and almost black colour of their heads. By way of illustrating the insidious and thorough destruction that is wrought to timber attacked by this species of Termite, the writer has in his possession the portion of a dead plank taken from an outhouse at the Derby Government Residence, from which the specimens here photographed were actually evicted. To all outward appearances the plank is perfectly sound and solid. It can, however, be easily crushed to pieces in the hand, the entire interior being eaten away, leaving nothing but an external shell or veneer of paper-like tenuity.

It is a common phenomenon to all dwellers in White Ant countries that, at that season of the year which, in the tropics, immediately precedes or ushers in the rain monsoon, swarms of the winged individuals make their appearance, and at night so crowd to all accessible house lights as to constitute a veritable nuisance. Of the American species, *Termes flavipes*, it has been recorded by Hagen that, at the swarming- season in Massachusetts, these winged Termites form a thick cloud, accompanied by no less than fifteen species of birds, which gorge upon the insects to such an extent that they are unable to close their beaks.

These swarming White Ants are in all cases the matured males and females developed from the previously described nymphs. Of the countless thousands, or it might be said millions, that emerge from the termitaries on these occasions, but a few stray units fulfil the main function of their existence and become the parents of new communities; by some authorities it is even contested whether there are any survivors. The advent of the emerging Termite swarms is a day of great rejoicing to the host of expectant banqueters that are awaiting to devour them. These include innumerable species of ordinary ants, carnivorous insects generally, spiders, birds, lizards, and other animals, and in some countries even man himself. In India, by way of example, as testified to in Smeathman's Memoir, hereafter quoted *in extenso*, the swarming

NEST MOUNDS OF WHITE ANTS, KIMBERLEY TYPE, DERBY, WESTERN AUSTRALIA. p. 119.

termites are systematically collected for the purposes of food immediately prior to their emergence from their nests. Mixed with flour, they are made into various descriptions of pastry, which is sold at a low rate to the poorer classes. It would appear that a too abundant use of this description of food occasions a form of colic and dysentery that is speedily fatal. In West Africa, Smeathman deposes, the natives content themselves with securing those of the swarming millions that have fallen upon the surface of neighbouring water. These they skim off with a calabash, carry to their habitation in kettles-full and parch them in iron pots over a gentle fire, stirring them continually after the manner of roasting coffee. Thus prepared, and without any additional materials, they regard them as the most delicious food, putting handsful into their mouths at a time, much as a European youth might devour comfits. Smeathman's personal attestation as to the edible properties of the West African Termites is highly favourable. He speaks to having frequently partaken of them prepared in the manner above described, and to having found them delicate, nourishing, and wholesome; something sweeter than, but not so fat and cloying as the grub of the Palm-tree beetle, *Curculio palmarum*. Other gentlemen of Smeathman's acquaintance acquiesced in their being most delicious eating, one comparing them to sugared marrow, and another to sugared cream and a paste of sweet almonds.

It is a remarkable fact that, so far as the author has been able to ascertain, Termites are not turned to account as an article of diet by the natives of Australia. On the other hand, it has been observed by him in the Kimberley district of Western Australia that the aborigines are in the habit of devouring large quantities of the earthy substance of the White Ant mounds or termitaria. Frequently, in the course of walking expeditions, the author has seen the natives step aside and break off and eat a handful of this substance, usually from a hillock which the termites had abandoned or from which they had died out, and which was consequently in a more or less disintegrated state. This earth-eating propensity was not analogous to that recorded by Humboldt of the Otomac tribe on the Orinoco, for the purpose of alleviating the pangs of hunger in the absence of better food. In this case the partakers were native retainers of Australian settlers, who had an abundance of provisions at command. As White Ant hillocks or termitaries contain a large amount of proteaceous matter in the form both of secretions by their constructors and of adventitiously growing microscopic fungi, it would appear probable that it is the presence of these materials that makes the component earth palatable to the native taste. A chemical analysis of different descriptions of termitaria with the object of ascertaining the correctness or otherwise of this anticipation might yield interesting results.

When, from among the emerging annual swarms, a matured pair of White Ants have run the gauntlet of their many foes and meet together, it usually happens that after a short preliminary toying in the air they alight upon the ground again, but before the marriage tie is consummated are discovered by the outlying workers of some other community, who, adopting them for their future king and queen, thereupon enclose them in a clay chamber around which a new nest is built. In this nest they are thenceforth immured, assiduously fed, and attended upon for life. It has been ably argued by Fritz Müller, that, notwithstanding the apparent waste of life associated with the periodical swarming of the Termites, it is only through such means, as is cogently advocated in accord with the tenets of the Darwinian theory, that the very desirable cross-fertilisation of these insect races can be effectually carried out.

White Ants possess, and correctly so far as a considerable number of species extend, a most unenviable notoriety for their terribly destructive, and more especially xylophagous, or wood-eating, propensities. Although the softer woods, such as European deal, are, where such choice exists, preferentially selected by them, scarcely any description of wood, not excepting even the world-famed Western Australian Jarrah, would appear to be absolutely proof against their depredations. According to the observations of Mr. R. C. Hare, sometime Government Resident at Wyndham, in Cambridge Gulf, in which township White Ants are conspicuously destructive, a considerable amount of difference is exhibited by jarrah wood grown in different neighbourhoods with regard to its being proof or otherwise against White Ant ravages. Only such wood as has been derived from the ironstone ranges, or from a mineral district is, it would appear, absolutely ant-proof. Blocks of the same wood grown in other localities and exposed by way of test beside it are speedily attacked. The author has been informed, on the other hand, by the above-named gentleman that the wood of the Cypress pine, indigenous to tropical Australia, resists the attacks of the White Ants for a considerable time, and is on this account chosen in the northern districts for the construction of survey posts and pegs. The fact that Termites can corrode metal and even glass by means of some special secretion has been attested to by Hagen, who correlates the property with glandular structures, situated near the insect's rectum. At Jamestown, in St. Helena, and also in the bonded stores at Wyndham, Western Australia, the metal capsules of wine and beer bottles have been perforated by White Ant agency, the corks destroyed and the liquid wasted. Where the covered passages of the Termites passed over the surfaces of the bottles it was observed by Mr. Hare that the surface was distinctly eroded as though it had been brought in contact with a revolving emery-wheel.

This phenomenon of the erosion of glass by the excretions of the Termites goes far towards the explanation of the remarkably indurated character of the termitaria of the majority of the mound-constructing species. As recorded further on, the food chambers of these termitaria are almost invariably found filled with short lengths of dried grasses, which apparently constitutes the main diet of these species. Those internal juices of the Termites, which on excretion erode a glass surface, are, it may be anticipated, capable of dissolving the silica contained so abundantly in the substance of the grass stalks eaten, and are, in a soluble state, ejected and mingled, as a temper, with the earthy material that constitutes the bulk of the hillocks.

The corrosive action of the fluid secretions of the Termites, as observed with reference to glass, is manifested in a yet more marked degree when the softer metals, such as tin and lead, are attacked. The circumstance of the metal capsules of wine bottles being perforated by these insects, apparently with the direct object of reaching and devouring the woody substance of the corks, has been already referred to. Instances fell within the author's observation at Port Darwin, in the Northern Territory of South Australia, in which lead sheeting, of considerable thickness, had been perforated by White Ants in order to enable them to gain access to a supply of coveted wood that lay beyond. To guard against White Ant depredations, the houses throughout tropical Australia, and in Queensland, even so far south as Brisbane, are to a large extent isolated on hard-wood or brick and cement pillars, having zinc or other metal cappings to prevent the Termites' ingress. Where these pillars are of wood a constant watch has to be kept to check the earliest indications of the insects' advent.

Should these precautions be neglected and the house perhaps be left tenantless for a few months' interval, the owner may on his return make a most unwelcome discovery. A clay-covered gallery half an inch or so in width up one of the supporting pillars is the only outward and visible sign of the enemy's presence. Appearances inside may at first sight be even less conspicuously premonitory of the ruin that has been accomplished. Stepping into the first room, however, the foot may crush through the flooring boards, the door posts and lintels may be found hollowed to a shell, and even the legs of the chairs and tables, while apparently sound to the external view, may be so tunnelled interiorly as to crumble into dust with the slightest pressure.

Australia is commonly alluded to as a land of "contrarieties." In the matter of Ants the scriptural proverb will allow of material modification. The sluggard is therein counselled to "go to the Ant." In Australia a slavish obedience of the prophet's injunction would be altogether a work of supererogation. The Ant, in that happy land, comes to, or goes for, the sluggard!

It is commonly supposed that it is the fabricators and inhabitants of the huge hillocks or termitaria, which are so conspicuous a feature in many tropical districts, that are the authors of the ravages that have been so abundantly attested to. So far, however, as the writer's experiences have extended, this by no means applies to the Australian species. These, as presently shown, differ very materially with relation to the dimensions and contours of the termitaria they construct, but in no instance among the many forms observed and here placed on record were they found to contain that specially destructive variety of Termite whose depredations are so conspicuous throughout the northern territories of Australia. This species, which appears to be closely allied to the Indian *Termes taprobanes*, is represented by the photograph of a living colony of some hundred individuals, previously referred to, and reproduced on Page 103. This type of White Ant is never found inhabitating any of the several descriptions of termitaries characteristic of the same districts, even though it may infest dead or decaying timber almost in contact with them. These notoriously destructive and essentially wood-eating Termites are, in point of fact, subterranean in their habits, living and breeding in underground galleries, whence they construct mines and tunnels to the immediate scene of their depredations. On the other hand, the huge mound constructors are in Australia, so far as observed by the author, exclusively graminivorous, collecting and hoarding up in their innumerable provision chambers vast stores of finely-cut up grass fibre.

It is a somewhat singular circumstance that the marked diversity of size, form, and correlated peculiarities that distinguish the habitations of the different species of White Ants throughout the tropical areas of the world's surface, have hitherto received such scant scientific attention. It will further surprise many readers, probably, to learn that up to the present date a paper contributed over a century ago (1781), by Henry Smeathman, to the Transactions of the English Royal Society,

with relation to certain nest or mound-building African species, constitutes the standard account, and that his illustrations are reproduced, with trivial variations, in most modern zoological textbooks and other popular narratives of White Ants and their fabrications.

This remarkable hiatus in our knowledge of a very important insect group is fortunately on the eve of being filled in by the labours of several independent workers. Dr. J. D. Haviland, more especially, has in preparation the account of his recent extensive investigations of the Termitidæ of Singapore and South Africa, from which widely-separated regions he obtained no less than eighty-one distinct specific types, a magnificent addition to the small list of less than one hundred previously recorded species. Dr. Haviland was further fortunate in bringing certain of these forms alive to England, and produced a notable sensation by exhibiting them in conjunction with his preserved examples at one of the past year's scientific meetings of the Linnæan Society of London. A reference to Dr. Haviland's original observations concerning certain of the species whose habits he investigated is included in a future page.

The Termite fauna of Australia will probably prove on examination to be especially rich in the number and variety of its specific types. Little is attempted in this Chapter beyond the portraiture and a descriptive outline of the leading modifications of the nest-mounds or termitaria of the species peculiar to the eastern, western and northern districts of the tropical regions of this Island-Continent. The systematic identification and classification of the Australian Termitidæ has been already commenced by Mr. W. W. Froggatt, of the Sydney Technological Museum, and as soon as his investigations are more advanced, it will be an easy matter to correlate the typical mound-forms figured in this volume with the technical nomenclature of the special White Ant types that build them. So far, as a preliminary instalment towards a comprehensive Monograph on this subject, Mr. Froggatt has contributed to the "Proceedings of the Linnæan Society of New South Wales" for the year 1895 a paper which constitutes a general survey of the distribution of the hitherto known species and a special reference to a mound-constructing type distinctive of New South Wales. No technical name, however, is associated with either this or any other Australian specific forms that are less extensively referred to. The communication by the author to Mr. Froggatt of notes and sketches relating to the termitaria here figured has served only to elicit the fact that their constructors are apparently in all instances specifically new and undescribed, and await the appearance of Mr. Froggatt's projected

Monograph for the relegation of individual titles. It is at the same time by no means improbable that some of the North Queensland types, more especially, may prove to be specifically identical with species obtained from Singapore and the Malayan region by Dr. Haviland. The transport of these insects in floating driftwood or even by indirect human agency on board intercommunicating trading vessels would be easy of accomplishment, and has, as hereafter recorded, actually occurred in the case of areas far more remotely separated.

Various authorities, including Lespes, Quatrefages, Fritz Müller, and, most recently, Drs. Battesta Grassi and Andrea Sandias, have done much towards elucidating the domestic economy, and developmental phenomena of the respectively indigenous and acclimatised European species *Termes lucifugus* and *Calotermes flavicollis*. These species, however, are not mound-constructors, but live and breed in subterranean chambers and passages, or in galleries excavated within the dead wood or standing timber upon which they feed. The destructive powers and propensities of these two species have so conspicuously manifested themselves in certain districts in Sardinia, Spain, and the South of France as to compel scientific attention, if only with the object of checking their depredations. Of the two species mentioned, *Calotermes flavicollis* would appear in these countries to concentrate its energies upon the destruction of olive and other valuable fruit trees, and *Termes lucifugus*, in like manner, to devour the oaks and fir trees. It is, however, the first-named of these two species that has won for itself so wide a notoriety on account of its destructive inroads upon human habitations. At Rochefort, La Rochelle, Sainte, and other townships in the department of the Lower Charente, the woodwork and furniture of public and private buildings have been invaded and destroyed to an enormous extent, and the utmost difficulty has been experienced in effectually contending against their ravages. According to the evidence submitted in M. Quatrefages' Memoir on the subject (*), this house-invading Termite appears to have been originally imported to La Rochelle with ship's cargo from Saint Domingo, in South America, so long since as the year 1780, and from that centre to have been distributed to the neighbouring townships. *Termes lucifugus*, on the other hand, appears to be an indigenous and, excepting for its attacks upon timber in the open country, a retiring and relatively harmless species. According to a more recent contribution upon this subject written by the late W. S. Dallas in Cassell's Natural

* "Rambles of a Naturalist," Vol. II., p. 346, 1857.

History, 1883, a third species, *Termes flavipes*, found in Portugal and the south of France, was originally introduced from North Africa.

Still more recently (1870), a very destructive species of White Ant, *Eutermes tenuis*, has been accidentally introduced into St. Helena with a captured slave ship from tropical America. Such were the ravages which this species committed on that Island, that in Jamestown, the capital, the larger number of the buildings were destroyed and had to be rebuilt, the damage to property being reckoned at the lowest estimate to amount to no less than £60,000. Some of the results of the destructive work of this termite are thus narrated by Mr. J. C. Melliss in his book "St. Helena," 1875. "It was a melancholy sight five years ago, to see the town desolated as by an earthquake or, as a visitor remarked, by a state of siege—the chief church in ruins, public buildings in a state of dilapidation, and private houses tottering and falling, with great timber props butting out and meeting the the eye at every turn. Books and records in the library and Government offices were destroyed, and merchandise of every consumable description devoured in the warehouses." Among other remarkable phenomena chronicled as having been brought about at Jamestown through this termite agency was the very singular spectacle of a large Margossa tree, *Melia azederach*, in full foliage, which, without any previous warning, and to the great discomforture of two native policemen who were standing near it, was seen to sway, totter, and suddenly fall to pieces. On examination it was found that the inside of the tree was completely eaten away, leaving only a thin external shell. *Eutermes tenuis* agrees with the majority of the more destructive species in constructing no conspicuous nest or mound, but lives and breeds in subterranean chambers and galleries.

It has been remarked on a previous page that the account and figures given by Henry Smeathman of the termitaria and habits of certain West African Termites so long ago as the year 1781, constitute up to the present date the most complete record extant concerning the tropical nest-building forms. As an indication of the good work in the same direction that awaits accomplishment with relation to the internal architectural details of the edifices, as well as the life phenomena and habits of the hitherto unstudied Australian species, Mr. Smeathman's Memoir may be advantageously quoted at some length. Briefly summarising the results of Mr. Smeathman's investigations, we find that they include the descriptions and figures of three nest-building Termites, and also a reference to a fourth variety, the Marching Termite, remarkable for the recorded circumstance that, unlike all other known species, it travels in processional order above ground and in broad daylight. Concerning the nest or termitaria-constructing species, the form he distinguished by the title of *Termes bellicosus* is undoubtedly the most remarkable. This Termite constructs clay hillocks ten or twelve feet high of an obtusely conical contour, supplemented by a variable number of smaller conical pinnacles or turrets which spring from around the base, and from the general surface-area of the central cone. According to Mr. Smeathman, these termitaries make their first appearance above ground in the form of one or two, or more, sugar-loaf-shaped turrets of a foot or so in height, which through constant increase in size and numbers, become finally amalgamated.

In vertical section the nest mound of *Termes bellicosus* is depicted as consisting of an outer shell averaging a foot in thickness, and traversed throughout its height by a wide spirally ascending tunnel. The Royal Chamber containing the abnormally inflated queen, which in this species, see p. 103, fig. 8, measures as much as three or four inches long, by one inch thick, and also the king, is situated in the immediate centre of the hillock, at the same level as the surface of the ground. Around this royal cell are clustered the innumerable waiting chambers, or so-called Royal Apartments occupied by the labourers and soldiers which wait upon and guard the king and queen. These form an intricate labyrinth which extends for a foot or more in diameter from the Royal Chamber on every side. Here the nurseries and magazines of provisions begin, which, intercommunicating by other empty chambers and galleries, are continued on all sides to the outer shell.

It is remarked by Smeathman that the dimensions and composition of the nurseries of this species, *Termes bellicosus*, differ materially from those of the other chambers, being not only much smaller, but composed of woody material in place of clay. In addition to their eggs or infant brood, they were always found to be slightly overgrown with minute fungoid growths, upon which, according to collateral observations made by M. König of an East Indian species, it would appear the infant Termites are fed. The contents of the magazine chambers proper were found to consist of what, at first sight, appeared to be raspings of the wood and plants which the Termites customarily destroy, but on nearer investigation, proved to be composed of various descriptions of gums or inspissated juices.

The complex mass of nurseries and provision magazines in the nest-mound of *Termes bellicosus* is described as extending vertically for some two-thirds or three-quarters of the entire interior space, leaving a more or less considerable hollow area between its roof and that of the dome-shaped outer shell for their extension and multiplication. This is referred to by Smeathman as somewhat resembling a cathedral nave, the similarity being further heightened by the presence of three or four large Gothic-shaped arches, sometimes two or three feet in height, which give additional support and strength to the dome-like roof. There is also a considerable interspace in the centre of the main mass of nurseries and provision magazines, with groined lateral walls and intercommunicating bridges, which appear to be for the purpose of shortening the distance through which the newly deposited ova have to be carried from the royal chamber to the upper nurseries. From the basement of the termitarium cylindrical subterranean passages, lined with the same clay of which the tenement is composed, run in every direction to vast distances, some of these at their commencement being found by Smeathman to measure no less than thirteen inches in diameter.

According to Smeathman's descriptions the soldier individuals of the communities of *Termes bellicosus* are by no means pleasant antagonists to encounter at close quarters. If, on breaking open the nest, the aggressor remains near enough for them to reach him, the soldiers will attack with such fury as to make their hooked jaws meet in the flesh, drawing as much blood as is equal in weight to their body and allowing themselves to be pulled to pieces rather than quit their hold. If left undisturbed, the soldiers shortly retire and the labourers then appear upon the scene in crowds to repair the breach. This they do by each bringing with it a load of tempered clay, which it immediately places in position then retiring for more. The hundreds, or, it might be said, thousands, who thus contribute towards the repairing work, soon, unless some of the main chambers or avenues have been exposed, close up and obliterate all traces of the passages laid open. During the progress of these repairs, the few soldiers that remained to supervise the work were observed at intervals of a minute or two to raise their heads and strike the walls of the building in such manner as to make a distinct vibrating sound. This noise was immediately answered by a loud hiss which apparently came from all the labourers, who immediately addressed themselves to their task with re-doubled energy. Should the onslaught on the termitary be renewed the workers all vanish again within a few seconds and the soldiers rush out in numbers to do battle with the enemy as valiantly and vindictively as before.

Within a comparatively recent date the account given by Smeathman of the nest mounds and habits of *Termes bellicosus* have been confirmed in all important points by Mr. T. J. Savage, who has recorded the results of his investigations of this species in the Annals and Magazine of Natural History for the year 1850. Savage,

P

however, considers Smeathman to have been mistaken in attributing to this form wood-destroying properties which he correlates with a smaller species more nearly resembling, if not identical with, *Eutermes arboreum*.

The second form of White Ant hills or termitaria described by Smeathman is referred to by him as being the joint work of two species of Termites, upon which Dr. Solander had conferred the respective titles of *Termes mordax* and *T. atrox*. In each instance the nest consists of perpendicular cylindrical columns about three quarters of a yard high, consisting of very firmly consolidated black brown earth or clay and surmounted by an overlapping conical roof of the same material, which imparts to them an aspect comparable to that of gigantic mushrooms. The substance of these termitaria is described as being of such rigid consistence that the whole structure can be more easily uprooted from its earth foundation than fractured across the centre of the cylindrical column. It was further observed of examples that had become thus accidentally overturned, that the inhabitants commenced the construction of a new column, vertically and at right angles from the prostrate one. The internal structure of these so-called turret-nests are described by Smeathman as presenting none of the systematic complexity possessed by the hillocks of *Termes bellicosus*, consisting entirely of innumerable cells of irregular shape, each of which possessed two or more entrances by which it communicated with its neighbours. After one of these turret-nests is finished, it is not further altered or enlarged, but another column is constructed within a few inches of the first. A group of half-a-dozen or more of these mushroom-like termitaria is described as being often seen at the foot of the trees in the thick woods.

The fourth variety of West African nest-constructing Termites, figured and described by Smeathman, is distinguished by the title of *Termes arboreum*. As its name implies, it builds its nest on the arms or stems of trees, sometimes at the considerable height of seventy or eighty feet from the ground. This nest or termitarium is usually spherical or oval in shape, and may be as large as a sugar cask. The substance out of which it is composed differs from that of the hillock and turret-shaped varieties. It consists, in place of clay, of minute particles of wood, combined with the gums and juices of trees, and built up into innumerable little cells of irregular shape and size. These nests are described as being so compact that there is no detaching them, except by cutting them in pieces or sawing off the branch upon which they are built, and to which, indeed, they are so firmly united that they will even resist the force of the tornadoes to which they are not unfrequently exposed, so long

PLATE XVIII.

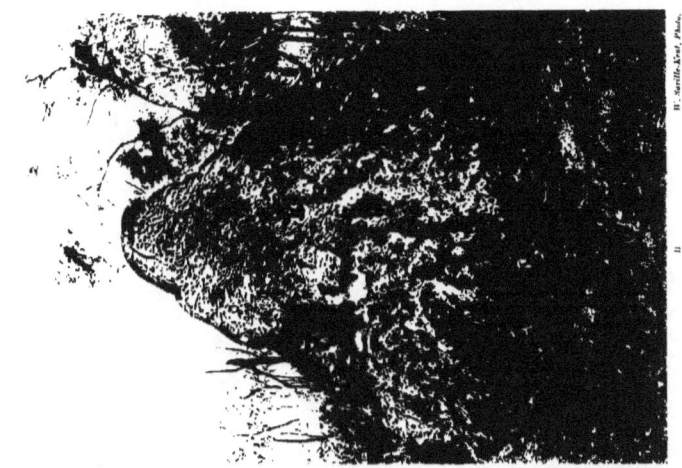

H. Saville-Kent, Photo.

as the tree stands upon which they are fixed. Covered galleries of the same woody composition lead from the nests of these tree-building species to the ground, and thence extend beneath its surface to considerable distances.

One remaining West African Termite recorded by Smeathman invites brief notice on account of the contrast its habits present in comparison with those of ordinary species. He confers upon this variety the title of the Marching Termite, *Termes viarum*. Smeathman's discovery of this species originated through his attention being arrested by a loud hiss at repeated intervals, which proceeded from a spot close to the pathway in the forest he was traversing. On tracing it to its source, he came suddenly on an army of Termites issuing from a hole in the ground about five inches in diameter. At a short distance from the hole they divided into two streams or columns of some twelve or fifteen a-breast, which united again and descended into other subterranean passages some few paces further on. For over an hour there was no diminishing in the numbers, or even pace, of the marching army. The bulk of the procession consisted of labourers, among which were distributed a few soldiers, while, as if for the column's protection, numerous soldiers occupied a position as sentinels on either side all along the line of march, having climbed up and stationed themselves for such purpose on the leaves of the neighbouring and often overhanging herbage. These marching termites were considerably larger in size than the hill-constructing *T. bellicosus*, and were especially remarkable for the circumstance that both soldiers and labourers possessed conspicuous eyes, indicating their essentially daylight proclivities. These individuals, in all the ordinary termite communities, are completely blind, and carry on their works at night, or under the shelter of artificially covered chambers and galleries. Although this particular species of Termite has not been re-discovered by subsequent observers, it is of interest to record that an eyed, daylight-working species is included among the many varieties recently obtained by Dr. Haviland in Natal.

The foregoing somewhat lengthy summary of Mr. Smeathman's observations on the West African Termites has been introduced with the hope that it will prove a stimulus towards the closer study of the Australian species. Many of these are shown in this Chapter to be the constructors of equally remarkable habitations, and while to the writer time allowed only the present record of the broad superficial characters of the termitaria, a wide and interesting field is left open for the more intimate investigation of their puny architects by those possessing leisure and the advantage of a permanent residence within tropical Australian districts.

116   THE NATURALIST IN AUSTRALIA.

Taking the Australian types of Termitaria observed and photographed by the author in consecutive order, reference may first be made to the form reproduced as

TERMITE NESTS, ALBANY PASS, NORTH QUEENSLAND.

W. Saville-Kent, Photo.

the corner illustration of the opening page of this Chapter, and also on the present page. This description of Ant-hill is very prevalent in Cape York Peninsula, in the North of Queensland, and more especially in the neighbourhood of the Albany Pass, where the photographs here reproduced were taken. These Termite mounds form so conspicuous a feature of the landscape on approaching the Pass from the south, that the name of Ant-Hill Point was conferred upon it by one of the earlier navigators, and is still retained upon the Admiralty Charts. To passengers

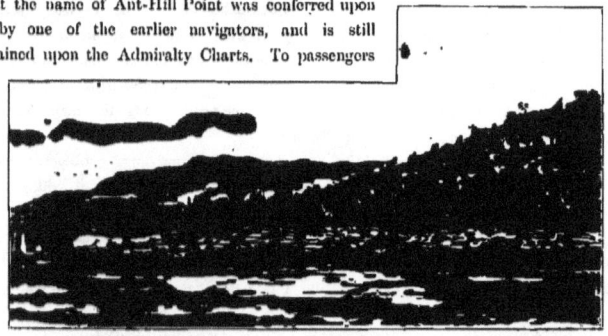

ANT-HILL POINT, ALBANY PASS, NORTH QUEENSLAND.   W. Saville-Kent, del.

making their first voyage through this picturesque strait, it is often suggested in jest, and accepted with exemplary faith, that this area is the site of a native burial ground, and that the ant-mounds represent monuments erected in loving memory of their deceased ancestors. To a new arrival from temperate climes, it is certainly difficult to realise that these remarkable structures are the edifices and tenements of puny insects.

Writing of this White Ant community in his interesting contribution towards a history of the Australian Termitidæ, Mr. W. Froggatt remarks: "At Somerset, Cape York, there is one of the most remarkable termite cities in the world. Viewed from the sea, and looking up beyond the old Government Residency, now occupied by Mr. Frank Jardine's homestead, it appears as if the plain for a mile or more in extent is covered with pointed pillars six or seven feet in height (*), broad at the base and tapering to the summit, forming rather symmetrical pyramids. They are thickly dotted over the plain, often only a few yards apart; the effect is much heightened if the grass has been recently burnt off, as it had been the first time I passed Somerset."

In the absence of a photograph of this remarkable termite community as seen from the most advantageous point, when passing on ship-board through the Albany Pass, the accompanying adaptation of a sketch of it, made by the author on making its first acquaintance in the year 1888, is reproduced in the lower of the two illustrations on the preceding page.

In contradistinction to the several varieties of White Ant habitations described and illustrated in this Chapter, the York Peninsula type of termitarium above referred to may be most appropriately styled the "Pyramidal." The mounds in this variety are always constructed with a wide base and taper gradually towards the apex. This is usually represented by a single one, but, in the largest hills, may be divided into two or three points. In some instances, as illustrated by the termitary to the right in the corner illustration of page 101, the single apex is prolonged in a columnar fashion. The sides of the larger hills, more especially, are usually very distinctly groined or buttressed, after the manner of the trunks of many of the forest trees in the adjacent scrub. The matured height of these pyramidal termitaries usually ranges from six or eight to ten or twelve feet. The author has, however, seen hills of apparently the same species on the bridle track between Somerset and the Patterson

---

* Many of these ant-mounds have an altitude of as much as ten or twelve feet.—S.-K.

telegraph station in York Peninsula that were at least fourteen feet high, and which towered above his head while passing them on horseback. Unfortunately, this happened in the author's pre-photographic days, or a pictorial record of these giants of their tribe would have been secured. The colour of these Cape York termitaries is usually a rich ferruginous red, corresponding in this respect with the character of the underlying soil.

A species of White Ant, apparently identical with this Cape York form, erects pyramidal termitaria throughout the Northern Territory of the Australian Continent. On the occasion of the writer's first visit to the neighbourhood of Wyndham, Cambridge Gulf, Western Australia, as the guest of Captain the Hon. H. P. Foley Vereker and the officers of H.M.S. "Myrmidon," during their survey of this coastal area, in the year 1888, he availed himself of the opportunity of examining many of these nests, and with the co-operation of Mr. R. C. Hare, then resident magistrate of Wyndham, made sections and excavations in selected hills, with the hope of discovering the royal chamber with its enclosed queen. None of these efforts, however, proved successful. Not being at the time cognisant of its approximate position, or of the rapidity, as attested to by Smeathman, with which the labourer termites wall up all the entrances to this chamber, it seems highly probable that as soon as the disturbing strokes of the pickaxes and shovels approached this royal cell, it was converted by their masonic dexterity into what was apparently a homogeneous lump of the component clay, and thus evaded notice. It is commonly stated in this district of North Australia that the subterranean excavations of the White Ants penetrate as deeply into the earth as the hillock is raised in altitude above it. This certainly appears to be the case with relation to many of the termitaria examined, which revealed beneath the surface of the ground as extensive and complex a mass of cells as was found in the previously demolished overlying hillock. The food material contained in the magazine cells of these termitaries consisted, so far as was observed, of finely chopped grass, such as is found in the habitations of the species next described. Concerning this food storage it is worthy of record that, in many instances, it was possible to trace the continuity of the clay-covered galleries for several hundred feet along the surface of the ground in the grassy plains near Wyndham. It would appear that the termites issue out of these galleries to mow down and garner in their harvest of cut-up grasses by innumerable temporarily opened breaches during the night hours only.

The second type of termitarium, with reference to the circumstance that it has been met with by the author only in the Kimberley district of Western Australia,

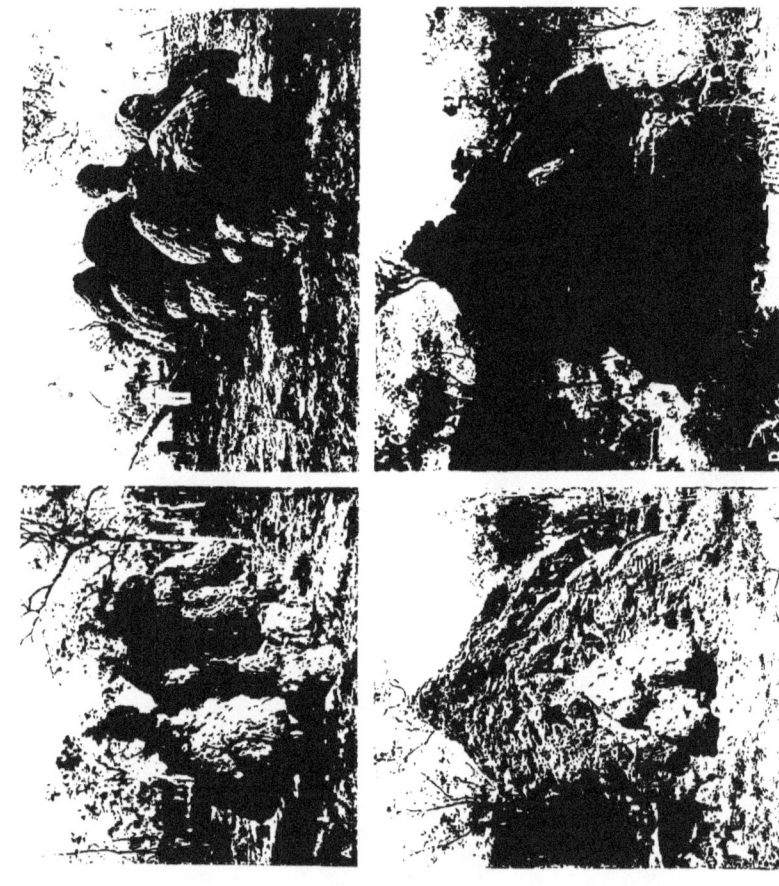

NEST MOUNDS OF WHITE ANTS, KIMBERLEY TYPE, ILLUSTRATING : A & B, ABNORMAL SHAPES, p. 119, AND C & D, PHASES OF RECONSTRUCTION, p. 120.

may be suitably referred to here as the Kimberley type. It is very abundant in the thinly-wooded scrubs, or so-called "pindan," adjacent to Roebuck Bay, but attains to its most luxuriant phase of development near the township of Derby at the head of King's Sound. The several photographs of these Kimberley termitaria reproduced in Plates XVI. to XIX. were obtained in this neighbourhood, and will suffice to show how exceeding variable the external contours of individual hillocks may be. Throughout the entire series, however, it will be recognised that one very peculiar and characteristic plan of construction is predominant. They all present the appearance of having been built up, as it were, by the superposition of consecutive bucketsful of half solidified mortar, which, before setting, has partly overflown and overlapped the preceding instalments. The greatest altitude of this description of termitary that has been observed by the author was fourteen feet, that being the measurement of the example photographically reproduced in Plate XVIII., fig. A. As clearly indicated in this figure, the individual termitarium had passed the zenith of its development, and was already much weathered and eroded on its upper surface.

The contours of some of the termitaria constructed by this Kimberley White Ant are most fantastic and grotesque. In Plate XVII., the summit of the larger fore-ground hillock presents a remarkable resemblance to the head of a lop-eared spaniel or retriever, with its tongue partially protruding. It is, moreover, worthy of remark that this singular resemblance was not noticed at the time of taking the photograph, though immediately recognised with the pulling of the first print. Plate XIX., figs. A and B are yet more grotesque in aspect, and, as a matter of fact, portray two different views of the same termitarium. In fig. A, the general contour is not unlike that of a primitive form of locomotive engine, which for some unexplained reason has become embedded in a thick coating of clay; the indication of the chimney, with its terminal spark-catcher, being particularly prominent. In the second view of the same termitary, fig. B, there is a ludicrous likeness to a group of human figures, clad in voluminous, fleece-like garments. A man resembling the stereotyped delineations of Father Christmas or Robinson Crusoe, with a pack on his back, leads the way, and is followed by what might be his better half, wearing what bears a suspicious resemblance to a divided skirt, combined with the very latest fashion in balloon sleeves. A matronly woollen cap, corresponding with the top of the engine chimney in the previous view, adorns her head. To her right there would appear to be the indication of a child similarly muffled up in shaggy clothing. In all of these forms, the same typical construction of rugged, overlapping clay-masses is distinctly apparent.

The colour of these Kimberley termitaries is usually a light-red or Indian-red hue, corresponding with that of the red sand-stone that predominates in the district where they occur. Not unfrequently, however, as when constructed on the alluvial flats of the Fitzroy river, they agree with the immediate subsoil in being of a similar light brown tint. The opportunity was afforded the author, through the assistance of Dr. Ernest Black, then Government Resident at Derby, of making sections through termitaries of this White Ant, and also of acquiring information concerning the rate at which such half-demolished nests are reconstructed. Plate XVIII., fig. B, represents one such typical section made by the aid of pickaxe and a cross-cut saw. The quest for the Queen Ant was unfortunately futile, and neither did the exposed chambers present anything approaching that diversity and symmetry of design originally credited by Smeathman to the West African *Termes bellicosus*. The larger but very irregular lacunæ, situated chiefly towards the centre of the termitary, do not appear to have been specially constructed, but to represent the interspaces that originally existed between the superimposed clay-layers. From the centre to the topmost and outer crust, the space is chiefly occupied with the closely crowded magazine chambers. There was, however, a smaller central nucleus, consisting of more diminutive cells, apparently the nurseries, but these were unoccupied when exposed to view. The magazine or provision chambers were, on the other hand, for the most part fully stocked with characteristic food material. This consisted exclusively of finely cut up leaves and stems of the grasses growing thickly in the vicinity of the nests. The presence of this material is plainly shown in many of the upper chambers of the section photographed, but still more conspicuously where it has fallen from the chambers, and lies scattered among the *débris* at the base of the hillock.

The period of time occupied in rebuilding the partially destroyed hillocks of this Kimberley Termite was accurately gauged by sections made through examples in the neighbourhood of Derby. On the writer's first visit in September, 1893, there happened to be a very characteristic example of one of these White Ant hillocks that had been accurately bisected a little over one year previously, when erecting the fence line of a new road to the Derby race-course. A photograph of this termitary with its reconstructed portions consisting of four over-lapping clay masses at its base, is reproduced in Plate XIX., fig. C. As will be recognised, the newly added portions cover somewhat less than one-fourth of the entire bisected surface. In March, 1895, the passage to England, via Singapore, gave the author an hour or two on shore at Derby, which opportunity was hastily utilised for securing a second photograph of this particular termitarium, which by this time had had another eighteen months for the work of reconstruction. The replica of this second photograph, Plate XIX., fig. D, shows that a very considerable advance had been accomplished. The newly added patches extend now to over two-thirds of the area originally laid bare, and it may be reasonably anticipated that another twelvemonths' uninterrupted work at the same ratio, will effect the covering in and complete obliteration of the original sectional scar. It is worthy of note that an examination made and a photograph taken on the same occasion of the more recently bisected termitary, Plate XVIII., fig. B, revealed the fact that it had accomplished an almost identical amount of progress in the direction of reconstruction as that exhibited after a similar interval by the one here made the subject of special notice. On the other hand, a second photograph taken at the same time of the undisturbed termitary having a dog's-head-like apex, Plate XVII., exhibited, after this intervening interval of eighteen months, scarcely any perceptible alteration of contour or increase in bulk. Examining the two photographs carefully with a hand-glass, it is difficult in point of fact to detect so much as even the difference of a wrinkle between them. This somewhat surprising fact can be explained by the supposition that this particular termitarium had attained to the full zenith of its development, that the queen probably was deceased, and that any future changes would be those of erosion and decay. That this process is already in progress is in fact distinctly visible, the external clay layer being worn and weathered in places as in the larger example portrayed in Plate XVIII.

The third type of termitarium constructed by the Australian White Ants inviting attention in this Chapter, is that illustrated by Plates XV. and XX. This specific form is not built to a remarkable height, eight feet representing the tallest observed. Its specially conspicuous features are the peculiarly elongate compressed contours of every individual hillock, supplemented by the circumstance that in every instance the long axis of the structure is in a precise line with the north and south poles of the magnetic compass. On this account the very appropriate titles of Magnetic, Meridian, or Compass Ant-hills have been conferred upon these special forms. The photographs here reproduced were taken by the writer in the valley of the Laura river some 60 miles inland from Cooktown, North Queensland.

122   THE NATURALIST IN AUSTRALIA.

In accord with the character of the peat-like deposits out of which they are excavated, these ant-hills are of a black or dark ash-grey tint. A very prominent feature of these Laura Valley Meridian Ant nests is their highly ornate structural type. They consist, as shown in the accompanying photographs, of a congeries of slender pinnacles erected in the same straight line close to one another, which are subsequently amalgamated. Upon these, again, numerous subsidiary pinnacles are usually constructed. Some of the larger and wider of these termitaries, when viewed end on, as shown in Plate XX., present, with their closely crowded spires and pinnacles, a not unremote resemblance to the architectural

BROADSIDE VIEW OF LAURA VALLEY MERIDIAN ANTS' NEST.

PLATE XX.

NEST MOUNDS OF WHITE ANTS, MERIDIAN VARIETY, LAURA VALLEY, NORTH QUEENSLAND;
END ON VIEW, p. 122.

mass of some grand cathedral. A broadside view of one of the most elaborately pinnacled examples of this description of termitarium, observed and photographed by the author in the Laura district, and closely corresponding with the one depicted in Plate XX., is reproduced on page 122.

Another more widely known variety of Meridian Ant nest occurs, and has been seen by the author, on the Victoria river plains, some forty miles from Port Darwin, in the Northern Territory of South Australia. To the kind courtesy of Mr. Paul Folsche, for many years superintendent of the Police Department at Palmerston, the writer is indebted for the very excellent photographs of these remarkable Port Darwin termitaria that illustrate this Chapter. The Meridian termitaria in this instance, as portrayed in Plate XXI., figs. A and B, are about the same height as the Laura river type last described, but differ from them in being much more solid and compact. There is also an absence of the ornately pinnacled plan of architecture notable in the Laura variety, their upper edge as shown in profile being nearly smooth or but slightly serrated. Another peculiarity of the Port Darwin variety, pointed out to the author by Mr. Folsche, is the circumstance that they are usually distinctly convex on the east, and concave on the west, of their opposing, broad lateral surfaces.

White Ant termitaria corresponding in aspect and structure with the Port Darwin type, have been also reported to the author by Dr. T. L. Bancroft, of Brisbane, as occurring in the neighbourhood of the Howard river, North Queensland. From a photograph of them placed at the writer's disposal by that authority, it would appear that they attain to a more considerable elevation, some of them, as shown by the figures in their vicinity, being not less than nine or ten feet high. It is further worthy of remark that, while in the Port Darwin district the Meridian Ant-hills are erected on a more or less open grassy plain, their co-types near the Howard river occur in the midst of a thickly timbered country. Possibly, on a nearer investigation, this Howard river form will prove to be a third distinctly differentiated type of the Meridian structural plan.

The *raison d'être* of the north and south directions of these Meridian Ant nests has given rise to much speculation and various interpretations. By some it is supposed to bear a direct relationship to the prevalent winds. As, however, these, in the districts where they occur, are chiefly, according to the seasonal monsoons, south-east or north-west, but predominantly the former, that theory would hardly appear to afford a satisfactory explanation. A more probable interpretation presents itself to the writer's mind, with reference to the circumstance that, by being constructed with this orientation, their larger surfaces present the least possible direct exposure to the meridianal rays of the tropical sun, and in consequence absorb and retain interiorly a minimum amount of solar heat.

The author is beholden to Mr. Paul Folsche for the photographs of yet another form of White Ant termitarium, reproduced in this volume. It occurs also in the neighbourhood of Port Darwin, and probably represents the most colossal example of these remarkable insect habitations that is to be found in the Australian Continent. The architectural plan of this variety may be most appropriately designated the "Columnar." The altitude of the highest of these termitaria observed, Plate XXII., fig. A, is no less than eighteen feet, as may be verified by the figures of the man, horses, and vehicle standing beside it. In common with the pyramidal type of termitarium belonging to the Cape York Peninsula district of Northern Queensland, previously described, this columnar form is associated with strong ridges or buttresses, which, extending throughout the length of the lofty column, must add materially to its rigidity and strength. It will be of interest to ascertain whether any remarkably large species of White Ant is the fabricator of these tower-like structures. Although unable to produce any direct evidence on this subject in the present volume, the writer has been informed by Mr. Froggatt that its constructors are most probably referable to the genus *Eutermes*.

In addition to the remarkable descriptions of White Ant habitations now enumerated, there are a number of Australian Termites awaiting identification and classification that either construct smaller and architecturally insignificant termitaria or which, erecting no mound whatever, are purely minors, living in decaying wood or subterranean chambers and galleries, whence they extend their depredations to remote distances. Among the former category, irregular mound-shaped termitaria some two or three feet high, probably constructed by distinct species, may be met with at a considerable distance south of the tropic of Capricorn, on both the Queensland and Western Australian sides of the Island-Continent. In the former instance they may be referred to as being abundant in the neighbourhood of Brisbane, and in Western Australia as extending as far south as Pinjarrah and Bunbury, in about latitude 33°. It frequently happens that birds of various species excavate holes and build their nests within these smaller termitaria. A White-breasted Kingfisher, probably *Halcyon sanctus*, thus constructs its nest in those in the southern districts of Western Australia. In similar irregular heap-like termitaria in the neighbourhood

NEST MOUNDS OF WHITE ANTS, MERIDIAN VARIETY, PORT DARWIN, NORTHERN TERRITORY OF SOUTH AUSTRALIA.

A.—Broadside. B.—Edgeways View. p. 123.

of Rockhampton, Central Queensland, the beautiful Parrakeet, *Posephotus pulcherrimus*, burrows and builds its nest, while the handsome White-tailed Kingfisher, *Tanysiptera sylvia*, an ally of the more remarkable Racket-tailed species, *T. microrhyncha*, burrows a hole and deposits its eggs in a small symmetrically ovate form of termitary that occurs in the thickly-wooded scrubs in the coastal ranges at Bloomfield, near Cooktown, North Queensland. For the information relating to these two species, as also for the characteristic photographs reproduced in the lower half of Plate XXII., portraying the termitaria with their associated nest burrows, the author is indebted to Mr. D. Le Souef, the Director of the Melbourne Zoological and Acclimatisation Society's Gardens.

On submitting outline sketches of these last-named types of termitaria to Mr. Walter Froggatt, the writer was informed by that authority that they most probably represent, in both instances, ground-constructed nests of *Eutermes funipennis*, a species which builds indifferently on the ground or in branches of trees, after the manner of the *Termes arboreum*, originally described by Smeathman, referred to on a previous page. The arboreal nests of this or an allied species have been observed by the author throughout North Queensland and are of common occurrence in Thursday Island, Torres Straits. The termitaria constructed under these conditions are dark coloured, usually sub-spheroidal or ovate in shape, about eighteen inches in diameter, and most frequently constructed at a height of some ten or twelve feet only from the ground, with which they are connected by clay-covered galleries. The substance of these nests, as in the African arboreal forms, consists almost exclusively of finely comminuted woody particles, cemented by the excreted juices of the termites into a homogeneous mass of remarkable density.

As is familiar to most Northern Australian settlers, snakes, lizards, rats, scorpions, and a variety of other animals take up their abode within the natural or artificially excavated fissures of the White Ants' nests, some little caution, on account of the first-named reptiles, having to be exercised when dismantling a hillock. Among the utilitarian uses to which the termitaries have been largely applied in the Kimberley district is that of road-making, a circumstance which is accountable for the disappearance of numbers of the largest size that formerly occupied a position close to the township of Derby. Used as a top layer, it binds down and hardens with the weather into a cement-like mass of great hardness and durability. For a like purpose termitary earth, damped and rammed, is highly esteemed as the top covering of the flooring of settlers' huts. As an extemporary oven again, a small ant-hill, scientifically excavated, may, in the hands of an experienced bushman, be made to accomplish undreamt-of culinary triumphs.

The author on one occasion witnessed a rather interesting episode that was enacted in connection with a small mound-shaped termitarium in the southern district of Western Australia. As a rule the Termites utilise the night hours only for the enlargement of their borders, their architectural work of the preceding night being clearly visible in the early morning in the form of a darker semi-dried clay patch on one or more areas of the termitary surface. To effect any such extension of their premises, they have to temporarily break passages through the original outer shell, a proceeding which, if done in broad daylight, would doubtless leave them open to the attacks of many enemies. An exceptional case to this general rule observed by the writer was on an afternoon immediately preceding a heavy rainfall, which proved to be the forerunner of continuous wet weather. The Termites apparently possessed an instinctive foreknowledge of the approaching meteorological change, and were determined to make the most of the little dry weather left. Small breaches were consequently being made at innumerable points of the termitary, and the heads of the working individuals were every now and then visible as they arrived with their loads of tempered clay.

There were other interested watchers of their operations. These were a colony of large red and black ants, from a neighbouring subterranean nest. A number of these insects were continually patrolling the surface of the White Ants' hillock, and every now and then one of them would make a plunge at a scarcely visible opening. Its head and powerful jaws disappeared for a second or two within the clayey matrix, and would then be withdrawn, on most occasions dragging forth in its tenacious grip an unfortunate termite, which was speedily dispatched and borne away by the red ant with as much ease as a terrier could carry off a rat. The true ants, *Formicidæ*, are as a rule very high appreciators of the succulent Termites, or White Ants, as an article of food, and where a colony of the destructive species is found making its depredations, there is scarcely a surer method of getting rid of these pests than by exposing their galleries and chambers to the inroads of the larger varieties of ordinary ants, which are almost always to be found in the vicinity. It not uncommonly happens, indeed, that termitaria are tenanted partly by termites and partly by ordinary ants. In such instances it may be assumed that the predatory ants have, in the first place, invaded and established a foothold in the White Ants' habitation, from which the latter, while securely barricading the rearward chambers, have been unable to more effectually dislodge their aggressors.

As an instance of the highly scientific tactics which the true ants will display when raiding a White Ant colony, Mr. R. C. Hare mentioned to the author that on one occasion, on disturbing a piece of timber infested by termites, so as to expose them to the attacks of the enemy, they proved to consist, to a large extent, of winged nymphs, just on the point of taking flight. Some large black ants that were reconnoitring in the neighbourhood immediately rushed in to take possession, but, instinctively recognising the position of affairs, fell to in the first instance to cut off the wings of all the matured nymphs before attempting to kill or secure a single victim.

White Ants, Termitidæ, of a harmless description, that feed exclusively upon dead or decaying wood, and construct no conspicuous nest, are found as far south as Tasmania in the Australasian system. In November, 1884, when residing at Hobart, the author communicated to the Proceedings of the Royal Society of that colony a brief account of the infusorial parasites that infest the intestines of this Termite. They belonged chiefly to the genera Trichonympha and Pyrsonympha, originally instituted by Dr. Leidy, of Philadelphia, for corresponding parasites of the North American White Ant, *Termes flavipes*, though differing essentially in their specific characters. Other infusorial forms found infesting the Tasmanian Termite were referable to the genus Lophomonas, that had hitherto been recorded as parasites only of the Orthopterous insects Blatta and Gryllotalpa. Figures and full descriptions of the American infusorial parasites are included in Volumes II. and III. of the author's "Manual of the Infusoria," published in 1881-2. An illustration of the more characteristic aspects of the Tasmanian species, upon which the author conferred the title of *Trichonympha Leidyi*, in honour of the founder of the genus, are figured for the first time as a tail-piece to this Chapter.

The conspicuous features of the species of Trichonympha, as compared with all other known Infusoria, are the remarkable abundance and length of the fine hair-like cilia with which their bodies are clothed, and the remarkable activity and Protean shapes they display or assume in the living state. In the case of the American type, the aspects of these animalcule, as they contort their bodies and fling around their cilia in whirling mazes, have been compared by Dr. Leidy to the flowing of a thin sheet of water over the brim of a fountain, swayed to one side or the other by contending wind currents; or, again, to dancing nymphs in a spectacular drama, wearing as their chief adornment a cincture of long cords suspended fringewise from their shoulders. The mazy evolutions and gyrations of the more modern skirt-dance would probably provide an equally apposite simile. A phenomenon observed by the author of the Tasmanian Trichonymphæ, but not recorded of the American type, was the circumstance that they customarily anchored themselves by the longer posterior cilia to organic *débris* or other convenient fulchra, and, thus attached, spun to and fro after the manner of extemporised roasting jacks, but with great rapidity. The writer was much disappointed in finding no Trichonymphidæ within the intestinal tracts of the many large nest-building tropical Australian termites examined. A distinct variety was, however, found associated with the large wood-destroying species figured on Page 103. Similar infusorial parasites have been reported, but not specifically described as infesting the European *Termes lucifugus*. M. C. Lespes has recorded that they constitute the greater portion of the brown pulp of which the contents of the intestine of this species are composed. Dr. Haviland also reports them as being very abundant in various Singapore species. In examining these parasitic Infusoria, Dr. Leidy found that they could be kept alive for some time by transferring them from the termite's intestine to a small drop of the white of egg. By similarly employing diluted milk the author was able to keep and examine the Tasmanian species under the microscope for lengthened periods.

Through the kind courtesy of Drs. D. Sharp, F.R.S., and J. D. Haviland, the writer has been afforded the opportunity, while concluding this Chapter, of glancing through the proof-sheets of the article on the Termitidæ, edited by Dr. Sharp, which will be included in the forthcoming Insecta volumes of the Cambridge Natural History.* It embraces the results of the most recent investigations of this remarkable insect group, including the record of extensive and valuable researches prosecuted by Dr. Haviland at Singapore and in South Africa. In this last-named region it is interesting to observe that Dr. Haviland was fortunate enough to discover a form closely identical in structure and habits with the remarkable Marching Termite, the *Termes viarum* of Smeathman, described over a century ago, but of which no recent knowledge had been recorded. Like *Termes viarum*, the workers and soldiers of this Natal species, which Dr. Sharp has appropriately named *Hodotermes Havilandi*, are possessed of facetted eyes, all of the ordinary forms being blind, and come abroad during the heat of the day to cut and carry the grass and foliage upon which they principally feed. The subterranean galleries of this Natal type measured one-third of an inch

* Now published. Macmillan & Co.

PLATE XXII.

A. TERMITE MOUND, PORT DARWIN COLUMNAR TYPE, p. 124.

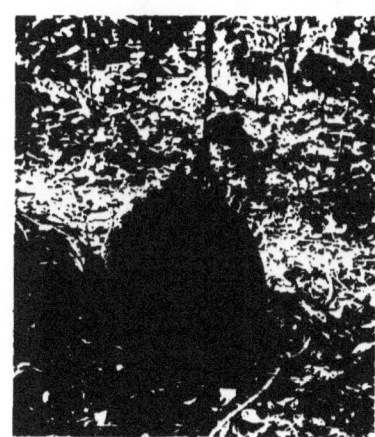

B. OVATE TERMITE MOUND, WITH NEST BURROW OF *Tanysiptera sylvia*, p. 125.

C. SMALL EXTRA-TROPICAL TERMITE MOUND, WITH NEST BURROW OF *Psephotus pulcherrimus*, p. 125.

in diameter, and were traced by Dr. Haviland for a distance of twenty feet, and to a depth of five and a half feet, but without coming to a main nest which probably exists. It would seem to be by no means improbable that a more searching investigation in tropical Australia will reveal the existence of some allied eyed, day-working, Termitidæ also in that Continent. Such a possibility is, at all events, well worth the quest.

The recent researches of Dr. Haviland, anticipated partly by Bates' exploration in the Amazons Valley, have established the fact that several species of Termites may be found inhabiting the same termitarium. Smeathman, in fact, has attested to two species inhabiting the characteristic turret nests of Senegal. In South Africa, Dr. Haviland reports having found as many as five species of Termitidæ and three species of true ants in a single mound. In this instance, while in close proximity, there appeared to be no absolute intimacy between the species. Other instances, however, are reported in which a second species lives as a guest on intimate terms with the original constructors of the mound. This tendency of the Termites to live in mixed communities considerably enhances the difficulty of accurately determining their specific range and identity, more especially having regard to the circumstance of the plurality of individual modifications in the same species. Supposing, for example, that the five consorting species above referred to presented from twelve to fifteen individual modifications; this would signify that from sixty to as many as seventy-five distinct types might be yielded for discriminative classification from a single termitarium.

As the result of the investigations of the morphological structure of the Termitidæ that have been conducted up to the present date, it has been found that the variations presented by the singularly modified heads and jaws of the soldier individuals yield the most reliable and readily recognisable characters for specific diagnosis. Consequently, in all collections made of these insects for classificatory purposes, special care should be taken to include samples of the soldier types. As an indication of the structural modifications that may obtain in this direction, a few of the more notable ones, as originally indicated by Hagen, are reproduced in Figs. 9 to 13, on page 103. The most remarkable deviation from the customary scissors-blade contour of the opposing jaws is presented by Fig. 13, that of *Termes armiger*, in which the formidable mandibles are replaced by a tubular pike-like prolongation of the front of the head, out of which a fluid secretion can be ejected for the purposes of either assisting in the nest construction, or for the discomforture of its enemies. With reference to this peculiarity, the suggestive title of "Nasuti" or "Snouted" Termites has been conferred upon these individuals.

In addition to the conformation of the jaws, the pattern of the veining or "neuration" of the wings in the matured flying individuals has been found to be of signal service for the scientific classification of the Termitidæ, and it is with reference more especially to such distinctions that the known species are at present relegated to the several genera *Termes, Eutermes, Hodotermes* and *Calotermes*.

The feeding habits of the Termitidæ have been proved by the recent investigations, more especially of Grassi and Sandias with relation to *Calotermes flavicollis*, to be very peculiar. No refuse matter of any kind is permitted to remain in the chambers or galleries, which are always remarkable clean. Whether cast skins, dead individuals, material regurgitated from the mouth or passed through the alimentary canal, such refuse is devoured or redevoured until no nutritive properties are left, when it is either built into the substance of the habitations or galleries, or cast outside them. It was observed of the above-named species that the so-called "proctodœal" food or alimentary matter in its semi-exhausted state as passed from the alimentary canal represented the favourite nutriment. One individual yields it to another on being stroked posteriorly by the antennæ and palpi of the latter, much in the same manner as the aphides yield their honey-like excretions to the true ants. This food-provision phenomenon is associated with the labourers or greater bulk of the community. In the case of the soldiers, it would appear that they are essentially cannibalistic, preying upon the sickly or purposely destroyed labouring units.

Among other data elicited through Professor Grassi's investigations, the fact was fully established—with regard to the European species—that the potentiality for increase of the termite community is by no means seriously impaired by the destruction of the original royal pair, or founders of the colony. Some one or more of the "neotcinic" or "substitution" kings and queens were always available for election to the vacant thrones. According, moreover, to the description of food supplied to them, it has been ascertained by Professor Grassi that any of the numerous individual termite modifications can be developed from the ordinary larvæ. A full translation of Grassi's important Memoir on the European Termitidæ is, while going to press, November, 1896, in course of publication in the "Quarterly Journal of Microscopical Science." Its perusal is strongly recommended to any who may contemplate the systematic investigation of the habits, metamorphoses and ætiological phenomena of the Australian or other exotic species.

Summarising the existing knowledge of the Termitidæ, Dr. Sharp reports that the number of described species does not exceed a hundred, and that very probably there are as many as one thousand in existence. Australia, while yielding probably for its area the largest and most diversified variety of specific forms and the most remarkable architectural modifications of termite habitations, has, so far, been practically neglected by scientific specialists, and consequently offers a most rich and almost virgin field to the Termitologist. It is hoped by the writer that the data concerning the more noteworthy habits and fabrications of this insect group embodied in this Chapter, will prove a stimulus to readers residing in tropical Australian districts to devote some attention to the study of such varieties as may be peculiar to their neighbourhoods. By so doing, and by combining with their recorded notes the conservation in alcohol or, possibly, more readily accessible whisky, of specimen samples of the variously modified individuals found in each specific form of mound or subterranean gallery, they will render valuable service to science. While assisting science they will also gain to themselves that newer interest and zest in life which is so often wanting to the isolated Australian settler. For the frequently enforced monotony of such a calling there is no panacea so sure as an intelligently bestridden natural history hobby.

Trichonympha Leidyi, S.-K., INFUSORIAL PARASITES OF THE TASMANIAN WHITE ANT, EXHIBITING DIVERSE PROTEAN MODIFICATIONS. MAGNIFIED 600 DIAMETERS. *p.* 127.

W. Saville-Kent, Photo.
WRECK POINT, PELSART ISLAND, p. 136.

CHAPTER V.

HOUTMAN'S ABROLHOS.

MUCH interest attaches itself in the minds of most biological students to the contemplation of the indigenous fauna and flora of islands occupying a more or less remote distance from the nearest mainland. Oftentimes, though the intervening space scarcely outdistances the range of human vision, the terrestrial inhabitants of the divided lands may be notably distinct, the distinction extending itself even to such migratory forms as birds and insects. This fact is especially familiar to all who have made themselves conversant with the contents of Dr. A. R. Wallace's fascinating volume, "Island Life."

The subject matter dealt with by that accomplished naturalist in the treatise quoted is limited exclusively to the consideration of the denizens of the terrestrial

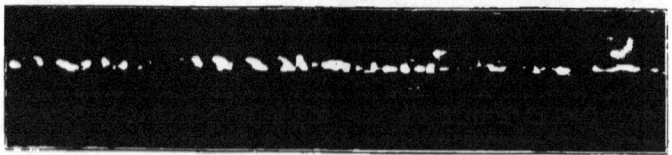

MIRAGE-ELEVATED BREAKERS ON OUTER BARRIER, PELSART ISLAND, p. 143. *W. Saville-Kent, Photo.*

and aerial elements. The assumption that corresponding strongly marked divergencies may possibly exist with respect to the fish and other inhabitants of the adjacent waters of similarly contiguous lands would seem, at first sight, to possess but little ground for serious entertainment. The main object of the present Chapter is to testify to a very conspicuous example of such marine faunal diversity that has fallen within the writer's purview. The geographical region which has furnished the testimony to be presently submitted is a small group of islands lying at a little distance off the coast of Western Australia, and diversely distinguished on maps and charts by the titles of "Houtman's Rocks" and "Houtman's Abrolhos."

From many points of view this island group yields data of high interest, and invites attention apart from the more especial subject matter of this Chapter. Houtman's Abrolhos takes its title in commemoration of the name of Frederic Houtman, one of the earliest Dutch explorers, who has been credited with their discovery, as well as in contradistinction to an island group of corresponding character, likewise named the Abrolhos, that is situated off the coast of Brazil. More particular interest attaches itself to Houtman's group with reference to the circumstance that it was the wreck there of the Dutch East Indian Company's ship, the "Batavia," in the command of Captain Francis Pelsart, in the year 1629, that led to the earliest recorded discovery of the great Island-Continent of Australia. As the islands lay but a little to the eastward of the course of the vessels trading round the Cape of Good Hope to the, at that time, extensive Dutch possessions, it is not a matter of surprise that in those early days, prior to the existence of reliable charts, numerous vessels suffered shipwreck on the low-lying reefs. The "Vergulde Draeck," the "Zeewyk," the "Ridderschap von Holland," the "Zuysdorp," and others, all belonging to the same nationality, may be named among those which, in addition to the "Batavia," came to an untimely end on the Abrolhos reefs. The last-named vessel, however, is the one around which pre-eminent notoriety will always cling, by virtue of the tragic events attending its wreck. The reader may be referred to that very excellent little booklet "Pinkerton's Early Australian Voyages," published by Messrs. Cassell and Co., for a succinct account of the events referred to, which are further embodied in a more artistically embellished form in Mr. W. J. Gordon's novel, "The Captain General." For the purposes of this Chapter the following brief epitomisation of the leading facts will suffice.

The "Batavia," in the command of Captain Francis Pelsart, was one of a fleet of eleven Dutch vessels which sailed from Texel for the East Indies, via the

Cape, on the 28th of October, 1628. On the night of the 4th of June in the following year, the ship, being separated from the fleet, was driven in a storm on to the Abrolhos shoals, and became a total wreck. The crew and passengers, to the number of over two hundred, effected a landing on three separate islands of the group. Owing to the shortness of water, Captain Pelsart, with some of the officers, proceeded in the skiff, at the request of the ship's company, to seek for further supplies in some of the adjacent islands. This search, so far as the islands were concerned, proved unavailing, and they extended their investigations to the adjacent mainland. After many, for the most part but partially successful, attempts to land, they were carried so far to the north-east by the prevailing currents that they decided to go on to Batavia, and to return thence with another vessel to the relief of the shipwrecked survivors on the Abrolhos Islands. This earliest recorded observation of a distinct northerly set of the ocean currents up the Western Coast of Australia is, as hereafter shown, of considerable interest.

During the protracted absence of Captain Pelsart and his officers in quest of relief, a grim tragedy was enacted at the Abrolhos. It would appear that previous to the catastrophe that befel the "Batavia," the supercargo, one Jerom Cornelis, in league with the pilot and some others, had plotted to run away with the vessel, carrying her either to the French port of Dunkirk, or turning pirates in her on their own account. Captain Pelsart's absence revived in Cornelis' mind the same design on a new basis. Being left practically in command of the situation, he determined, in company with his elected associates, to make himself possessed of the very considerable treasures which had been saved from the wreck, and, further, to seize and appropriate any vessel which Captain Pelsart might return in to their succour.

In order to get rid of the large number of the lost ship's company who were likely to oppose, or to prove an incumbrance to the realization of, his plans, he decided to put them all to death in cold blood, and actually carried this murderous design into partial fulfilment. A portion of the intended victims, however, managed to escape to one of the adjacent islands, on which another colony, in charge of Lieutenant Weybheys, had been sentenced to destruction by the mutineers. He, with the united forces, successfully repelled the several attacks which Cornelis subsequently made upon them with the object of their extermination. In the course of these encounters Cornelis was taken prisoner by Lieutenant Weybheys, who succeeded in giving timely advice to Captain Pelsart, on his

PLATE XXIII

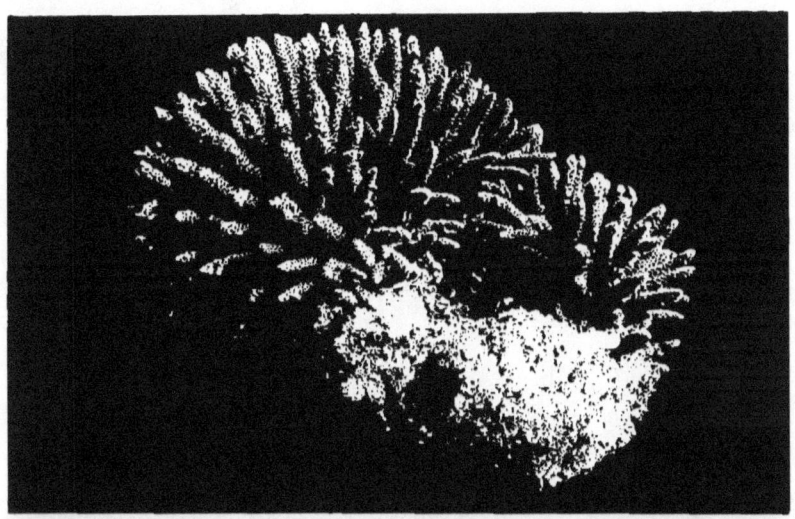

ABROLHOS CORALS, GENUS MADREPORA, p. 140. Half natural size.

arrival shortly after from Batavia in the frigate "Sardam," that the conspirators designed to seize his ship. Two boat-loads of armed men, in fact, put off with this intention, but on being threatened by Captain Pelsart that he would sink them with his big guns if they did not immediately throw their arms overboard and surrender, they gave in. They and the remainder of the mutineers were captured, and all the participators in the previous massacres were summarily executed.

A very considerable amount of treasure was on board the "Batavia" when she was lost. Captain Pelsart succeeded in recovering all the jewellery and other valuables which had been appropriated by the mutineers, and likewise in raising from the wreck five out of the six chests of silver coin that were being brought out to Batavia. The sixth one is supposed to be still somewhere immersed among the coral reefs Numberless relics from the numerous wrecks that have occurred on Houtman's Abrolhos, including a gun, cannon shot, coins, pipes, glass and earthenware, etc., have been already discovered, more especially on Gun and Rat Islands, during the process of excavating the guano which has accumulated there in large quantities, and a large number of these are now on view in the Perth Museum. Possibly the missing chest, or the bulk of its contents, may yet reward a persevering search.

To proceed with the more legitimate subject of this Chapter, it is desirable in the first instance to give a brief account of the precise geographical position and other essential details concerning the island group under discussion. Topographically defined, Houtman's Rocks or Houtman's Abrolhos consists of a little archipelago, for the most part of coral formation, situated between latitudes 28° 15" and 29° S., some thirty miles off the mainland coast of Western Australia and immediately opposite Champion Bay and the thriving port of Geraldton. More closely examined, the Abrolhos archipelago is found to be separable into four secondary groups, characterised in order from north to south as the North Island, Wallaby, Easter, and Pelsart groups. With the exception of the Wallaby group, which contains plutonic rocks corresponding in character with those of the mainland, and having an elevation of some thirty or forty feet, the larger residue is entirely of coral formation, while reefs of considerable extent also encircle the Wallaby series. Their composition, as manifested, more particularly in the islets of the Easter and Pelsart groups, consists of hard coral limestone conglomerate, undermined and weathered on its exposed aspects into low overhanging cliffs and promontories often of the most fantastic shape, which frequently show embedded in their eroded surfaces but slightly altered coralla of the Madreporidæ of which they are principally composed. A characteristic illustration of the aspect of one of these overhanging limestone promontories is afforded by the photograph reproduced as the heading to this Chapter. This view represents the extreme end of Wreck Point, in Pelsart Island, and is generally supposed to be the site of the stranding of the "Batavia" in 1629.

From time immemorial, as testified to by the deep guano deposits, Houtman's Abrolhos has been the home or breeding centre of countless hosts of sea-birds, which still resort thither in enormous quantities in the breeding season. At the dates of the writer's visits to these islands, July and August, 1894, the nesting time for the birds had not arrived. During the last visit in August, however, the "Wideawakes," or Sooty Terns, absent during the daytime, commenced to assemble over their breeding grounds on Rat Island as soon as it became dark, and, flying to and fro in increasing numbers throughout the night, filled the air with their discordant noise. The first-named title borne by these birds is supposed to be descriptive of their cry. To the writer their note agreed, phonetically, much more nearly with the four syllables, "Come-out-of-that," but it is probably open to various interpretations. There can be no question, however, as to the fact that to light sleepers the burden of this sea-gull's song must be eminently conducive to wakefulness. An investigation into a report upon the number of varieties and breeding seasons and other data of interest concerning the sea-birds frequenting the Abrolhos, has been recently carried out by Mr. A. J. Campbell, F.L.S., the well-known Victorian ornithologist, and is published in Vol. II., 1890, of the Reports of the Australian Association for the Advancement of Science. The following is an abbreviated list of the sea-birds recorded in Mr. Campbell's paper as permanently frequenting or resorting to the Abrolhos Islands at their breeding seasons:—

LIST OF SEA AND SHORE BIRDS FREQUENTING HOUTMAN'S ABROLHOS RECORDED BY Mr. A. J. CAMPBELL, F.L.S.

Halietus leucogaster, White-bellied Sea-eagle.

Pandion leucocephalus, White-headed Osprey.

Hæmatopus longirostris, White-breasted Oyster-catcher.

Hæmatopus unicolor, Sooty Oyster-catcher.

Ægialitis ruficapilla, Red-capped Dottrel.

Tringa albescens, Little Sandpiper.

Tringa subarquata, Curlew Sandpiper.

Strepsilas interpres, Turnstone.
Numenius cyanopus, Australian Curlew.
Numenius uropygialis, Wimbrel.
Demiegretta sacra, Reef Heron.
Hypotœnidia philipensis, Pectoral Rail.
Porzana tabuensis, Tabuan Crake.
Anas castanea, Australian Teal.
Larus pacificus, Pacific Gull.
Larus longirostris, Long-billed Gull.
Sterna caspia, Caspian Tern.
Sterna bergii, Common Tern.
Sterna dougalli, Graceful Tern.
Sterna anæstheta, Panayan Tern.
Sterna fuliginosa, Sooty Tern or Wide-awake.
Sternula nereis, Little Tern.
Sternula inconspicua, Doubtful Tern.
Anous stolidus, Noddy Tern.
Anous tenuirostris, Lesser Noddy.
Puffinus nugax, Allied Petrel.
Puffinus sphenurus, Wedge-tailed Petrel.
Procellaria fregata, White-faced Storm Petrel.
Phaëton candidus, White-Tailed Tropic Bird.
Phaëton rubricauda, Red-tailed Tropic Bird.
Graculus varius, Pied Cormorant.
Pelecanus conspicillatus, Australian Pelican.

On account of the vast accumulations of guano resulting from the sea-birds having so long made the Abrolhos their headquarters, this island group possesses a considerable commercial value, and has been leased for some years past by the Western Australian Government to the enterprising firm of Messrs. Broadhurst and McNeil for the exclusive right to collect and export this valuable product. The record of the facts that up to 1894 no less a quantity than 46,000 tons of guano was excavated and exported by the firm, and that the royalty that accrued to the Government thereon amounted to £16,000, will suffice to indicate the important position which Houtman's Abrolhos Islands occupy as a source of income and revenue. In order to profitably work the guano deposits, tramways for land carriage and jetties for its convenient shipment have been constructed on several of the larger islands, and the hundred or more labourers, chiefly Malays, usually employed in delving for and transporting the excavated mould in bags, barrows, baskets, and every available receptacle, make a most animated spectacle. It is interesting to observe that this Abrolhos guano, notwithstanding that it is among the richest known in phosphates and other most highly prized constituents, is, in its virgin state, absolutely devoid of smell, and presents the aspect of a by no means extra rich, light-coloured, garden mould. The chemist consequently has to specially concoct and add an appropriate stink to this raw material, for guano without an odour would be to the agriculturist a veritable case of

"Hamlet" minus the Prince of Denmark. In short, the mightier the perfume the more potent the supposed fertilising properties.

At the instance of Mr. Thomas Broadhurst, the present representative of the firm, who takes a warm interest in all matters connected with the actual or potential commercial capabilities of these islands, the author was deputed by the Western Australian Government, in his capacity of Commissioner of Fisheries, to examine and report upon their eligibility for the establishment thereon of oyster, mother-of-pearl shell, and other profitable fisheries. The investigations made with this special object led to the discovery of a very unexpected constitution of the marine fauna of these islands, and at the same time permitted the writer to make a favourable report to the Government in directions which had not been anticipated. The ordinary Australian Rock Oyster, *Ostræa glomerata*, occurs in such abundance and under such conditions on several of the islands, that it could no doubt be made a subject of remunerative cultivation. The smaller West Australian variety of mother-of-pearl shell, identical with the *Meleagrina imbricata* of Reeve, was found growing very sparingly among the reefs, and although it could no doubt be abundantly propagated there by recourse to scientific methods, it did not seem worthy of attention in comparison with the unexpectedly favourable conditions which the author found to obtain at the Abrolhos for the introduction and acclimatisation of the larger and far more valuable tropical species, *Meleagrina margaritifera*.

A separate Chapter being devoted later on to fuller details concerning the various species of Australian mother-of-pearl shell, it will suffice here to remark that the large commercial species, *M. margaritifera*, is an essentially tropical type, not found growing indigenously below the parallel of $23\frac{1}{2}°$ S. The smaller shell, *M. imbricata*, while a native of the tropics, attains to its most prolific and vigorous growth several degrees south of the Tropic of Capricorn, being most abundant in Shark's Bay, on the Western Australian coast, and in Wide Bay and Moreton Bay, on the Eastern or Queensland sea-board. By direct experiment the writer had proved, a short while previously, that the larger tropical species might be artificially transported to, and would thrive and propagate in, the extra-tropical waters of Shark's Bay; and his investigations of the Abrolhos reefs led him to anticipate that in their vicinity this valuable tropical species would meet with even more favourable growth conditions. The specially propitious conditions noted were intimately associated with the correlated marine fauna and environments of the respective districts. In Shark's Bay, about 25° 55" S., where the shell was experimentally

ABROLHOS CORAL. *Montipora verrucosa.* Nat. size, p. 146.

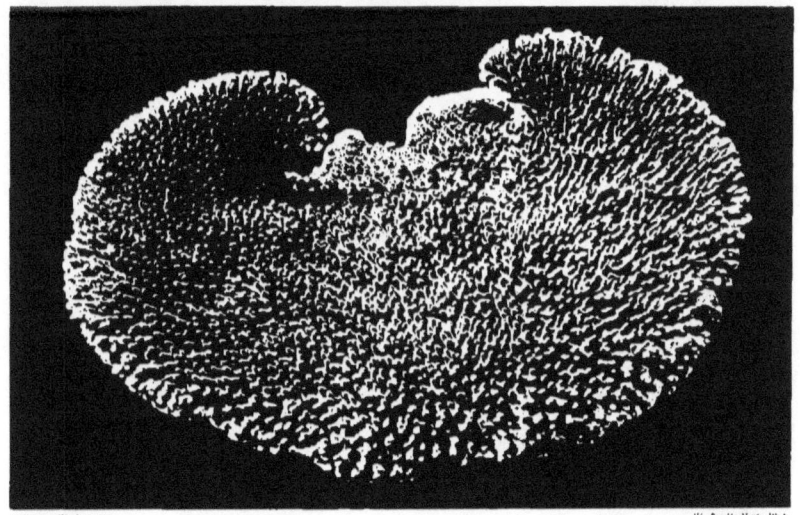

ABROLHOS CORAL. *Montipora corymbosa.* One-fifth nat. size, p. 140.

laid down, the sea-bottom was covered by coral banks of considerable extent, which were composed exclusively of representatives of the genus Turbinaria. *Turbinaria conspicua*, a most luxuriant foliaceous species, and irregular dome-shaped masses of *T. peltata*, represented the dominant forms upon these banks. Specimens of these Madrepores, no less than five feet in diameter and some fifteen feet in circumference, were secured by the author from this locality and are now on exhibition in the Coral Galleries at the Natural History Museum, South Kensington. They constitute, up to the present date, the largest examples of *Madreporidæ* yet brought to England, or, indeed, to any European museum.

It is worthy of record in this connection that the genus Turbinaria has, as the result of a somewhat extensive investigation of the coral reefs around the Australian coast, been found by the author to enter most extensively into reef-composition in the colder, or extra-tropical, areas within Australian waters. They have, in this manner, been observed by the writer to predominate in the reefs in Wide Bay, Queensland, on the southern outskirts of the Great Barrier Reef; in the colder, though more northern, inter-tropical waters of the Gulf of Carpentaria; and, finally, in the Shark's Bay district of Western Australia. It should be further mentioned that these Turbinaria reefs, while attaining to within a short distance of the surface of the water, are never exposed above it by the falling tide, as commonly happens in reefs of mixed *Madreporidæ* growing within the tropics. A fuller reference to these Shark's Bay Turbinaria, together with illustrations of the more prominent species, will be found in that portion of Chapter VIII. which specially deals with Corals and Sea Anemones.

The most prominent testimony indicating that the Houtman's Abrolhos reefs were better suited even than Shark's Bay for the introduction and cultivation of the large tropical Pearl Shell, was afforded by the composition of the reefs. These, in place of being built up of the extra-tropical or cold water Turbinariæ that flourish in Shark's Bay, were composed of numerous varieties of branching *Madrepore*, or so-called Stag's Horn Corals, with which were intermingled many species of *Porites*, *Montipora*, *Pocillopora*, *Seriatopora*, *Cæloria*, *Goniastræa*, *Mussa*, *Symphyllia* and other essentially tropical generic types. Typical examples of all these genera are, in fact, included in the collection from these reefs presented by the writer to the British Museum.

The reef corals of Houtman's Abrolhos, while growing very near to the surface level of the water at low ebb tide, very rarely appear above it, and then to the extent of a few inches only. It was observed by the author on one such occasion during his residence on Pelsart Island, that the polypes pertaining to the projecting extremities of the branching Madrepore, which were thus laid bare during the early hours of a winter's morning, were killed by the brief exposure to the chilly atmosphere, leaving, after speedy decomposition, the coralla extremities bleached perfectly white. Within the moderate depths of from just beneath low water mark downwards to as much as ten or fifteen fathoms, the corals upon the Abrolhos reefs and within their contained lagoons attain to a luxuriance of growth that is not surpassed even upon the Great Barrier of Queensland. Nowhere else, indeed, has the author met with such extensive sheets, as it were, of one and the same species spread over continuous areas. This was especially notable of certain species of Madrepore or Stag's Horn Corals. One of these, nearly allied to *Madrepora hebes*, but having a more luxuriant growth form, covered acres of the sea bottom in the enclosed lagoon of Pelsart Island. Being of a dark purple or purplish brown hue, with the tips and all the growing points brilliant mauve or violet, it produced, with the intervening patches of pure white sand, a most fascinating scene. In other areas a finer branching species allied to, or identical with, *Madrepora syringoides*, of a dark seal-brown tint with creamy-white tips, and growing points, formed equally extensive patches. A cluster of this Madrepora is photographically reproduced in the lower figure of Plate XXIII. Seen in bulk, this delicately branching white-tipped species was highly suggestive of a free blossoming white-flowered heath, the contour of the tubular terminal calicles viewed from a little distance wonderfully resembling Epacris blooms. In a third locality a more robust light buff-coloured species with heliotrope tips, near *Madrepora pulchra*, but with closely radiating, less divided, branches, represented the dominant type. A double or twin corallum of this variety is depicted in the upper half of the Plate last quoted.

A large crateriform or vase-shaped species, closely resembling *Madrepora corymbosa*, of a golden-brown or brown-pink hue with cream-white edges, undoubtedly represents one of the most conspicuous members of the genus indigenous to the Abrolhos group. A single corallum of this type is figured in the lower part of Plate XXIV. The specimen here delineated measured close upon three feet across. Larger matured coralla, however, commonly attain to a diameter of five or six feet and upwards, or it may happen that two or three specimens become joined laterally to one another, producing an extensive horizontal sheet. The species grows under manifold conditions, forming independent islets, clothing exclusively the precipitous sides of the submarine cliffs with tier upon tier of its horizontal fan-shaped coralla, or constituting but one among the many varieties entering into the composition of the larger reefs. It is as seen growing on the perpendicular walls of the deeper reef edges that it furnishes the most remarkable spectacle, often occurring under these conditions in such luxuriance as to be visible in superimposed tiers or terraces as far down through the clear water as the eye can penetrate. A specially favourable opportunity of studying this coral growth was afforded the author on one of his return voyages from the Abrolhos Islands, when owing to lack of wind the lugger drifted and grounded for some little time, fortunately at low water, on the margin of one of these hanging coral gardens. Shortly after the vessel drifted loose again with the rising tide, a unique sight presented itself by the appearance of a company of sharks, no less than fourteen, ranging from eight to ten or twelve feet in length, being discernible at one time from the boat's deck. Without wishing to cultivate their nearer acquaintanceship, one could not withhold admiration of the leisurely grace of their motion as seen at various depths in the emerald clear water, and of the amazing swiftness, like the release of an arrow from a bow, with which they would abruptly dart away in pursuit of some passing fish, or in a race for an attractive lure thrown to them from the boat.

To return to the subject of the Abrolhos corals, there was one form developed under conditions closely corresponding with the last-mentioned type, but notable for its usually pale sage-green hue, and the remarkable diversity in the contour of its coralla. In some instances, it formed robust, shortly branching, bush-like growths; in others, fan-like expansions, or even encrusting masses, or it might be a combination of these types. Two characteristic illustrations of this coral, which is apparently an undescribed species, but allied to *Madrepora sarmentosa*, also a new species collected by the writer on the Great Barrier Reef, are given in Plate XXV. In the upper of these two, the encrusting character of the initial or basal growth-stage of the species is very obvious. A third and remarkably fine example of this coral, obtained in the south entrance to Shark's Bay, which measures over three feet across, figures as a tailpiece to this Chapter. It combines in its individual corallum so many noteworthy variations that detached fragments from separated areas might be readily mistaken for distinct species. All of these specimens figured, together with several additional ones, have been contributed by the author to the British Museum Coral Galleries. They there constitute a most instructive object lesson for the benefit of those systematic zoologists with whom small or fragmentary specimens alone have been hitherto available for classificatory purposes. In recognition of the multitudinous growth-forms exhibited by this particular coral, it is proposed here to provisionally distinguish it by the suggestive title of *Madrepora protæiformis*.

Many of the areas of the Abrolhos reefs were characterised by an interblending of all of the various species of Madrepore enumerated in the foregoing paragraphs. The author's attention was, however, particularly impressed by the very definite border-like plan of their growth upon each side of the numerous river-like channels which circulate through the Pelsart Island Lagoon. In the regularity of their development and blended tints they vied, on colossal lines, with the artificially laid out flower-parterres of a well-appointed garden.

The endeavour has been made by the author in Chromo-Plate IV., facing page 144, to commit to paper a faint and necessarily very inadequate idea of the unique and remarkably beautiful spectacle yielded by an area occupied chiefly by the violet-tinted Madrepora, as seen and roughly sketched while standing in the bows of a small row boat, and drifting down one of the intersecting channels. Through the glass-clear water in the immediate foreground every coral branch was distinctly visible, the clustered coralla in many instances constituting harbours of refuge to parrot and other fishes of the most brilliant hues, which would dart to and fro across the intervening spaces as the boat approached.

The horizon line, looking oceanwards, as shown in this sketch, was, in its way, almost equally remarkable. The boundary in this direction is represented by the lovelly raised surface of the rocky platform, which constitutes a massive breakwater between the placid waters of the lagoon and the tumultuous billows, which break unceasingly, and with a sustained roar mightier than that of Niagara, upon the precipitous edge of the outer barrier. An attempt has been made to portray the singular appearance of the rebounding columns of water thrown up against the horizon-line by the breaking waves, the altitude of which is optically greatly enhanced by mirage. This atmospheric phenomenon is here, as throughout tropical seas, of general occurrence. As an illustration of its prevalence, the very low points of the archipelago of islands that enter into the composition of the Houtman's Abrolhos commonly appear to be elevated to an abnormal height above the horizon. A little further north, on the adjacent Western Australian coast, it is by no means unusual, as the effect of mirage, for the passenger steamers to discover the boats of the pearling fleets, or it may be of other denominations, some hours before they are

PLATE XXV.

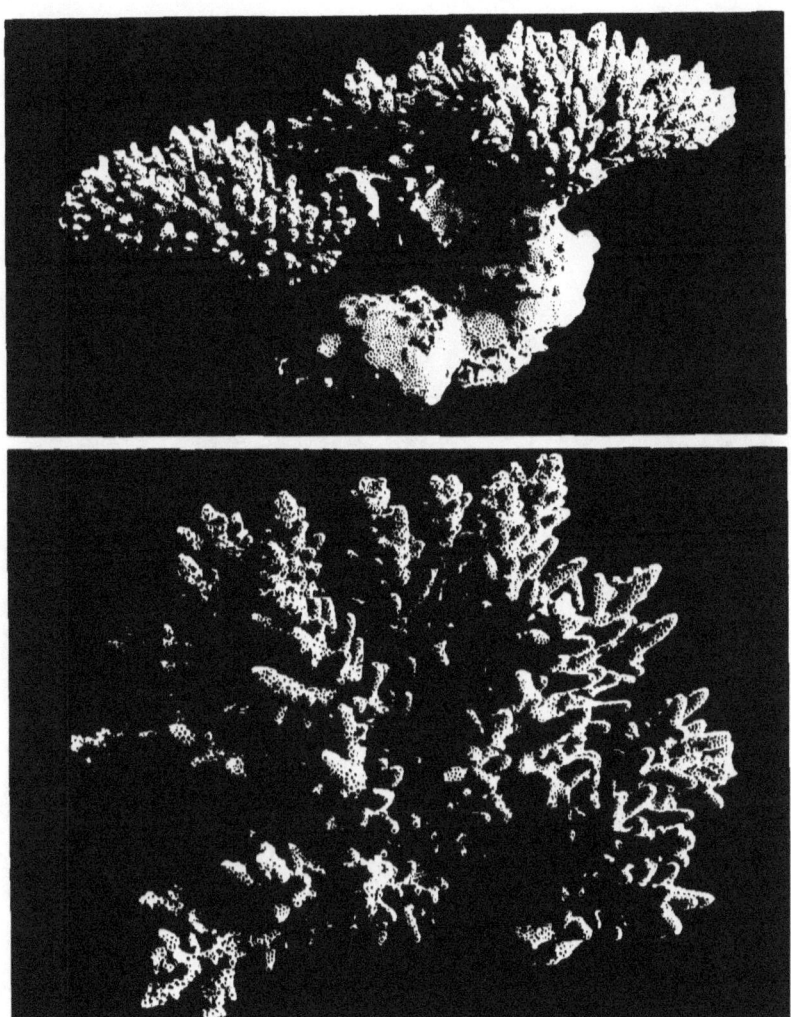

ABROLHOS CORALS. *Madrepora prostrata*. One-fourth and one-third nat. size, p. 111.

actually in sight, and the boats may indeed disappear from view again before they come within the range of ordinary vision. On the reefs of the Great Barrier system, on the Queensland coast, the phenomenon of mirage exemplified by the abnormal elevation of the breaking waves may be very generally observed, the detached rock-masses, lying on the outer margins, black through exposure of the weather, and popularly known as "nigger-heads," standing out in marked contrast to the snowy whiteness of the breakers. In the neighbourhood of Torres Straits and the Warrior Reefs, white pelicans commonly take up their station on these reefs at low tide, and, approached under conditions favourable to the manifestation of mirage phenomena, seem to stand high in the air and to be of colossal size.

An additional illustration of the aspect of the mirage-elevated breakers on the Pelsart barrier is afforded by the photograph reproduced at the foot of page 132. In this photograph, which was taken with the camera erected on a partly submerged rocky ledge within the lagoon, several distinctive features are clearly defined. While the breakers extend throughout the entire length of the horizon, the middle distance is occupied by the level platform reef, upon which are resting several huge masses of rock that have been torn off the outer margin of the reef and hurled to their present position during a storm of unusual violence. On the rock-mass to the left, one of the shore-birds popularly called Red-bills, Hæmatopus, is distinctly visible, and in the original negative there are others also on the farther rocks. A very characteristic photograph of similar storm-stranded coral-rock masses, taken at close quarters on the Capricorn Island reefs, on the Queensland coast, is furnished by Plate XXX. of the author's recently published work "The Great Barrier Reef of Australia." A near approach to the masses on the Abrolhos barrier was not practicable during the seasons of the author's visits. The water, even at lowest tide, was continually surging over from the outer ocean, and though a foothold for a brief interval with the water knee-deep might be retained, a succeeding wave would bring a flood against which it was impossible to stand, much more to effectually erect and work the camera. In the photograph here reproduced, one exhausted wave is delivering its spent energies into the lagoon, while a second one overflowing from the outer waters has advanced about half-way across the platform reef.

With reference to the violet-tinted branching Madrepora that enters so extensively into the composition of the Coloured Reef View illustrated by Chromo-Plate IV., it may be mentioned that in the deeper and stiller areas of Pelsart Island lagoon, with the water from five to ten fathoms deep, masses of this same Madrepora cover

144 THE NATURALIST IN AUSTRALIA.

W. Saville-Kent, Photo.
STAG'SHORN CORAL GROWTH, *Madrepora hebes*, GREAT BARRIER REEF, QUEENSLAND.

acres in extent, and attain to a tall coppice-like growth of five or six, or more, feet in height. The dead and decaying coralla of this species that also bestrew the sea-bottom in portions of these areas present a remarkable resemblance to cut and prostrate brushwood. A very interesting demonstration of the reparative powers of the Madreporidæ is furnished by the accompanying illustration. It represents a branchlet of the purple species, which, having been broken off at some previous period,

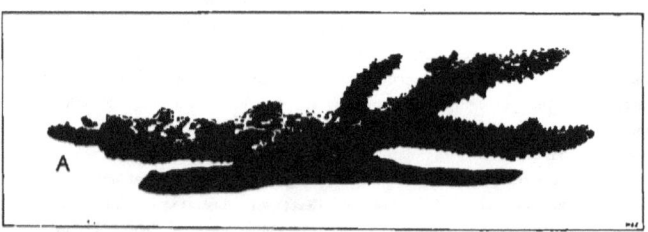

W. Saville-Kent, Photo.
MADREPORA BRANCHLET, SHOWING RE-GROWTH AT BROKEN END, MARKED A. TWO-THIRDS NATURAL SIZE.

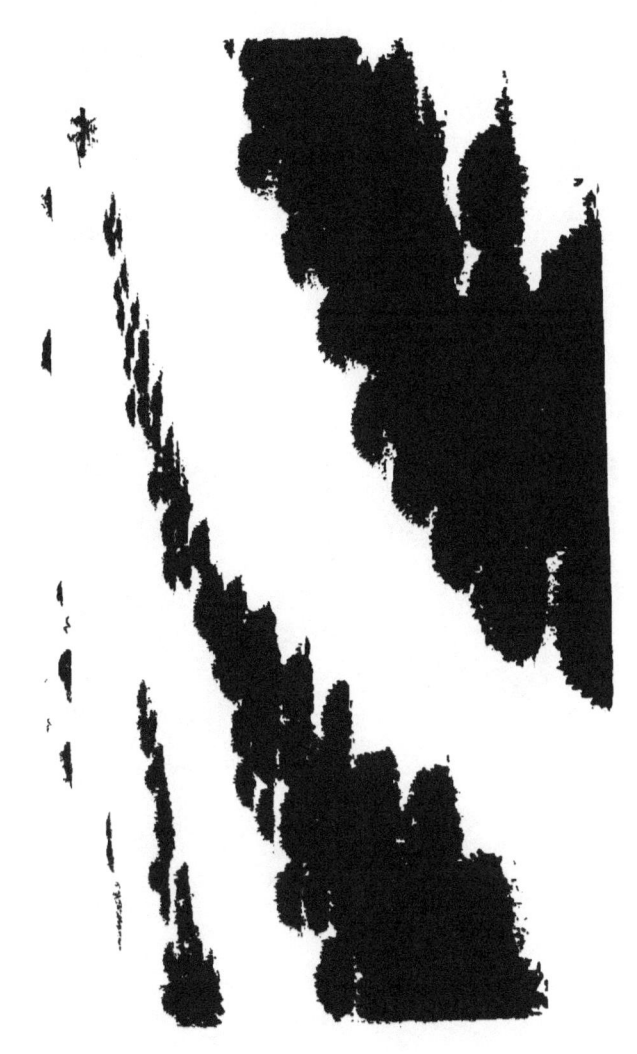

has commenced to grow retrogressively from its broken end. The original axial corallite has in this manner developed a shoot about three-quarters of an inch long, from which numerous lateral corallites have also commenced to bud.

Mention has been previously made of the close resemblance of this purple-tinted Stag's-Horn Coral to the *Madrepora hebes* of the Queensland coast. As an illustration of the correspondence that may be observed in their respective growth plans, a photograph of a tidally exposed reef area, consisting almost exclusively of that variety taken by the author in the vicinity of Lark Passage on the Great Barrier, is reproduced on page 144. Additional reef views embodying the same type will be found in the writer's volume especially descriptive of the Great Barrier Reef products. The coral on this Queensland reef lacked the brilliant tints of the Abrolhos colony-stocks, being for the most part of a warm brown hue with whitish tips. Some few of the coralla among the mass were, however, brilliant green or lilac, while in other localities the same species was met with in which the greater portion of the coralla was bright grass-green, but every branch tip to the extent of about half-an-inch, an intense violet. As recorded by the writer in the volume above quoted, the colour characters of the Madreporidæ are not only exceedingly variable in the same species, but are even unstable in individual coralla. Thus a colony-stock of the type under notice, *Madrepora hebes*, growing in a pool at Thursday Island, Torres Straits, was, when first observed, made up of pinkish-brown stems and branches with greenish-white tips, while the polypes were all of a light emerald-green tint. On examining the same growths two years later, it was observed that the branches and main stems were now a dark seal-brown and their tips, to a large extent, a pale lilac-blue tint, while the polypes had assumed, for the most part, a clear red-brown hue. It is consequently quite possible that the reef areas in Pelsart Island Lagoon here portrayed, may at some future date be found to have exchanged the purple tints recorded in 1894 for more sombre brown or brilliant green.

Next to Madrepora, the genus Montipora builds up the most conspicuous coral developments in Pelsart Island Lagoon. Its members are chiefly of either an encrusting or a foliaceous character, and in many instances of the most brilliant violet or even magenta hue. Selected specimens, quickly dried in a breeze, still in the writer's possession, have retained for over two years much of their pristine splendour, and exhibit sufficient of their original tints to convey to the untravelled a faint idea of their living beauty. One species of this genus Montipora, apparently new to science, but identical with a type collected by the writer also at the Palm Islands on the Queensland coast was notable

T

for the erect foliaceous character of its coralla, which in many instances took a very elegant and symmetrically convoluted scroll-like contour, and has on this account been associated by the writer with the title of *Montipora circinata*. This species was not so remarkable for its colour, being usually of a pale pinkish-red hue, with lighter edges, but was at the same time conspicuous through the circumstance of its growing in deep water in distinctly isolated patches of many square yards in extent. The more matured coralla not unfrequently attain to a height of as much as three feet, and, seen in masses in the crystal water beneath the boat, might be readily mistaken for beds of broadly foliaceous plants. A photograph of an example of this type, viewed vertically to illustrate its spirally convoluted growth pattern, is given life size in the upper half of Plate XXIV. A lateral view of the same specimen is also reproduced on a smaller scale in the accompanying figure.

In connection with the subject of the Houtman's Abrolhos Madreporidæ, it is worthy of remark that nowhere perhaps do such facilities exist for making a systematic record of the growth rate of tropical species as here. Many of the component reefs and lagoons are so sheltered and so readily accessible that the necessary measurements, observations, and experimental cultivations could be carried on at all seasons. In support of this suggestion, it may be mentioned that the coralla of various species of Madrepora have already established themselves upon the basement of the stone jetty constructed by Messrs. Broadhurst and Co. on Rat Island, in the Easter group, for the shipment of guano. Certain of these taking the form of cauliflower-like clusters, and having a diameter of some nine or ten inches, represented, the writer was informed, a growth of less than two years; while smaller examples, with a diameter of three or four inches only, had been built up within the interval of a twelvemonth. It is noteworthy to observe that these Madrepora clusters were

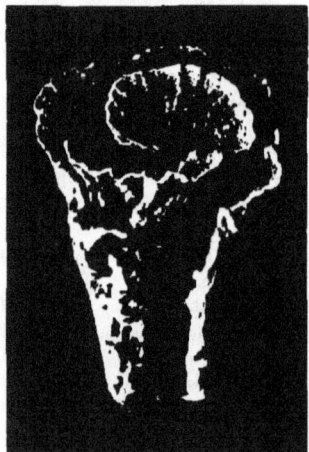

PELSART ISLAND SCROLL CORAL, *Montipora circinata*,
TWO-FIFTHS NATURAL SIZE.

in most instances the initial growths of the handsome violet-tinted species growing on the adjacent reefs, which has been referred to as attaining to such a luxuriant growth in the Pelsart Island Lagoon.

At some future date, when the Colony of Western Australia shall have passed its present lusty adolescence, and arrived at that maturer age when it shall possess its own University and Chairs of Natural History, it may be safely prophesied that these Abrolhos reefs, within a twenty-four hours' journey from the metropolis of Perth, or but three or four hours' sail from Geraldton, will constitute one of the happiest and most productive hunting grounds and fields for biological investigation to the associated students of and graduates in Natural Science. In addition to the unprecedented facilities here offered for the most exhaustive study of living Stony Corals or Madreporaria, either individually or in the bulk, abundant material is also to hand for the observation and record of the numerous phenomena of wider scope relating to the formation and growth of the reefs, to their environments and food supply, and also to the complex questions of their rise or subsidence.

As a characteristic illustration of the earliest recognisable phase of a coral island, attention may here be directed to the photograph reproduced in this page taken by the writer in close vicinity to Gun Island, in the Pelsart group. It

"THE BIRTH OF A CORAL ISLAND," ABROLHOS ARCHIPELAGO, WESTERN AUSTRALIA.

represents the first accumulation of loose coral fragments raised above high water mark, which, by continued accretion and solidification, becomes fashioned into a typical coral island upon which herbage and terrestrial organisms may eventually become established. This photograph, which has been suggestively designated "The Birth of a Coral Island," was taken, as will be recognised, at low water, the weather-bleached central accumulation of coral fragments alone showing above the surface when the tide is up. This infant coral islet, which measures some twenty feet only in length, is the direct produce of the swirling currents that circulate at high water over the surface of the flat, lifeless expanse of platform reef. During some abnormal tide or storm this heap of loose fragments was first swept into its present position and constitutes the nucleus of an islet that will probably hereafter become amalgamated with Gun Island. The living corals which are to be found growing only around the circumference of this platform reef, beneath low water mark, are thus shown to take no immediate part in the building up of the coral island, which is the direct outcome of the winds and currents acting on the previously detached reef *débris*.

The power and selective force of the waves and currents in determining the character of the beaches and the subsequent formation of coral islands, such as the Houtman's Abrolhos, is very instructively illustrated by the two photographic views reproduced in Plate XXVI. The lower of these illustrations depicts an area of the open beach of Pelsart Island, a little to the east of Wreck Point, which is composed almost entirely of the eroded crateriform corallæ of the Madrepores that originally grew upon and have become detached from the steep escarpments of the outer reef. The slightly raised coral limestone cliffs at Wreck Point, figured in the Chapter heading, are also composed for the most part of aggregated coralla of the same description, the separately embedded coralla being discernible at various points in the original negative. A striking contrast to the picture last described, with its ponderous elementary components, is afforded by the second one, immediately above it. Here, in place of massive coralla, the beach, for a long stretch, is composed almost exclusively of the pearly shells of the univalve Molluscs *Turbo margaritaceus* and *Trochus cærulescens*. This wonderfully prolific shell beach also belongs to Pelsart Island, but is situated a little away inside the lagoon, where comparatively quiet water prevails.

Passing on to the consideration of zoological groups, other than the Madreporaria, which bear testimony to the essentially tropical character of a large portion of the marine fauna of Houtman's Abrolhos, we find that some of the most remarkable evidence is yielded by that group of the Echinodermata, distinguished by the title of the Holothuridæ, which comprises the so-called Sea-Cucumbers or Trepang and Bêche-de-Mer of commerce. Torres Straits and the Northern moiety of the Queensland Great Barrier Reef represent the regions on the Australian coast which have alone, so far, yielded the most valuable varieties of the last-named marine

PLATE XXVI.

SHELL AND CORAL BEACHES, PELSART ISLAND, HOUTMAN'S ABROLHOS, p. 148.

commodity; supplies to the value of from £15,000 to £20,000 being annually exported thence to the Chinese market. From the tropical areas of the Western Australian coast-line, and more particularly from the neighbourhood of King's Sound, a single entirely distinct, and, as compared with the choicer Queensland species, an inferior commercial variety of Bêche-de-Mer, nearly allied to, if not identical with what is known in Queensland as "Surf-Red," *Actinopyga mauritiana*, is collected for export. Neither has a diligent investigation, made with the express object of discovering the presence of other more valuable species, so far proved successful. It was consequently a most unexpected surprise to the writer, in the course of his explorations of the Pelsart Island reefs in Houtman's Abrolhos, to meet with not only one, but no less than three of the most esteemed Torres Straits and North Queensland varieties. These several species are commercially known in the Queensland market as "Black Fish," "Red Fish," and "Teat Fish," or, in the Chinese vernacular as "Woo-Sum," "Hung-Hur" and "See-Ok-Sum." On spirit-preserved specimens of these varieties being submitted to the British Museum Echinoderm specialist, Professor Jeffry Bell, that authority decided that only one of these, the "Red Fish," *Actinopyga obesa*, had hitherto been associated with a scientific name, and the writer has accordingly portrayed and describe the remaining two, "Black Fish" and "Teat Fish," in his treatise on the Great Barrier Reef, under the respective titles of *Actinopyga polymorpha* and *Holothuria mammifera*.

The relatively short time available for the exploration of favourable Bêche-de-Mer feeding grounds among the Abrolhos reefs revealed the presence of all of the three above-mentioned valuable commercial varieties in tolerable abundance, and there is little doubt that, in combination with the development of other fishery potentialities of the district, the systematic conservation and export of Abrolhos Bêche-de-Mer would constitute a remunerative source of income. One other essentially North Australian representative of the Holothuridæ, though having no intrinsic commercial value, was observed by the writer at the Houtman's Abrolhos. This was the *Synapta Beselli*, of Semper, a species first discovered by that naturalist in the Philippine Islands, and remarkable for its extraordinary length, no less than five or six feet when fully extended, and the nodulated, quadrangular contour of its transparent pink, or more or less mottled brown, integument. This species occurred in considerable abundance on the reefs in the neighbourhood of Warrior Island, Torres Straits, but, so far as the writer's investigations have extended, does not occur on the mainland coral reefs of Western Australia.

The fish fauna of Houtman's Abrolhos was found, as might be anticipated in virtue of its essentially migratory constituents and its proximity to areas of relatively cool water, an interesting intermixture of both tropical and temperate species. Conspicuous among the fishes indigenous to the temperate Australian sea-board may be mentioned such species as the Snapper, *Pagrus major*; the Sergeant Baker, *Aulopus purpurissatus*; Australian Whiting, *Sillago ciliata*; Yellow-tail, *Seriola gigas*; and a species of what in the Sydney market would be designated a Morwong, Chilodactylus. Characteristic tropical fish were, on the other hand, specially represented by innumerable varieties of Parrot-fishes, Labridæ and Scaridæ. Many of these, it is interesting to observe, such as species of Julis and Pseudoscarus, had not been met with by the writer further north on the Western Australian coast, but were familiar to him, as in the case of the Holothuridæ, as inhabitants of Torres Straits and the Queensland Great Barrier region. Such species, again, as *Platax orbicularis* and *Mesoprion Johni*, the Golden Snapper of Thursday Island, Torres Straits, may be mentioned among the essentially tropical forms that were found frequenting the Abrolhos reefs.

No zoological groups in addition to the foregoing yielded such substantial evidence of tropical affinities, though the investigations prosecuted were necessarily of too short duration to allow of any very positive deductions in this direction. Among the Crustacean Class Crawfish, Palinuri, were exceedingly abundant and identical with the ordinary Perth and Fremantle market type. Crabs referable to genus Grapsus, also similar to those frequenting Fremantle harbour, swarmed at the bases of the more sheltered coral limestone cliffs. There was missing, however, from this locality the toothsome Blue Crab, *Neptunus pelagicus*, that abounds along the Western Australian coast from Shark's Bay southwards. It was here replaced by another smaller and more essentially tropical representative of the same genus.

Apart from considerations of tropical or temperate affinities, reference may be here made to a most magnificent Nudibranchiate Mollusc obtained by the author in the neighbourhood of Rat Island, in the Eastern Group of the Abrolhos. The species, of which a coloured illustration is reproduced in Chromo-Plate V., is, as shown by the conformation of its gills or branchial plume, referable to the genus Doris, of which it represents a new and, so far as ascertainable, the largest discovered species. In its condition of full extension, as delineated in the illustration, it occupies a quadrangular area of no less than nine or ten inches long with an inch or two shorter diameter, but it is at the same time of so flattened a shape as to more closely resemble a huge complanate Planarian akin to the genus

Pseudoceros. One of the most noteworthy features of this grand Doris was its remarkably brilliant colour. This differed somewhat among individual specimens observed. Red was always the predominating hue, but this varied from a dull orange or Indian red in some examples to the most brilliant vermilion tint characteristic of the specimen figured. This gorgeous tinting of the general surface of the body is, by contrast, considerably enhanced by the elegant scarlet and white frill-like border which encompasses the entire periphery. Such was the abnormal size and brilliancy of the first example of this Nudibranch that fell under the writer's observation, floating in the water at a little distance from the Rat Island jetty, that the possibility of its being a living organism did not immediately present itself. It was, in fact, passed by at first sight, under the impression that it was simply a portion of the lung of a freshly-killed sheep that had been immolated that morning for the use of the station. When it was seen a second time, it had drifted nearer to the jetty, and the recognition of the splendid prize awaiting appropriation dawned upon one's vision as a revelation. With reference to its magnificent proportions and brilliant hue, the writer has provisionally conferred upon this fine Nudibranch the regal title of *Doris imperialis*.

To conclude with a consideration of the most salient facts chronicled in this Chapter—namely, the very remarkable interblending of both tropical and temperate marine organisms, and more especially the phenomenon of species occurring in the Houtman's Abrolhos which, while indigenous to Torres Straits and the north Queensland coast, are not inhabitants of the adjacent sea-board of Western Australia—but one interpretation appears to be permissible. This is that an ocean current setting in from the equatorial area of the Indian Ocean penetrates as far south as this island group without impinging on the adjacent mainland, and that that stream is the medium that has conveyed thither the floating germs of the Cœlenterates and Holothuridæ, which possess such essentially tropical affinities. The presence here of migratory tropical fishes admits also of the explanation that they would very naturally follow a warm stream bringing them to such congenial conditions and environments as exist in and among the Houtman's Abrolhos. That this interpretation is the correct one is substantially supported by a reference to the Admiralty Charts denoting the courses of the ocean currents in this region, that have been kindly placed at the writer's disposal by the Hydrographic Department of the Admiralty. In one of these, No. 2640, there is, in fact, clearly indicated a prevailing northerly drift of the ocean currents along the Western Coast of Australia, but at the same time there is a distinct southerly intrusion of the equatorial waters of the Indian Ocean, at some distance off shore, down towards, though not absolutely reaching, the Abrolhos Islands.

Further substantial evidence supporting the correctness of the interpretation here submitted, is afforded by the results of synchronous readings of the sea temperature obtained by the writer with the assistance of Mr. G. Beddoes, Mr. Broadhurst's managing partner, in the vicinity respectively of the Abrolhos Reefs and at Champion Bay. These taken in the early morning of July, midwinter, 1894, at three feet below the surface of the water exhibited a difference of from ten to fourteen degrees Fahrenheit; readings at the Abrolhos averaging 69° and 70°, while those at the Geraldton Pier registered as low as 56°.

Many of the zoological data recorded in this Chapter were, it may be mentioned, embodied in a paper contributed to the 1893 Norwich Meeting of the British Association, and are published in abstract in the Proceedings for that year. Figures and descriptions of a lizard species, *Egernia Stokesii*, which is particularly abundant on Gun Island, in the Abrolhos group, will be found in Chapter III.

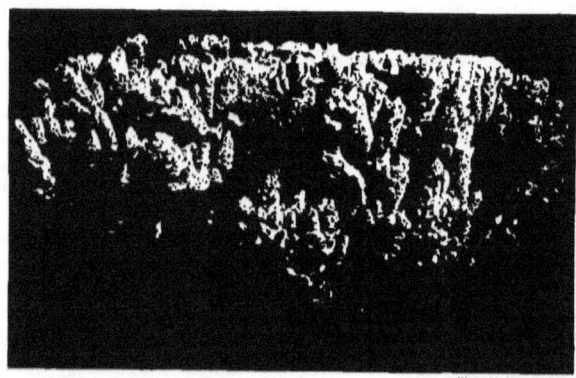

ABROLHOS CORAL, *Madrepora protei formis*, p. 142. ONE-SEVENTH NATURAL SIZE.
W. Saville-Kent, Photo.

Chicago Plate. VI

PLUMED TREVALLY, *Caranz gallus*, p. 169.

PHENOMENAL AND ECONOMICAL.

THE marvellously varied fish fauna of the Australian seas and rivers would require a book, not to say volumes, to do it justice. Within the brief space at present at disposal, little can be attempted beyond an outline sketch of its leading features, and an indication of those among its multifarious representatives which, with regard to their economic utility, bizarre contour, or remarkable habits or alliances, stand out pre-eminently among the fishes of the world.

Viewed in its entirety, the fish fauna of Australia is found to allow of a fairly equal separation into two primary regional subdivisions. The first of these, embracing the denizens of the tropical sea-board and estuaries of the Australian

Continent, possesses a marked affinity, and even to a large extent identity, with the species characteristic of the Indo-Malay region. The second one constitutes in itself a highly characteristic south-temperate fish fauna which possesses many points of correspondence with that of the littoral regions of both the North Atlantic and the North Pacific. In some few instances, the correspondence of the types inhabiting respectively these widely separated oceans is so remarkable that it has not been found possible to associate them with independent specific diagnoses. It becomes consequently incumbent to conclude that these in all recognisable respects homoplastic types have been either arrived at by heterogeneous processes of evolution, or that they represent one and the same species, whose members have become widely separated as the result of some marked change in the relative positions of, and intercommunicating currents between, the larger oceanic areas since their earliest appearance under their existing specific forms.

As examples of fish which are bracketed with a corresponding synonomy, and occur as indistinguishable types in such remotely situated areas as the temperate Australian and European seas, reference may be made to the John Dory, *Zeus faber*, purchasable in either the London or Sydney markets; the Sprat, *Clupea sprattus*; the Conger Eel, *Conger vulgaris*; the Frost Fish of Australian and New Zealand seas, *Lepidopus caudatus*, synonymous with the Scabbard Fish of European seas; the Tunny, *Thynnus thynnus*; the Horse Mackerel, *Trachurus trachurus*; the Skipjack of Australia, *Temnodon saltator*, known on the East Coast of North America as the Blue-fish; and the little Bellows-fish, *Centriscus scolopax*. Many species of Sharks and Rays, such as the Blue Shark, *Carcharias glaucus*; the Porbeagle, *Lamna cornubica*; the Fox Shark, *Alopecias vulpes*; the Spiny Dog-fish, *Acanthias vulgaris*; the Angel-fish, *Rhina squatina*; and the Eagle Ray, *Myliobatis aquila*, are also common to the European and Australian waters, but, as they are for the most part ocean rovers with an almost cosmopolitan distribution, their presence in these widely separated areas does not possess the same significance.

In addition to the list just enumerated, a number of species might be mentioned, concerning which, while they occupy such widely separated habitats, the recognisable points of distinction are of so trifling a description that it is a matter of dispute among ichthyological authorities whether they are to be regarded as identical or independent species. The Australian Jew-fish, *Sciæna antarctica*, is thus regarded by some as differing in no essential points from the European Maigre, *Sciæna aquila*. One of the Australian Mackerels, the *Scomber antarcticus* of

Castlenau, is declared by Professor McCoy to possess no distinctive characters that he can detect when compared with the Common Mackerel, *Scomber scomber*, of the European seas. There is likewise an Anchovy abundant on the Southern Coast-lines of Australia that so closely resembles the much-esteemed European species that it is indifferently described in ichthyological works as *Engraulis encrasicholus*, *var. antipodum*, or as *Engraulis antarcticus*.

Apart from the close affinity subsisting between the specific types enumerated in the foregoing paragraphs, an interesting parallelism obtains between certain more abnormal genera which possess characteristic representatives in both the North Temperate and South Temperate hemispheres, separated by a wide tropical area from which they are absent. In this manner, that most singular type, the Northern Chimæra, *Chimæra monstrosa*, possesses its counterpart in the so-called Elephant-fish, *Callorhynchus antarcticus*, of the New Zealand and Southern Australian seas, while the peculiarly modified sucking fishes, Gobiesocidæ, represented in British waters by several species of Lepidogaster, find their co-types on the Australian Coast-line in several species of the genus Crepidogaster.

The Cod family, Gadidæ, yields a remarkable instance of a group of littoral fishes with allied but diverse genera, subsisting in Australian and other South Temperate waters, and in those of the North Temperate zone, with an at present impassable barrier of tropical ocean between them. Gadus proper, embracing eighteen species, includes such well-known forms as the Common Cod, *G. morrhua*; the Haddock, *G. æglefinus*; the Whiting, *G. merlangus*; the Pollack, *G. pollachius*; the Saith or Coal Fish, *G. virens*, and other European market species, which are entirely confined to the Arctic and North Temperate zones. The Hake, *Merlucius vulgaris*, however, has its counterpart in *M. gayi* of the New Zealand and Magellan seas; and the Rocklings, genus Motella, numbering three British species, have also representatives on the Coast of New Zealand and at the Cape of Good Hope. There are but two genera, Lotella and Pseudophycis, belonging to the true Cod family, which inhabit New Zealand and Southern Australian waters and are almost entirely restricted to this area of distribution. Both of them are very nearly allied to the European genus Phycis, including the so-called Forkbeards, *P. blennioides* and *breviusculus*, of British waters, from which they differ only in the relatively less reduced condition of development of the ventral fins. The largest and commonest form, *Pseudophycis barbatus*, is known as the Rock Cod in Tasmania and New Zealand. It grows from a more general average of two or three to a weight of eight or nine pounds, is taken in large quantities with hook and line, and constitutes one of the most abundantly represented species in the Hobart fish market. As many as four species of the allied genus Lotella, one of which, *L. callarias*, is the Cod of the Melbourne fishermen, and another, *L. marginata*, the "Beardie" of the Sydney fish market, extend as far north in their distribution as Port Jackson, in New South Wales. The last-named species has been also obtained by the writer in Tasmania. In face of these facts, the statement in Dr. Gunther's "Introduction to the Study of Fishes," p. 284, that the fish fauna of Australia, as compared with that of New Zealand, is characterised by "an apparent total absence of all Gadoids," requires modification.

It is a noteworthy circumstance that the rivers of Europe and North America yield a single fresh-water representative of the Cod family. This is the well-known Burbot or Eel-pout, *Lota vulgaris*, limited in the British islands to the streams of Yorkshire, Durham, Norfolk, Lincolnshire, and Cambridgeshire, but is by no means plentiful in either of these counties. In shape and aspect the Burbot very closely resembles the Australian Lotellæ, and grows to an attested length of over three feet, and a weight of from six or eight to as much as twenty pounds. In English waters, however, a weight of two or three pounds is the more common calibre. While no true fresh-water Gadoid has as yet been discovered in Australia—the so-called Murray Cod, as hereafter explained, being a perch—one exceedingly interesting fish that constitutes the sole type of a family most nearly related to the Gadidæ is abundant in certain of the rivers of Tasmania and the Southern Australian colonies. This is the fresh-water Black-fish, of the Australian colonists, *Gadopsis marmoratus*, an excellent table fish, which, in the Ringarooma in North Tasmania, not unfrequently attains to a weight of ten pounds and nearly three feet in length.[*]

The general contour of the Australian Gadopsis is very much that of such gadoids as Phycis or Pseudophycis, and, like the former, it has filamentous, bifurcated, jugular fins. Owing to the fact, however, that a small portion of the membranes of the elongate dorsal and anal fins are supported by slender spines in place of flexible rays, ichthyologists have decided upon the relegation of this type to an independent

---

[*] Dr. Gunther, Catalogue of Fishes, Vol. IV., p. 318, refers to an example of this species in the British Museum Collection as a *fine* specimen, which is only four inches in length, with the added note that it attains to twice those dimensions. A correct systematic record of the adult dimensions attained by the many species described would be a valuable addition to a subsequent edition of this Catalogue.

group just without the pale of the true Gadidæ, with which it would otherwise be most naturally associated. With respect to the rudimentary development of the spinous rays, Gadopsis, in addition to its other points of interest, may be regarded as a form that bridges over, in a natural manner, the hiatus that is supposed to subsist between the ordinary spine-finned (Acanthopterygii), and spineless-finned (Anacanthini) fishes. In this direction it would appear to the writer that additional transitional forms, apparently now extinct, probably united this abnormal Anacanthinoid with such Percoids as Oligorus and other allied Australian genera characteristic of the Murray river and its tributaries. Several of these share with Gadopsis a peculiar excavated contour of the frontal region of the head; the ventral fins, though not jugular, are set far forward, and have two filamentous prolongations, as commonly obtains among the true Gadidæ. The peculiar pattern markings of Gadopsis are, it may be further remarked, wonderfully similar to those characteristic of the Murray Cod, *Oligorus macquariensis*, consisting of a series of darker scribblings and reticulations on a ground colour of olive or golden green, but which are scarcely, if ever, precisely alike in two specimens. This colour pattern, in Gadopsis, becomes speedily obliterated after death, leaving the fish a uniform black hue, from which it takes its characteristic popular title of the "Black-fish." In the case of the Murray Cod, Oligorus, the circumstance that the foregoing vernacular name should have been relegated to it serves to accentuate the fact that, setting aside the presence of its spinous fins, it presents to the eye of the uninitiated much in common with the external aspect and configuration of the common Cod. This last suggestion is necessarily not adduced as a serious argument respecting the possible affinities of the fish under discussion, but merely as a floating straw indicative of the existence of a probable strong current of positive evidence that may be found flowing underneath.

It is a peculiarity of the distribution of *Gadopsis marmoratus* in Tasmania that, in common with the large fresh-water lobster, *Astacopsis Franklinii*, it is indigenous only to those rivers which flow northwards and discharge themselves into Bass's Straits and does not occur naturally in all of these. This fish has been artificially introduced from the St. Patrick's to the South Esk, one of the northern rivers hitherto deficient in that species, and has since multiplied and thriven therein, showing that there were no special conditions existing that were inimical to its previous establishment in that river. Still more recently, the writer, acting in his capacity of Superintendent and Inspector of Fisheries to the Government of Tasmania, successfully transported to and established the same species in, the River Derwent, whence it may now be easily distributed to other of the more important rivers comprised in the southern water-shed of that island-colony.

The Percidæ, or Perch family proper, which is among the largest and most widely distributed groups of the spined-finned or Acanthoid fishes, is very extensively represented in Australian waters, both salt and fresh. The Murray Cod, *Oligorus macquariensis*, Plate XXXI., fig. C, which represents the largest fresh-water member of the group, has already been briefly referred to. The facts that the fish commonly attains to a weight of 60 or 70 pounds, and has been taken over one hundred-weight, and also that as large an average quantity as ten tons are supplied weekly to Adelaide for several months in the year, and about half that proportion during the residue, corresponding quantities going to Melbourne, Ballarat, and other Australian cities, will give some idea of the abundance and commercial importance of the species. The Murray river, traced from the source of its largest affluent, the Darling, originating in Queensland, and flowing through the Colonies of New South Wales, Victoria, and South Australia, describes a course of but little less than two thousand miles. Added to this are the main river and numerous primary and secondary tributaries, seven or eight of which have independent courses ranging in length from 350 to 800 miles. All these streams and their affluents teem more or less throughout their courses with the Murray Cod and several other allied commercially valuable members of the Perch family, which, as will be readily understood, provide among themselves the materials for a highly important fishing industry.

The Golden Perch, *Ctenolates ambiguus*; the Silver Perch, *Therapon Richardsoni*; Macquarie's Perch, *Macquaria australasica*; and various species of Murrayia, the majority of which attain, in their adult state, to weights ranging from three to five or six pounds, contribute substantially towards swelling the supply, consisting chiefly of Murray Cod, that is perennially transported from the Murray river to the larger Australian cities. While the Murray Cod is commonly regarded as being limited in its distribution to the Murray river system, it occurs also in several of the rivers of Queensland and New South Wales that debouch upon the eastern coast-line.

Western Australia possessing many rivers in its southern district which have hitherto been devoid of any fish of economic value, one of the writer's latest professional undertakings in that colony was the transportation thither from the Murray river of a stock of young Murray Cod and Golden Perch, which should, a few years hence, increase to such an extent as to be available as a food supply. The fish transported, numbering several hundred, averaged from a quarter to half a pound in weight. After porterage by train from Morgan, on the Murray river, to Adelaide, they were there stored for a while in a pond in the Botanic Gardens of that city. Thence they were shipped in batches by the Orient and P. and O. mail steamers to Albany, a distance of over one thousand miles, and from there conveyed by rail again to their ultimate destinations. The majority of the fish were liberated in the waters of the Upper Swan or Avon, in the neighbourhoods of York and Beverley, and the residue were turned into a lake, receiving constant accessions of fresh water, some ten miles out of Albany. It will be interesting to note the results of these acclimatisation processes a few years hence, and should the fish have commenced to multiply, steps should be taken to distribute them to other waters suitable for their reception, where their presence will be of public utility. A year previously (1893) a small tentative consignment of the Murray river Golden Perch, *Ctenolates ambiguus*, and also the Victorian Silver Eel, *Anguilla australis*, was transported by the writer *vid* Adelaide to the Upper Swan river, and according to latest accounts are doing well there. The last-named type has, in fact, already commenced to multiply.

Before dismissing the subject of the Murray Cod, it is worthy of record that several huge members of the same genus, *Oligorus*, frequent the Australian sea coasts and estuaries, being most abundantly represented among the coral reefs and in the estuaries of the tropical districts. One of the best known of these species, *Oligorus gigas*, first reported from New Zealand, attains to a weight of three or four hundredweight. Two distinct species, *O. Goliath* and *O. terra-reginæ* from Queensland and other North Australian waters, rival it in dimensions. The popular title attached to these huge fish in Australia is that of Gropers, a name, however, which must not be confused with the several varieties of fish, Cossyphus, Chironemus and Chilodactylus, locally bearing the same name, which belong to the southern or temperate Australian waters. The so-called Rock Cods, including the allied genera, Serranus and Plectropoma, are additional representatives of the Perch family, which enter very extensively into the commercially important fish fauna of Australia, being especially abundant, in both numbers and varieties, in the sub-tropical and tropical districts. Many of these fish are remarkable for their brilliant colouration, being variously ornamented with stripes or bars or spots. In one especially handsome species, *Plectropoma Richardsoni*, not unfrequently exposed for sale in the Fremantle fish market, Western Australia, the ground colour of the body is a most brilliant carmine with a tendency to yellow beneath, diversified throughout the back and sides with ultramarine spots of an almost sapphire-like intensity.

Among the Australian members of the Perch family, the Gippsland Perch of the Victorian markets, *Lates* or *Percolates colonorum*, is worthy of brief mention, on account of the excellent sport it will afford, fished for with rod, line and artificial fly, after the manner of the lordly salmon. As its popular name implies, it is specially abundant in the fresh-water lakes of Gippsland, Victoria, and their tributary rivers. The distribution of the species is, however, tolerably extensive, it frequenting the majority of the river estuaries throughout the Victorian, New South Wales, and South Australian coasts, and also certain of those on the north coast of Tasmania. In aspect and habits of feeding from the surface, the Gippsland Perch suggests points of comparison with the English Bass, *Labrax lupus*, also a percoid, which likewise yields good sport with the salmon rod. *Lates colonorum* grows to a weight of some six or seven pounds, and is, in this respect, thrown altogether into the shade by its near ally, *Lates calcarifer*, which is exclusively a denizen of tropical waters.

This magnificent species, most appropriately designated the Giant Perch, frequents the estuaries of all the inter-tropical Australian rivers from the Fitzroy in Queensland around the northern sea-board to the Ashburton in Western Australia. It attains to a length of four or five feet, and a weight of over sixty pounds. This fine fish, known in India as the Cock-up or Nair-fish, occurs also in China, and has been observed by the writer in the Singapore fish market. At Rockhampton and Cooktown on the Queensland coast, the Giant Perch is most familiarly known by the native name of Barramundi, a title which is rather misleading, it being applied indifferently to other large fresh-water fish, including the Lung Fish, Ceratodus, and also to Osteoglossum. Some surprise was experienced by the writer on hearing the term of Barramundi applied to the fish in the Fitzroy district of Western Australia, but the mystery was solved on finding that it had been instituted by a recently imported Queensland native. A characteristic photograph from life of this Giant Percoid is included in the series of illustrations of Queensland fish embodied in the author's "Great Barrier" book.

Next in order to the Perch family, that of the Sea-Breams, Sparidæ, demands brief notice. This also is a most cosmopolitan group, its members being distributed throughout the world, and including some half-a-dozen British species. That Australian representative of the tribe, however, which above all others is held in highest repute, both for sport and on gastronomic grounds, is the so-called Snapper,

PLATE XXVII

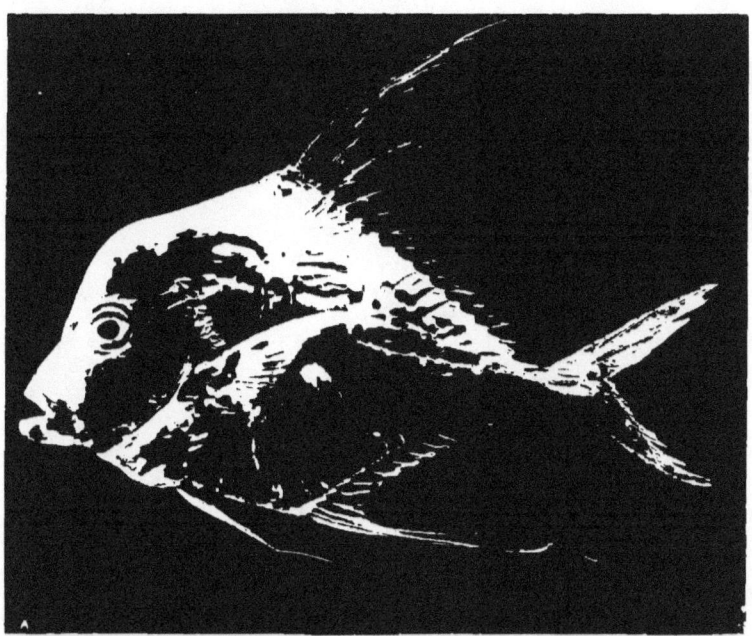

A. DIAMOND TREVALLY, *Caranx gallus*.
One-third

B. WESTERN AUSTRALIAN SNAPPER.
One-sixth

Pagrus unicolor. This fine species is abundant all round the Southern Coast-line, extending northwards on the Eastern Coast to Mackay, in Queensland, and to Adelaide, in South Australia. What would appear to be a racial variety only of the same fish occurs also on the Coast of Western Australia, as far north as the Abrolhos Islands and Shark's Bay. Adult individuals of this magnificent Bream not unfrequently attain to a weight of as much as twenty or even thirty pounds. A specimen, indeed, weighing twenty-nine and a half pounds, taken in Hobson's Bay, passed through the writer's hands, when making a model of it for inclusion with a series of typical Victorian fish for the Melbourne Centennial Exhibition in 1888. A peculiarity of the Australian Snapper, shared with some few other species of fish, is the circumstance that the adult males are distinguished by the development of a large bony knob in the fore part of the head, and which in very old specimens is accompanied by a fleshy excrescence on the snout. The profile outline of these individuals so strongly resembles that of a human face that they are popularly distinguished in the Sydney, Melbourne and Adelaide fish markets by the title of the "Old Man" Snapper. As previously remarked, the Snapper taken on the Coast of Western Australia presents some slight variations from that of the Southern and Eastern Colonies, and it is for that reason correlated by many ichthyologists with the distinctive title of *Pagrus major.* The technical differences between the two, as hitherto recognised, are, however, very obscure and difficult to define. Among other points, a slightly larger number of scales goes to form the lateral line in the last-named species, and a greater number to the transverse series in the former. Again, whereas in *Pagrus unicolor* the second anal spine is described as being longer but not stronger than the third, in *P. major* the same spine is stronger but not longer than its successor.

In connection with the considerable number of examples of the Western Australian Snapper that have recently fallen within the writer's observation, a point has been noted that may furnish an additional clue to the distinction of the two species under discussion. It relates to the respective contours of the heads of the adult males. Although extensively familiar with the typical "Old Man" individuals of the Southern and Eastern Colonies, the writer has not seen among them that particular modification that obtains at Fremantle, or *vice versâ.* In these "Westralian" examples there is no abnormal development of the occiput, but on the other hand the nasal protuberance takes an even more exaggerated form, giving a most bibulous expression to the fish's countenance. Should this recorded distinction prove to be a constant one, the title of the "Bottle-nosed" Snapper might be appropriately

x

conferred upon it. Photographic presentments of both the "bottle-nosed" male and normal female individuals of the Western Australian Snapper are reproduced in Plates XXVII. and XXIX. respectively.

Brief consideration may now be given to those fishes frequenting the Australian seas and rivers, which, with regard to the area of their distribution, are essentially Australasian, Indo-Pacific, or at all events unrepresented in European waters. A front position among this considerable assemblage must undoubtedly be given to that family group, the Cirrhitidæ, which includes, *par excellence*, that most justly celebrated fish, the Hobart or Tasmanian Trumpeter, *Latris hecateia*. The known representatives of the genus Latris are entirely limited in their distribution to the seas of New Zealand, Tasmania, and the southern shores of the Australian Continent, while the particular species named is almost exclusively confined to the colder waters of Tasmania and New Zealand. The Hobart Trumpeter, in both contour and colour, is a most handsome fish. Like the Snapper of the adjacent colonies, it affords excellent sport with the hook and line, and in deep water, fifty or sixty fathoms, using a crawfish bait, is not uncommonly taken weighing as many pounds. In addition to the considerable quantities of this fish that are disposed of in the Hobart market, a large number are dispatched in both the fresh and smoked conditions to Sydney and Melbourne.

The admirable practice is followed by the Hobart fishermen, as at Grimsby in England, of storing their catches in well-boats, and of transferring them on arrival in port to floating trunks. The Trumpeter, in common with many other species well adapted to this method of conservation, are in this manner kept alive to meet the fluctuating requirements of the local market; or any customer, if he should so desire, may avail himself of the opportunity of picking and choosing his own living fish. The life colours and distribution of the markings of the Trumpeter are very characteristic and attractive. In the diagnoses given in technical works on Ichthyology, the colour patterns as recorded of spirit-preserved examples are described as consisting of "four whitish longitudinal bands on a brown ground." In life, the paler hues predominate, the ground colour of the body being a most delicate opaline tint, silvery white ventrally, and with shades of palest green and blue upon the sides and back. Three dark olive-green longitudinal bands, originating immediately behind the head, traverse the upper part of the body. The top and bottom bands terminate independently at the base of the caudal fin, but the central one fuses with the lowermost one at a point corresponding with three-quarters of the total length of the band it joins. Beneath this lowest band, which is straight and divides the fish's side into two subequal moieties, there is usually a small linear patch of similar olive green, and narrow irregular lines of the same tint are produced from the eye to the end of the snout, and on the cheeks. The fins are more or less golden yellow, suffused with olive green, this latter tint being generally developed in regular striations on the membrane of the spinous dorsal, while the tips and edge of the caudal fin are frequently almost black. These details have been described at some length, on account, not only of their not having previously been correctly chronicled, but with reference to a singular phenomenon observed and recorded by the writer a few years since.

A number of these Trumpeters were kept alive for a considerable time in the sea-water tanks and ponds of the experimental fishery establishment inaugurated by the writer at Battery Point, Hobart. While in the daytime the fish were notable only for the colour pattern above described, it was observed on visiting the tanks at night, with the light of a lantern, that the fish had undergone a very distinct metamorphosis. The longitudinal colour bands were still conspicuous, but in addition to them, and as though belonging to a deeper sub-stratum of the integument, there were now distinctly visible five broad equidistant transverse blackish bands extending from the back to a little below the lowermost of the olive-green longitudinal stripes. It was observed at the same time that examples of the Silver or Bastard Trumpeter, *Latris Forsteri*, which, while of a different colour, is likewise longitudinally striped, exhibited at night similar dark transverse band-like markings. Later on, a specimen of *Latris hecateia*, in one of these fishery tanks, was attacked, apparently by a cat, or possibly a musk rat, and so injured that it lost the sight of both its eyes, though in other respects it recovered its health and appetite. The remarkable circumstance concerning this fish now, however, was that, being blind, it habitually exhibited the dark transverse markings that were only visible at night under normal conditions. It would appear that in ordinary daylight the stimulus upon the optic nerves and correlated system is such as to keep the pigment cells or chromatophores entering into the composition of these transverse bands in such a state of contraction that they are not visible. With the removal of the stimulating light or power of appreciating it, the tension is relaxed, and the expanded chromatophores become visible.

The existence of these supplementary colour-bands is of interest from another point of view. A further investigation revealed their constant though latent presence in many other unrelated species of fish. It was also observed by the writer that a very large number of sea fish exhibited these cross-bands constantly when in their young condition, though they gradually lose them, except in a latent state, as they attain maturity. These several facts all seem to favour the assumption that these cross-bands represent the residual traces of what constituted a conspicuous and permanent character in their ancestral forms. The familiar "parr" markings of the young of the majority, if not all, of the known Salmonidæ, appear in a similar manner to indicate that their ancestral archetypes were likewise permanently cross-barred. This subject was dealt with at some length in a paper communicated by the writer to the Hobart, 1892, Meeting of the Australian Association for the Advancement of Science. It is one worthy of further investigation and of discussion in a future edition of such a work as Mr. F. Beddard's interesting and instructive book on "Animal Coloration."

A fair idea of the general contour of the form and distribution of the colour markings of the Hobart Trumpeter will be afforded by a reference to Plate XXVIII., in which are depicted the photographic presentments of some two dozen species of Tasmanian sea fishes taken from a series of coloured plaster casts executed by the writer a few years since, which were presented by him to the Tasmanian Museum. The species under notice, *Latris hecateia*, is represented by two figures in this Plate, the one of a small example occupying the first place to the left of the third row in the series from the top, and that of a larger individual, which in life measured two feet and weighed eight or ten pounds, being situated to the extreme right in the second row below it.

The special facilities with which fish lend themselves to reproduction in Plaster of Paris was first recognised by the late Mr. Frank Buckland, whose personally executed casts of British fish, and notably Salmonidæ, on exhibition in the Buckland Museum, South Kensington, constitute probably the most extensive series of such models that has been brought together. Many of these plaster casts have been coloured from life by the late noted fish artist, Mr. H. L. Rolfe, "the Landseer among fishes," as he was justly and familiarly styled, and supply a far truer presentment of the living originals than is obtainable from the most carefully preserved specimens in which the softer parts become invariably more or less shrunk or otherwise distorted, and the natural colours completely obliterated. Similar, but yet more artistically finished, life-coloured models of various of the softer skinned lizards and snakes have been recently executed with great success under the auspices of the United

PLATE XXVIII

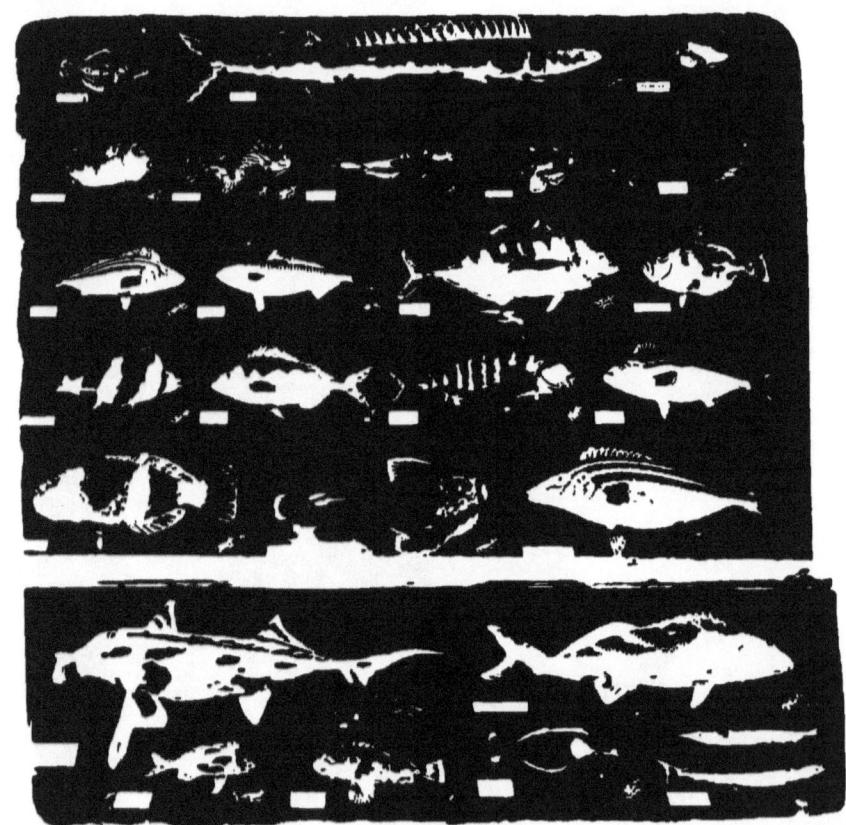

PLASTER CASTS OF TASMANIAN FISH.
Executed for and Presented by the Author to the Hobart Museum.
One-eighth of natural size. Index p. 165.

States National Museum authorities, to whom we are indebted for the several perfected examples of the art now on view in the Central Hall of our own Natural History Museum. As a recent exponent of the perfection with which certain animal forms may be thus immortalised in paint and plaster, it may be mentioned that Professor E. Ray Lankester's exhibit at last year's Conversazione of the Royal Society consisted of a couple of finely executed models of the New Zealand Lizard, *Sphenodon punctatus*, to which, on account of the soft, wrinkled character of its skin, it is almost impossible to render justice by any ordinary process of taxidermy. A diagrammatically arranged index which will allow of the ready identification of the many fish forms reproduced in this Plate, is herewith appended. The circumstance that the Barracouta, No. 2, occupying the central position in the top row, was, in the flesh, precisely three feet long, will afford a convenient key to the dimensions of the various species there portrayed.

In addition to the Hobart and Silver Trumpeters, *Latris hecateia* and *L. Forsteri*, there are two or three other species of the same genus, one of which, *L. ciliaris*, is common in New Zealand, and is met with as far north as Sydney. A large but rare form that is occasionally taken in Tasmanian waters has been

INDEX TO COLOURED PLASTER CASTS OF TASMANIAN FISH PHOTOGRAPHICALLY REPRODUCED IN PLATE XXVIII.

1 LEATHER JACKET. *Monacanthus sp.*	2 BARRACOUTA. *Thyrsites atun.*	3 LEATHER JACKET. *Monacanthus sp.*	
4 PORCUPINE FISH. *Chilomycterus jaculiferus.*	5 VELVET FISH. *Halsueuna cutaneus.*	6 ROCK COD. *Pseudophycis barbatus.*	7 FLATHEAD. *Platycephalus bassensis.*	8 FLOUNDER. *Rhombosolea monopus.*
9 HOBART TRUMPETER. *Latris hecateia.*	10 COLONIAL SALMON. *Arripis salar.*	11 SILVER TREVALLY. *Caranx georgianus.*	12 AUS. DOREY. *Cyttus australis.*
13 MAGPIE PERCH. *Chilodactylus bizonarius.*	14 SEA PERCH. *Chilodactylus macropterus.*	15 SEA CARP. *Chilodactylus Allporti.*	16 SILVER TRUMPETER. *Latris Forsteri.*
17 BLUE HEAD PARROT FISH. *Labrichthys cæruleus.*	18 CAT FISH. *Kathetostoma lævis.*	19 HOBART TRUMPETER. *Latris hecateia.*
20 ELEPHANT FISH AND EGG CASE. *Callorhynchus antarcticus.*	21 BUTTER FISH. *Chilodactylus spectabilis.*
22 RED PERCH. *Anthias rasor.*	23 ROCK GURNET. *Sebastes percoides.*	24 PARROT FISH. *Labrichthys laticlavius.*	25 GARFISH. *Hemirhamphus intermedius.*

described by the writer under the title of *Latris Mortoni*, and two small and somewhat doubtful species, *L. bilineata* and *L. inornata*, have been reported by Count Castlenau from the Victorian coast-line. The genus Chilodactylus, belonging also to the Trumpeter family, that of the Cirrhitidæ, possesses representatives which are much more extensively distributed. It includes the so-called Sea Carp, *Chilodactylus Allporti*; Sea Perch, *C. macropterus*; Magpie Perch, *C. bizonarius* and Butter-fish, *C. nigricans*, of the Tasmanian and Victorian markets. Two species, *Chilodactylus nigrescens*, the so-called Groper, and *C. carponemus*, the Leather Mouth, are not unfrequently exposed for sale in the Fremantle market, while a small form, *C. gibbosus*, which is also met with in Western Australia, occurs likewise on the eastern coast-line as far north as Moreton Bay. The several species of Chilodactylus of economic value that frequent the coast of New South Wales are distinguished in the Sydney market by the aboriginal title of "Morwongs."

The portraits of coloured plaster models of the majority of the species of the genus Chilodactylus above enumerated will be found among the series illustrated by Plate XXVIII. The four fish occupying the fourth row from the top are all Cirrhitidæ, and with one exception referable to the aforesaid genus. The fish to the extreme left, characterised by its boldly defined black and white bands, is the so-called Magpie Perch. In captivity it proved to be specially susceptible of domestication. One example was, in fact, so tame that it was accustomed to thrust its head out of water and permit itself to be stroked and fed from the hand with its favourite food, which consisted of the small round Pea-crabs, Pinnotheres, which abounded as mess-mates or commensals in the common mussels that were opened and cut up for the general commissariat. This Magpie Perch, and also a Tasmanian Flounder, which manifested similar sociable proclivities, constituted, it may be remarked, special objects of admiration and attraction to the late Admiral Sir George Tryon, who, when visiting Hobart with the Australian squadron, was a frequent visitor to the writer's Fisheries establishment.

The fish next on the line to the Magpie Perch is the so-called Sea or Black or Silver Perch of the Tasmanian market, and is especially notable for the abnormal length of one of the free rays of its pectoral fin. The larger, transversely striped fish next to this is "the Carp" of the Hobart market, *Chilodactylus Allporti*, notable in life for its handsome colouration, in which alternate bands of white and vermilion red predominate. Soon after removal from the water the whole body becomes suffused with the vermilion tint, but that again speedily gives place to a more uniform hue of dusky brown. The fourth fish to the extreme right in this series is the Silver Trumpeter, *Latris Forsteri*, having much the same pattern of colour markings as the "Real" or Hobart Trumpeter, *L. hecateia*, immediately beneath it, but entirely wanting the finer lines and symmetry of form of that species. The tints of the colour bands in this species are also very distinct, varying in individuals from a golden brown to a more or less definite red-brown or light red hue. The remaining example of the Cirrhitidae represented in this plate is the large fish to the right in the second row from the bottom. This is the "Butter-fish," *Chilodactylus nigricans*, of the Hobart and Melbourne markets, a somewhat rare visitor to the southern island colony, but plentiful on the Victorian coast-line. Extending westward, it constitutes one of the commoner market fishes at Adelaide and Fremantle. The colours of this species are somewhat attractive, consisting of a blue-grey ground, variegated with more or less well-defined oblique bands of blackish spots. A large fish is often over three feet in length and weighs ten or twelve pounds.

The yet larger massive fish, *Callorhynchus antarcticus*, on the same line as *Chilodactylus nigricans*, is, as will be recognised by the formation of its tail, a member of the shark tribe. As such it is appropriately referred to among other of its more immediate congeners at the end of this Chapter.

The genus Beryx, of the family of the Berycidae, has several representatives in Australian waters that are greatly esteemed for food. One of these, *Beryx affinis*, is known in Sydney as the "Nannegai." The same species and an allied form, *B. Mulleri*, are supplied to both the Adelaide and Fremantle markets, and are there distinguished by the title of "King Snappers." The fishes of this genus are among the most resplendent of known species. Both of the two forms named are, when freshly caught, the most brilliant scarlet carmine with various opalescent tints, chiefly blue and lilac reflections. Added to this, their abnormally large and lustrous eyes confer upon them a most distinguished appearance. The chief interest attached to these Berycidae is, however, the circumstance that they represent the oldest known group of spine-finned or Acanthopterygian fishes. They were most abundantly represented in the cretaceous epoch, and their known living representatives are mostly confined to deep water. The occurrence of the few existing littoral species which still linger around the Australian coast-line is of high significance, taken into account with the many other ancient forms of animal and vegetable life that so notably distinguish this Island-Continent. An illustration of *Beryx Mulleri*, the most abundantly represented Western Australian member of this interesting genus, is given in Plate XXIX.

The Tassel fishes, *Polynemidæ*, constitute one of the leading groups of economic value indigenous to Australian waters, though in this instance most abundantly represented in the essentially tropical rivers and estuaries. The distinguishing features of the most characteristic species of the single genus, Polynemus, is the almost salmon-like contour of their shapely bodies, correlated with which is the peculiar free style-like or filamentous character of a certain number of the rays of the pectoral fins. The most highly esteemed esculent species, *Polynemus tetradactylus*, is distinguished in the northern districts of Queensland and Western Australia with reference to its shape as the northern salmon. As its technical name implies, its free pectoral rays are only four in number, and these do not exceed in length the adjacent membrane-united rays. In the majority of other known species there are five free pectoral, and in one small but new form discovered by the writer in the Ord river, Western Australia, in association with the surveying expedition (1888) of H.M.S. "Myrmidon," there are seven such rays, four of which may extend backwards beyond the extremity of the very elongate lobes of the caudal fin. By way of compliment to Captain the Hon. H. P. Foley Vereker, in command of the above-named vessel, as whose guest the writer first visited Cambridge Gulf, the name of *Polynemus Verekeri* has been conferred upon this species.* This phenomenal fish not having been previously figured, except in diagrammatic outline, its aspect as drawn and coloured from life is herewith reproduced in monochrome.

SEVEN-RAYED TASSEL FISH, *Polynemus Verekeri*, ORD RIVER, CAMBRIDGE GULF, W.A. NATURAL SIZE.

The life tints of this newly discovered species are very distinctive, the ground colour being a prominent chrome yellow with darker shadings, the pectoral and

* Proc. Royal Soc., Queensland. Vol. VI. part V., 1889.

long-forked caudal fins bright orange, and the remaining fins a lighter shade of the same tint. The long filamentous free rays forming a fringe beneath the pectoral fins were, by way of contrast, a bright vermilion red.

Several of the Australian Polynemi attain to a considerable size. One species, *P. Sheridani*, first recorded from the Mary and Burnett rivers in Queensland, and there locally known as the King-fish, has been taken up to a weight of one hundred pounds. *Polynemus tetradactylus*, which occurs also in India, and is locally known there as the "Bahmeen" or "Pamban Salmon," is recorded in Dr. Day's "Fishes of India" as attaining to a weight of over three hundred pounds. In conjunction with the Giant Perch, or "Nair Fish," *Lates calcarifer*, previously referred to, it affords most excellent sport with rod and line, using a spinning bait, and as such is the subject of special mention in Mr. H. S. Thomas' angling work, "The Rod in India." All of the larger Indian Polynemi are further noted on account of the excellent isinglass that is manufactured from their sounds or swimming bladders.

The family of the Carangidæ or Horse Mackerels, while embracing many species that are practically cosmopolitan in their distribution, includes also numerous essentially Australian or Indo-Pacific types. The genus Caranx, which represents one of the first-named groups, comprises a species, *C. trachurus*, which is regarded as identical with the British Horse Mackerel. What are known as Trevallies on the North and Eastern, and Skipjacks on the Western, Australian Coast-lines, are also members of the same genus, but mostly of a much deeper and more compressed shape than the familiar British species. In the form *Caranx gallus*, selected for illustration, Plate XXVII., fig. A, both the head and body are remarkable for their angular contours, and the dorsal, anal, and pectoral fins for the filamentous elongations of their primary rays. In *C. radiatus*, there is a more rigid, fringe-like development of the second dorsal and anal rays, but the body does not depart from the ordinary ovate type. Both of these two last-named species belong to the tropical Australian Coast-line, and have been observed by the writer on the Queensland and Western Australian sea-boards. The first-named species is locally known, with reference to its shape and appendages, as the Diamond-fish or Plumed Trevally, while that of the Fringed Trevally has been conferred on *Caranx radiatus*.

The photograph of a young individual of the first-named species, *Caranx gallus*, in which the filamentous appendages are relatively longer than in the adult fish,

Y

has been utilised as a head-piece to this Chapter. A Tasmanian species, *Caranx georgianus*, known as the Silver Trevally, which is much esteemed for the table, is represented by No. 11 in the series of Plaster Casts photographically reproduced in Plate XXVIII.

The giants of their tribe in this group of the Carangidæ are the several members of the genus Seriola, which are extensively represented throughout the temperate Australian waters, and are known locally as Yellow-tails, Samson-fish, and King-fish. Two species, *Seriola grandis* and *S. Lalandii*, are found as far south as Tasmania; while *S. gigas*, of the Western Australian Coast, is especially plentiful in the neighbourhood of Houtman's Abrolhos. This species attains to a weight of over one hundred-weight, and is locally known there as the King-fish. The Tailor-fish, *Temnodon saltator*, while a near ally of Seriola, is much more cosmopolitan in its distribution. It abounds on the Eastern, Southern, and Western Coasts of Australia, and, ranging throughout nearly all the tropical and sub-tropical seas, constitutes an important fishery on the Atlantic Coast of the United States, where it is distinguished by the title of the Blue-fish.

The occurrence of the Frost Fish or Scabbard-fish, *Lepidopus caudatus*, belonging to the family of the Trichiuridæ, in New Zealand and Tasmanian waters, has already been referred to among the cosmopolitan species that are indigenous to European as well as to Australian seas. A more essentially Australian type, belonging to the same family, and of a correspondingly elongate band-like form, is the well-known Barracouta, of the Adelaide, Melbourne, Sydney and Hobart markets. Even this species, however, while not represented in the Northern Hemisphere, is abundant upon the Coast of Chili, and also at the Cape of Good Hope, where it is known by the popular title of the "Snook."

The photograph of a coloured cast of a fine specimen of the Barracouta occupies the greater portion of the top line of the series represented in Plate XXVIII. Its length in the flesh was just three feet. Much difficulty was experienced in securing a specimen of this fish for modelling in which the elongate but deeply indented dorsal fin was uninjured, this structure in the majority of the examples brought to market being much mutilated as a consequence of the rough conditions that attend their capture. The habits of the Barracouta are essentially gregarious, the species assembling in large numbers, and pursuing and preying upon the shoals of smaller fish of every description at the surface of the sea. It is in fact the presence of these marauding shoals of Barracouta that to a large measure renders it impracticable to prosecute fishing in Southern Australian waters by means of the drift net, which is so extensively used in the fishery for Mackerel and other surface species in the British seas.

The most approved method of capturing the Barracouta on the Tasmanian Coast is by means of the "Maorie jig," so-called since it is supposed to have been first employed by the natives of New Zealand. This fishing apparatus consists simply of a strong pointed but unbarbed hook, fastened to a small block of wood, preferably cedar. This again is attached by a strong line, a yard or two long only, to a stout staff. A bait, consisting of a piece of coloured cloth or of shark's-skin, is sometimes added, but is not necessary. With this very simple equipment, and the boat at full sail, the fish, when abundant, are hauled in as fast as the fishermen can throw over and recover their baited or unbaited hooks. The deck at such times speedily becomes one struggling mass of snapping, slippery monsters, and the fun and excitement, as the writer can testify, is most exhilarating for those partial to this class of sport. Off the Victorian Coast-line, and more especially in the neighbourhood of Warnambool, the Barracouta is more commonly captured from sailing boats with a long hand line and a glittering metal lure and hook trailed astern, much after the method practised in England for the capture of Mackerel, known as "whiffling" or "railing."

From a utilitarian point of view, the Barracouta is undoubtedly one of the, if not the, most important of the Tasmanian food fishes. While the Real or Hobart Trumpeter may be said to typify the species fitted, like the Turbot, to grace the table of the wealthy, the Barracouta may be as essentially styled the "poor man's fish." It takes in Tasmania the place that is occupied by the modest Herring or the Haddock in the English market. The fact, indeed, that a six or eight pound Barracouta of the best quality can at most times be bought for sixpence, or at less than half that price if any number are taken, places this fish at the command of the very poorest. Notwithstanding its cheapness, the Barracouta as a food fish is by no means to be despised. Carefully smoked, after the manner of the familiar Findon Haddock, it constitutes a no less toothsome adjunct to the breakfast table. Barracouta pie, again, the invention of a Tasmanian culinary genius, is a savoury dish that would tickle the palate of the most fastidious.

There is a second species of fish closely resembling the Barracouta, and belonging to the same genus, which, while somewhat uncertain in its appearance, occasionally visits the Tasmanian coast in vast shoals. This is the so-called Tasmanian

"King-fish," *Thyrsites Solandri*. It considerably exceeds the Barracouta in size and may weigh as much as from twelve to over twenty pounds. This King-fish is caught in the same manner as the last-named species, and, in seasons when it is plentiful, "three men in a boat" have been known to capture forty dozen, or over two tons weight, in a single night. Not many years ago this King-fish appeared in the bays and inlets adjacent to the Derwent estuary in such remarkable numbers that tons were stranded on the beaches. Such large supplies were at the same time taken by the fishermen that, in the absence of a sufficient market demand or ready means for their conservation, vast quantities of them were used for manuring the hop gardens and fruit orchards for which Tasmania is so justly famous. As though by way of retribution for the previous lavish waste of this magnificent food species, this King-fish has since been an almost entire stranger to the Tasmanian coast.

Another group of elongate predaceous fishes, having, with regard to both aspect and habits, much in common with the Barracoutas, Thyrsites, is that of the Sphyrænidæ or Sea-pikes, represented by two genera, Sphyræna and Lanioperca, and many species in both the temperate and tropical Australian waters. In the higher latitudes, beyond the range of distribution of Thyrsites, one species, *Sphyræna langsar*, is often locally distinguished by the corresponding title of the Barracouta. Large examples of this species have been known to attain to a length of seven or eight feet and to weigh as much as forty pounds. One of the most widely distributed of the Australian Sea-Pikes, *Sphyræna obtusata*, is photographically portrayed in Fig. C of Plate XXIX. The specimen figured was obtained at Fremantle, Western Australia. The "Barracuda" is the common name given to tropical Atlantic representatives of the genus Sphyræna in the West Indies.

The family of the Trachinidæ is deserving of brief notice, since it includes the several greatly esteemed species of so-called Australian Whitings, referable to the genera Sillago, Isosillago and Neosillago, but having necessarily no relationship to the European Whitings, which are true Gadidæ. Although there are no European representatives of these generic groups, the Weever fishes of the British seas, *Trachinus draco* and *T. vipera*, as also the Star-gazer, *Uranoscopus scaber*, of the Mediterranean, are usually assigned by ichthyologists to the same family. The association of these respective generic groups by no means recommends itself as a natural system of classification. In Dr. Gunther's "Introduction to the Study of Fishes" all the members of the family Trachinidæ are referred to as bad swimmers, whose locomotive powers are confined to moving along the bottom in small depths. This description applies

PLATE XXIX

A.—BOTTLE-NOSED SNAPPER, *Pagrus major*, p. 161. B.—KING SNAPPER OR "NANNEGAI," *Beryx Mulleri*, p. 167.
C.—SEA PIKE, *Sphyræna obtusata*, p. 172.

correctly to the more considerable number of genera referred to this family, which possess the physiological distinction of being devoid of an air-bladder. To that anatomical fact they owe the circumstance of being sluggish, bottom-reposing fish. The Australian Whitings, Sillago and its allies, on the other hand, possess a well-developed air-bladder, and are of the most active habits, cruising about in shoals, after the manner of Red Mullets, over the surfaces of the sandy banks upon which they feed. Different species of these Whitings abound throughout Australian waters from Tasmania to Torres Straits, and are all held in high esteem for the lightness and delicacy of flavour and whiteness of their flesh. It is upon this account that they have received the title that has been popularly awarded them.

The Flathead, genus Platycephalus, is another essentially characteristic Australian group that, like Sillago, possesses specific representatives throughout the seas of that Island-Continent, all of which are more or less esteemed for food. A characteristic representative of the commonest Tasmanian species, *Platycephalus bassensis*, occupies the second position from the right immediately beneath the Barracouta in Plate XXVIII. While unrepresented in European waters by any specific type, the genus Platycephalus is closely related to the Bull-heads and Gurnards, and is on such account relegated to the same family of the Cottidæ. The true Gurnards or Gurnets, belonging to the genus Trigla, are typified by several Australian species, many of them being remarkable for the brilliant, butterfly-wing-like aspect of their large pectoral fins. None of them, however, attain to a large size or are sufficiently plentiful to constitute a marketable species, as in British waters. A fish of somewhat rare occurrence that is occasionally taken on the Tasmanian coast, and is also referable to the Cottidæ family, is the singular Velvet Fish, *Holoxenus cutaneus*. A photograph of a coloured plaster cast of this species occupies the second position from the left in the same serial line that contains the Flathead. The most remarkable feature of this species is not so much its form as its wonderful colouring, which may be a brilliant scarlet vermilion throughout, or a varied mixture of vermilion and orange. Added to this, the skin, which is devoid of scales, is very soft and loose and of a granular or pilose texture suggestive of the surface of wet flannel or velvet.

The Grey Mullet family, Mugilidæ, is represented so abundantly, numerically and specifically, in the Australian seas and rivers as to constitute one of the leading economic groups. Its members agree, however, so nearly in all essential characters with their British, or it might be said, cosmopolitan, congeners as to dispense with the necessity of elaborate notice.

The Labridæ or Wrasse and Parrot-fish tribe embraces a multitude of forms, mostly of phenomenally brilliant colouration, but also including many species that are esteemed for the table. Among these, *Platychœrops Gouldii*, the so-called Blue Groper, of the Sydney fishermen, is one of the largest and most important representatives, growing to a length of three or four feet and a weight of over forty pounds. In its adult state it is of a uniform dark purplish-blue hue, but is variegated in its earlier phases of development with orange bands and spots. This species is met with round the entire southern coast-line to King George's Sound, in Western Australia, and is also reported from Tasmania. The head and shoulders of the Blue Groper boiled, with egg sauce, has been pronounced by Australian epicures to surpass the Cod and to closely approach the English Turbot in its delicacy of flavour. The so-called Pig-fish, *Cossyphus unimaculatus*, of a brilliant scarlet-vermilion hue, with a single large blue-black spot near the centre of the spinous dorsal fin, is also highly esteemed in the Sydney market. The Blue-head, *Labricthys ceruleus*, of Tasmania, and several species of Odax, known locally as "Strangers," Ground Mullets, or Rock Whitings, indigenous to all of the southern or temperate Australian Colonies, represent the leading remaining forms of the family group of the Labridæ that occupy a recognised position in the Australian markets. A cast of a fine example of the Tasmanian Blue-head, executed by the writer, occupies a position to the extreme left in the series of three fishes that constitutes the fifth row from the top in Plate XXVIII.

Apart from the foregoing there are numberless other Labridæ, chiefly inhabitants of the tropical coral reefs, that are much appreciated as a source of food supply by those engaged more especially in the Bêche-de-Mer and Mother-of-Pearl shell fisheries. They embrace fish of the most brilliant hues, commonly known as "Parrot-fishes," while some of the more sober-tinted types, belonging to the genus Chærops, are locally known as "Gropers." One of the commonest of these, *C. cyanodon*, is of a uniform greenish grey hue, having its dorsal, anal, ventral, and caudal fins edged with a fine line of turquoise blue, and also prominent canine teeth of the same tint. Coloured illustrations of some of the more phenomenally brilliant-hued of these Australian Parrot-fishes are depicted in Chromo-Plate XV. of the author's recently published work on "The Great Barrier Reef."

Among the few remaining fish groups that demand brief attention with reference to the phenomenal as well as economical character of their component members, that of the Herring tribe, or Clupeidæ, must be included. The presence in Australian waters of species regarded as identical with the English Sprat, *Clupea sprattus*, and others, such as *C. sagax*, very closely resembling the English Pilchard, has been recorded in a previous page. The comparisons then under discussion did not, however, permit of a reference to certain members of the same family which are essentially Australian or Indo-Pacific in their distribution and veritable giants of their tribe. These giant herrings, represented more especially by the two specific types *Chanos salmoneus* and *Megalops cyprinoides*, are restricted in their distribution to tropical and sub-tropical waters, and within these limits have been met with by the writer at the greater portion of the coastal and estuarine districts visited, from Moreton Bay on the east coast to the Gascoigne River in Western Australia. Both of these two species are most excellent eating. They ascend the rivers into fresh water in the young condition, and grow to lengths of four and five feet. In India and the Malay Peninsula, where both species are also found and much esteemed, they are made the subjects of cultivation in suitable tanks.

While of these two last-named giant members of the Herring family, *Chanos salmoneus* is the more prized for the table, the second form, *Megalops cyprinoides*, would appear to possess the most conspicuous potentialities for sport. This species is, in point of fact, very closely allied to the North American Tarpon, a species which, while long known to ichthyologists, has quite recently sprung into notoriety with regard to the magnificent sport it has been found capable of affording. The American Tarpon, *Megalops thryssoides*, attains to no less a length than six feet, with an accompanying weight of over 200 lbs., and is now systematically fished for, with a salmon rod and a live fish bait, in the tropical estuaries of Florida. The fish, when hooked, is a most game one, surpassing the salmon in the vigour of its leaps and in its rushes to every point of the compass to free itself from the restraining hook, and will frequently show fight for the space of several hours before yielding itself to the hands of its captor. A most fascinating account of Tarpon fishing, with illustrations and technical instructions as to the *modus operandi*, were published in the August and October (1895) numbers of the "Badminton Magazine." Enthusiastic anglers disposed to initiate similar exciting sport in Australian waters may be profitably referred to those pages. There can be no doubt, in the writer's opinion, that in addition to the Ox-eye Herring, *Megalops cyprinoides*, there are half-a-dozen other tropical estuarine-frequenting Australian fish that would yield equally exciting sport on the same lines. Included in such list would be the Giant Perch, *Lates calcarifer*; at least two of the Tassel fishes, genus Polynemus; the Gropers,

Oligorus gigas or *Terra-reginæ*; the Jew-fish, *Sciæna antarctica*; the Giant Mackerel, *Cybium Commersoni*; and several of the Yellow-tails, referable to the genus Seriola.

The extensive family of the Siluridæ or Cat-fishes, represented by but a single fresh-water species, the Wels or Sheet-fish, *Silurus glanis*, in European waters, is a conspicuous group in Australia, numbering many species that inhabit both salt and fresh water. The detached island-colony of Tasmania is, in fact, the only province which is destitute of members. Only one species, the large eel-like Cat-fish, or so-called "Tandan," of the River Murray, *Copidoglanis tandanus*, possesses a commercial status, and is made the subject of an export trade to the Melbourne and other markets. Several allied forms, including other species of Copidoglanis and representatives of the genus Plotosus, are nevertheless, and notwithstanding their repulsive aspect and formidable spinous armature, excellent eating. In the tropical districts, the large sea and estuarine Cat-fishes of the genus Arius, differing from the preceding in having forked instead of eel-like tails, are by no means to be despised for the table when better fish are scarce. One species of this group, identical apparently with *Arius thalassinus*, is not unfrequently taken in the neighbourhood of Wyndham, Cambridge Gulf, weighing as much as fifty or sixty pounds. All the Cat-fishes are distinguished, as their name implies, by the tentacular appendages or barbels which adorn their snouts, and which are suggestive of the whiskers, or more correctly the moustaches, of a cat. One of the commoner northern species of Plotosus or Cat-fish Eels is represented by Fig. C. of Plate XXX.

There are two fish represented in the Plate last quoted that are very essentially Australian. The first of these is *Aulopus purpurissatus*, Plate XXX., fig. A., known to the Sydney fishermen as the "Sergeant Baker." This local appellation appears to have been conferred upon the species with reference to its having been first taken in the early days in Sydney Harbour by a sergeant of that ilk. In form, and in its brilliant livery of scarlet and purple, the fish is, to a considerable extent, suggestive of a gurnet, and is in consequence sometimes invested with that name. As a matter of fact, it belongs to the characteristic family of the Scopelidæ, a group which includes the much-esteemed "Bummuloh" or "Bombay Duck," *Harpodon nehereus*, the deep sea *Plagiodus ferox*, and many singular pelagic and abyssal forms, armed with, in relation to their size, most formidable teeth and, in many instances, brilliant phosphoric organs. A diagnostic family character, clearly discernible in the photograph of Aulopus here reproduced, is the presence of the small dead or adipose fin, like that of a trout, situated immediately behind the long dorsal one. The male individual of the Sergeant

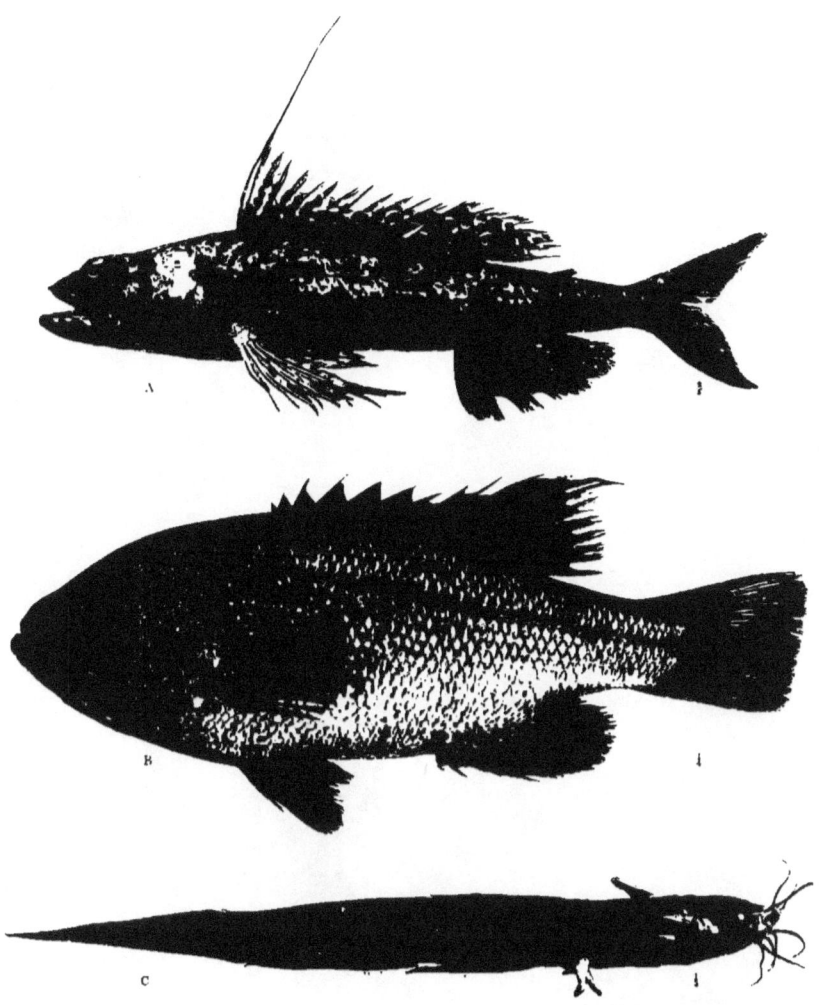

A. SERGEANT BAKER, *Aulopus purpurissatus*, p. 176.   B. W.A. JEW-FISH, *Glaucosoma hebraicum*, p. 177.
C. CAT-FISH EEL, *Plotosus sp.*, p. 176.

Baker, of which the specimen figured is an example, is readily distinguished by the long filamentous development of the united second and third rays of the dorsal fin. The species, while hitherto recorded only from the New South Wales and Victorian coasts, has been obtained by the writer at Fremantle and the Abrolhos Islands, Western Australia. The illustration here reproduced represents a specimen from the last-named habitat.

The remaining fish, Fig. B. in Plate XXX., is *Glaucosoma hebraicum*, the so-called Jew-fish of the Fremantle market, and appears, so far as is known, to be restricted to the Western Australian coast-line. It is taken with hook and line on the Snapper Grounds, attains to as large, or even a larger, size than *Pagrus unicolor*, and is esteemed by many to be superior to it from a gastronomic point. The colours in life, which are represented by opalescently interblending tints of purple, red, and gold, correspond to a marked extent with those of the last-named, and much more familiar, species. A second representative of the genus, *Glaucosoma scapulare*, occurs on the coasts of Queensland and New South Wales. It, also, is highly esteemed for the table, but is of such rare occurrence as to be but seldom seen in the market. The distinctive titles of Epaulette-fish and Jew-fish have been respectively conferred in Queensland and New South Wales upon this species. It differs most conspicuously from the Jew-fish, *par excellence*, *G. hebraicum* of Western Australia, in the abnormal size and smoothness of its scapular or shoulder bone and in the black tint of the integument that covers it.

This point is a suitable one for drawing attention to the utter confusion that prevails with regard to the nomenclature of the Australian market-fishes. The Jew-fish is an example in point. The name, as applied to the Western Australian *Glaucosoma hebraicum*, might be appropriately retained for that species. In Sydney and Brisbane, however, *Sciæna antarctica*, which is, to all intents and purposes, identical with the European "Maigre," is invested with this title, while the same name is also applied almost indiscriminately throughout the Colonies to the Cat-fish eels of the genus Plotosus. The title of the "King-fish" is even more heavily weighted. In Victoria and in the southern districts of Western Australia, this is the name by which *Sciæna antarctica* is distinguished, while the same form in Adelaide is known as the "Mulloway." The King-fish in tropical Queensland waters, identical with the "Seer-fish" of India, is a large Mackerel, *Cybium Commersoni*. In the same colony, one of the Tassel-fishes, *Polynemus indicus*, bears the same title, while in Tasmania it is a fish, *Thyrsites Solandri*, closely allied to the Barracouta. Finally, the King-fish of the Sydney

z

market is a species of Yellow-tail, *Seriola Lalandii*. Examples might be multiplied, but the intricacies involved by the plurality of application of the two names above quoted will suffice, perhaps, as an object lesson to the lay mind in demonstration of the fact that there is some virtue attachable to the intelligent application of technical terminology. King-fish, as obtained in Tasmania, the most toothsome of its race and a most delicious fish, would undoubtedly under any other name taste as sweet, but its fair title and reputation is open to serious disparagement at the hands of connoisseurs who, say at Sydney or at another of the Australian capitals, ordered for their feast the like-named fish.

Especially must the traveller be warned against expecting, on investing in a so-called Colonial Salmon, *Arripis salar*, anything approaching the British ·"King of fishes." It attains to a weight of ten or twelve pounds and upwards, possesses fine, almost salmon-like lines, and, fished for with rod and spoon bait, will yield magnificent sport. Placed upon the table, however, it proves to be one of the poorest and coarsest of the Australian market species. A specimen of this fish, which is a member of the Perch family, and occurs in vast shoals throughout the southern Australian sea-board, is portrayed in No. 10, or the second fish from the left of the series of casts of Tasmanian fishes represented in Plate XXVIII. A smaller, closely allied species, *Arripis georgianus*, is much superior from a gastronomic standpoint. It constitutes the so-called "Roughy" of the Melbourne market, but in Fremantle, Western Australia, is popularly known as Herring, and yields abundant sport to line fishermen throughout the year from the fine jetty of that sea-port. Large quantities of this fish are also kippered, and, so treated, bear a by no means inconsiderable resemblance, in both aspect and flavour, to the familiar kippered Herring of English celebrity.

Flat-fishes, Pleuronectidæ, are relatively scarce in Australia, though there are some excellent Flounders, *Rhombsolea monopus*, Plate XXVIII., No. 8, in Tasmanian and Victorian waters, and Soles, Plagusia or Synaptura, that are distributed throughout the coast-line. None of these latter forms, however, occur of sufficient size or in sufficient abundance to constitute an important item in the colonial fish markets. A few years since the writer discovered in the estuary of the Endeavour river, North Queensland, the presence of a larger member of this family group than had been hitherto recorded from Australian waters. This was the *Psettodes erumei* of the Red and Indian Seas, a form attaining to a weight of several pounds, and so much resembling the Halibut, *Hippoglossus vulgaris*, of the North Atlantic, that it has been referred to the same genus by some authorities. It is an excellent table fish, and if procurable in any quantity would be a valuable addition to the Australian market list.

The fish species illustrated by two examples, No. 25, at the base and to the extreme right in the series of casts figured in Plate XXVIII., is, although of small size, highly esteemed for the table in Australia. This is the Gar-fish, *Hemirhamphus intermedius*, a slender, cylindrical type, rarely exceeding a foot in length, which occurs in vast shoals throughout the temperate Australian seas. From the English Gar-fish, genus Belone, of which there are also Australian representatives, it may be readily distinguished by the circumstance that it is only the lower jaw, in place of both upper and lower jaws, that is developed in a characteristic beak-like manner. In the lower of the two figures given the small rudimentary character of the elevated upper jaw is clearly shown. Over half-a-dozen Australian species of this genus Hemirhamphus have been described, several of which are exclusively denizens of the tropical sea-board. In an allied Australian form, Arhamphus, the beak, as the name betokens, is entirely suppressed.

The fresh waters of Australia yield some fish of more phenomenal interest than the few already referred to. In *Ceratodus Forsteri*, the Lung Fish or Burnet and Mary River Salmon, of the Queensland colonists, is presented the only known surviving member of a genus which, with numerous allies, was abundantly represented in the Triassic and Jurassic formations of Europe, India, and America. Of its peculiar sub-order—the Dipnoi or lung-breathing fishes—*Lepidosiren paradoxus* of the Amazons in South America, and *Protopterus annectens* of tropical Africa, are again the only other known living types. In common with such other large-scaled fresh-water fishes as the Giant Perch, *Lates calcarifer* and *Osteoglossum Leichardti, Ceratodus Forsteri* is commonly associated by the Queensland natives with the name of the "Barramundi." This title, in its restricted sense, is now, however, exclusively attached to Osteoglossum.

This last-named form is almost as remarkable as is Ceratodus, with reference to the present geographical distribution of its nearest allies. A second species, *Osteoglossum Jardinei*, has been recorded by the writer from North Queensland; otherwise, the only known additional specific forms of the same genus, *Osteoglossum bicirrhosum* and *O. formosum*, are inhabitants respectively of the fresh-water rivers of Brazil, and of Borneo and Sumatra, while two allied genera *Arapaima* and *Heterotis* are indigenous, like the previously recorded Dipnoi, to the rivers of tropical South

America and Africa. One of these *Arapaima gigas*, of the rivers of Brazil, and the Guayanas, attains to the phenomenal dimensions of fifteen feet in length, with a weight of four hundred pounds, and is the largest living representative of the Teleostian or bony-skeletoned fish, as distinguished from the Chondrostian or Cartilaginous-skeletoned group, of which the Sharks and Sturgeons are familiar examples. The true Australian Barramundi, *Osteoglossum Leichardti*, has been obtained chiefly from the Fitzroy, Dawson, and other inter-tropical rivers debouching on the Eastern Coast of Queensland; while the second species, *O. Jardinei*, belongs to the river systems that flow into the Gulf of Carpentaria. Both species attain a length of two or three feet, and are most excellent eating, their flesh being even pronounced by connoisseurs to compare favourably with that of the European salmon. It remains yet to be discovered whether or not representatives of these two last-mentioned interesting genera, *Ceratodus* and *Osteoglossum*, have not living representatives in the few but as yet little explored rivers of the Northern districts of Southern and Western Australia. The typical illustrations of these two forms that are reproduced in Plate XXXI., figs. A and B, will assist towards the recognition of any near allies of these fish by those whose avocations may place them on a familiar footing with the indigenous fish fauna of these remote regions.

Among other members of the fresh-water fish fauna of Australia that invite brief notice is the small family of the Galaxiidæ, embracing many of the so-called species of the Native Trout, genus *Galaxias*, briefly referred to in Chapter I., which, in addition to many species distributed throughout all of the Southern and temperate colonies of Australia, has representatives in New Zealand and the southern districts of South America. All of the members of this family are of small size, not exceeding a few inches in length, and derive their popular name from their somewhat trout-like contour and spotted markings. In *Prototroctes marœna*, common to the rivers of Victoria and Tasmania, Australia possesses a fish very nearly akin, superficially, in contour and habits, to the English Grayling, Thymallus. It is an excellent sporting fish, and, possessing a corresponding cucumber-like odour, is commonly known as the "Cucumber Mullet." In this instance also it is of interest to note that its nearest generic and specific allies are found in the rivers of New Zealand, the Falkland Islands, and the extreme south of South America.

This brief record of phenomenal Australian fresh-water fish would be very incomplete without some reference to the remarkably fine races of English Trout and Salmon trout, *Salmo fario* and *S. trutta*, which for many years past have

PLATE XXXI.

A. LUNG FISH, *Ceratodus Forsteri*, quarter natural size, p. 179.

B. YORK PENINSULA BARRAMUNDI, *Osteoglossum Jardinei*, quarter natural size, p. 179.

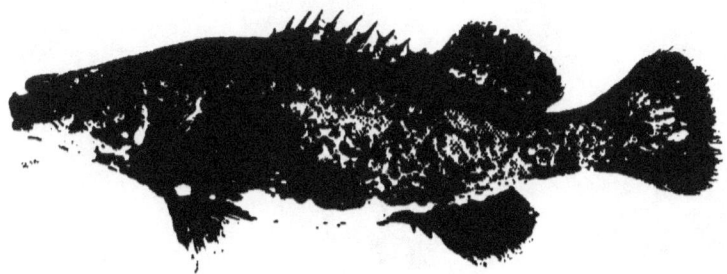

C. MURRAY COD, *Oligorus macquariensis*, quarter natural size, p. 158.

been thoroughly acclimatised in the rivers, lakes, and estuaries, more especially of Tasmania and Victoria. In the first-named Colony, the trout introduced have tended to grow to prodigious size, taking on all the characters of the English breed which is now known to be only a variety, but was formerly regarded as a distinct species and distinguished by the title of *Salmo ferox*, or the Great Lake Trout. An example referable to this racial variation was taken in the Huon river by the late Sir Robert Hamilton, formerly governor of Tasmania, which weighed no less than twenty-nine and a half pounds, while fish weighing twenty pounds and upwards are of common occurrence. Some of the large inland lakes of Tasmania, and notably the Great Lake, are famous for the size and abundance of the Trout they produce. The following, epitomised from a report which appeared recently in the columns of the "Field," will convey some idea of the sport that Tasmania can place at the disposal of the enthusiastic angler. These larger fish, it is scarcely necessary to say, will not rise to the fly, but may be successfully laid siege to with a variety of real or artificial spinning baits.

The magnificent bag referred to below and reported in the "Field" was made by three anglers who camped for a few days at the Great Lake, a sheet of water no less than 100 miles in circumference. The total number of fish taken was only 82. These, however, weighed collectively 759 lbs., yielding an average of $9\frac{1}{4}$ lbs. per fish, though, as a matter of fact, they varied in weight from $2\frac{1}{2}$ up to as much as 20 lbs. It is also recorded in this account that a fish had, a short while previously, been found dead on the shore of this lake which possessed an estimated weight of no less than 35 lbs. That such a phenomenal specimen may have actually been seen is by no means incredible. There is a remarkably large race of the ordinary brown trout indigenous to certain lakes in the Orkney Islands, and it has on that account been denominated *Salmo fario var. orcadensis*. As heavy a weight as 36 lbs. has been authentically recorded of one colossal individual belonging to these veritable Tritons of their tribe.

A second typical species of the British Salmonidæ has been permanently established in both Victorian and Tasmanian waters. This is *Salmo trutta*, or the so-called Salmon-trout, Sea-trout, or Sewin of the English, Scottish and Welsh fish markets. It is an essentially migratory form, going down to the sea and returning to the rivers to spawn. Its finer varieties very closely resemble the true salmon in shape and aspect, but it is a relatively smaller fish, rarely exceeding ten or twelve pounds in weight.

Much time, labour and treasure has been expended on the very laudable object of establishing the true salmon, *Salmo salar*, at the Antipodes; but, unfortunately, these attempts have not been rewarded with the same measure of success that has attended the experiments made with the Trout and the Salmon-trout. This subject of the establishment or otherwise of real salmon, in Tasmania more particularly, is one which has to be approached with some degree of diffidence. There are those who, not having the courage of their convictions, have acquiesced with and encouraged Tasmanians in the belief that their superb, but in comparison ungainly, trout are no other than lordly salmon. It has been a sore point, too, and a sincere source of regret to the writer, that when called upon to invest this or that fish which has been submitted to him, including the magnificent 29½ lb. Huon fish, with the aristocratic title of *Salmo salar*, he has been unable, conscientiously, to pronounce the verdict that was expected.

Prophets have even been summoned, like Balak, from afar to prophesy pleasant things concerning them. The only, and to most minds very cold, comfort accruing from such a course, however, has been the oracular announcement that certain of these fish would be sold "as salmon" in the English market. Now, according to the English Salmon Fisheries Act, 1861, all migratory species of salmonidæ, including the salmon-trout or the brown trout that frequent the river estuaries and run up the rivers to spawn, are "salmon." In the eye of the English law, consequently, the Hobart Museum Curator is perfectly justified in labelling the Huon trout a salmon, or any other authorities in declaring that "salmon" are established in Tasmania. If, however, sportsmen repair to Tasmania with the express object of enjoying salmon fishing as understood in the British Islands, and as correlated with the capture of *Salmo salar*, they are undoubtedly doomed to grievous disappointment.

A brief account of what has actually been accomplished in the direction of acclimatising the "King of fishes" in Tasmania may be acceptable. From the year 1852 down to within quite a recent date, the attempt to establish salmon in that Colony has been the main object of operations prosecuted by the Government, aided by a Board of Salmon Commissioners and liberal private assistance. The introduction of Trout and Salmon-trout, while conducted concurrently, was held to be of secondary importance. For several years the writer enjoyed the privilege of officially participating in these acclimatisation operations, and he was thus afforded the opportunity of making himself acquainted with all the attendant circumstances, and of arriving at what would appear to be the only logical explanation of the results obtained.

Since the date of the earliest successful consignment of salmon ova to Tasmania, by the ship "Norfolk" in the year 1864, literally millions of salmon ova have been similarly imported, their fry being hatched out at the Salmon Ponds, near New Norfolk, and subsequently liberated in the Derwent and other apparently suitable rivers. Sometimes these young salmon were turned loose into the rivers in the earlier or parr form, and at other times in the silvery "smolt" condition when they were ready to make their first migration to salt water. In no instances, however, have these salmon smolts been authentically known to return to the rivers the following year as "grilse" or in subsequent years as full grown salmon. Some have suggested that the young salmon on arriving at the sea were invariably devoured by sharks, barracouta, or other predatory fish. In that case, however, the salmon-trout that descended as smolts with them at the same time would have shared their fate. That, however, by no means happened, numbers returning to the rivers. It may be added that the active young salmon were far better qualified to take care of themselves in the salt water than many of the species of fish indigenous to Tasmania.

It occurred to the writer that possibly the question of temperature was an essential element in the refusal of the salmon to adapt themselves to their new environments, and a series of thermometrical observations was inaugurated. From these experiments it was ascertained that the sea water on the Tasmanian coast averaged some ten degrees higher than that of the British seas, and corresponded more nearly with that of the Mediterranean on the South Coast of France.

This fact was in itself a sufficient explanation of the disappointing results so repeatedly obtained. The salmon, unlike its hardier relation the trout, is known to be particularly sensitive to any changes of external conditions, and more especially to those of temperature. Consequently, the millions of young salmon fry and smolts that have been liberated in the rivers of Tasmania and made their way to the ocean, finding the temperature of the sea water in the vicinity of Tasmania too warm for their comfort, have, it may be predicted, wandered away, presumably in the direction of the Antarctic Ocean, in search of colder water. Interesting evidence in support of the correctness of this interpretation is afforded by the fact that precisely similar results have attended the attempts to establish salmon in rivers of the South of France which debouch upon the correspondingly warm waters of the Mediterranean. The young salmon have thriven as fry, parr, and smolts in the French rivers, but, on migrating to the sea, they wandered away and did not return to their native streams to spawn, as they would have done under their normal conditions.

That this temperature interpretation is the only one that will logically account for the decided objections manifested by *Salmo salar* to establish itself in Tasmanian waters—all other conditions being so favourable—is now pretty generally recognised. Tasmania, nevertheless, if not possessing salmon, may rest contentedly on her laurels as being *facile princeps* among the surrounding colonies in the production of Brobdingnagian trout. At the world-famous Trout-Hatchery on the River Plenty, of which the writer is proud of having been entrusted with the enlargement and re-organisation on modern principles, trout have been bred and the ova dispatched far and wide for stocking the streams of New Zealand, Victoria, New South Wales, South Australia, and, in these latter days, Western Australia. The latest operations undertaken by the writer in the last-named colony included the establishment of a trout-hatchery in the neighbourhood of Bunbury, and the distribution of fry raised there from Tasmanian ova in various of the more suitable rivers in the southern district of that colony. These included the Rivers Preston, Collie, Capel, Blackwood, Hervey, and Murray, from some at least of which streams satisfactory bags of trout are being hopefully anticipated in a few years' time.

English Perch, *Perca fluviatilis*, the Tench, *Tinca vulgaris*, and also the Prussian Carp, *Carassius vulgaris*, have long since been successfully introduced into Australian waters, the first-named species thriving most conspicuously in the artificial lakes and reservoirs in the vicinity of Ballarat, Victoria. In the writer's humble opinion there are, however, a number of indigenous fresh-water species more worthy of attention than the several last-named European types. Silver Eels, *Anguilla australis*, of high gastronomic excellence, and very nearly resembling the much esteemed English species, *A. vulgaris*, abound in the Southern Colonies, and about two years ago were experimentally introduced by the writer into rivers in the southern district of Western Australia, where they were hitherto unrepresented. According to recent reports, they have already commenced to multiply in the Upper Avon, where the majority were liberated, and, being distributed from this centre, should hereafter prove a welcome addition to the riverine settlers' commissariat.

Some little space may now be devoted to those representatives of the Australian fish fauna which are of an essentially phenomenal but non-economical character. Among these, no more appropriate group can be selected for illustration than the Lophobranchiate family of the Syngnathidæ, which embraces such types as the Pipe-fishes, Sea-horses and Sea-dragons. A few of the more notable of these are accordingly included in the coloured drawing, Chromo-Plate VI., which faces the commencement of this Chapter. No other part of the world, it may be asserted, produces so varied and remarkable an assemblage of members of this group as are to be found in Australian waters, and more especially within its southern or temperate limits. In this association it may be mentioned that the little European Sea-horse, *Hippocampus antiquorum*, scarce in British seas but common in the Mediterranean, and rarely exceeding a length of three inches, is represented in Tasmanian waters by a form, *Hippocampus abdominalis*, fully four times that size. Such a species as this would create a sensation at Brighton or other of the large public aquaria, and with suitable precautions could be successfully transported to England from the Antipodes. The colours of this Tasmanian Hippocampus vary considerably, ranging through all the gradations of creamy white, with usually a pale yellow chest and sparsely dark-spotted back, to examples in which a dark golden or more sombre brown predominate. In this item of colour the Mediterranean species exhibits even greater variability, the writer having had specimens in his possession, supplied to him by Mr. G. Hoadley King, the Aquarium naturalist, of Great Portland Street, London, in which some individuals were, while living, red, others pale yellow, or white, or brown with many intermediate tints. Two of the smaller forms of Australian Sea-horses or Hippocampi are included in the coloured plate previously referred to.

If the large Tasmanian Sea-horses are calculated to create a sensation at a public aquarium, the addition thereto of living examples of their relatives the Australian Sea-dragons, of the genus Phyllopteryx, might be expected to inaugurate a decided boom in favour of these aquatic institutions. This generic group is most essentially and peculiarly Australian, numbering three or four species, the two most remarkable of which are illustrated in Chromo-Plate VI. One of these, *Phyllopteryx foliatus*, is most abundant on the Tasmanian coast, and the second one, *P. eques*, is frequently taken in their dredges by the oystermen of Port Lincoln, in South Australia. While of somewhat smaller size and considerably less elaborate in contour and in the number and dimensions of its membranous appendages, the Tasmanian species is much the more brightly coloured. As shown in the illustration above quoted, its general ground tint is a brilliant scarlet carmine, spotted on the sides of the body and head with white. On the sides of the neck are seven or eight transverse stripes of violet or cobalt blue, while the narrow ventral surface of both neck and abdomen is bright yellow. The leaf-like appendages of both body and tail are deep crimson, with a yet darker linear outer border. Large examples measure as much as fourteen or

AA

fifteen inches in length. The aspect of this species when swimming in a semi-vertical position, as shown in the illustration, is remarkably suggestive of a hopping kangaroo, the two ventral appendages corresponding with the marsupial's fore-paws.

The South Australian Sea-dragon, *Phyllopteryx eques*, which occupies the central and most extensive area of our coloured plate, attains to a greater length than the Tasmanian type by several inches. Its most remarkable peculiarity, however, as indicated in the figure given, is the more angular contour of the body and the wonderfully luxuriant development upon it of outstanding spines and membranous leaf-like appendages. These structures bear a striking resemblance to the fronds of the seaweeds among which it commonly takes up its abode, and may be regarded as a special adaptation to ensure the animal's effectual concealment amid its thoroughly harmonious environment. In this manner bunches of a finer species of seaweed appear to spring from the head, while the broader leaf-like fronds of a larger variety grow seemingly from all of the more prominent angles of the creature's body. In addition to these there is a series of flattened plumeless spines along both the dorsal and ventral surfaces, which present a strong general resemblance to newly sprouting algal growths. Writing of this species of Sea-dragon, as illustrated in Dr. Gunther's "Introduction to the Study of Fishes," page 682, the Rev. Tennison Woods remarks, " It is the ghost of a sea-horse with its winding sheet in ribbons around it, and, as a ghost, it seems in the very last stage of emaciation, literally all skin and grief." In the figure alluded to, drawn from a specimen abnormally shrivelled up by conservation in strong alcohol, the spectre-like comparisons are by no means inappropriate. In its native element, however, resplendent in its natural colouring, in which various shades of light crimson or lilac predominate, bravely bedecked with flowing frills and furbelows, and with resplendent jewel-like eyes of sapphire blue, this fish constitutes one of nature's most exquisitely wonderful productions, capable of and distinctly manifesting the fullest capacity for life's enjoyment.

The fish at the bottom of Chromo-Plate VI. is a type, *Solenognathus spinosissimus*, that is intermediate between the slender Pipe-fishes—represented by a species of Syngnathus towards the left-hand top corner—and the Sea-horses, Hippocampi. Like the latter, it has a prehensile tail, with at the same time a gorgeous scarlet and yellow livery resembling that of the Sea-dragons. The form figured is occasionally taken with *Phyllopteryx foliatus* in Tasmanian waters, while a second species, *S. Hardwickii*, occurs as far north as Houtman's Abrolhos on the west, and Moreton Bay on the eastern Australian sea-board.

In *Gastrotokeus biaculeatus*, an elongate, grass-green species, a yet nearer approach is made to the ordinary Pipe-fishes. The tail, however, possesses the same prehensile character as that of the Sea-horses. Gastrotokeus is an essentially tropical type, occurring on the North Queensland coast and on the north-west coast of Western Australia. In common with Phyllopteryx and Solengognathus, the male individual of Gastrotokeus fulfils the *rôle* of foster parent to the infant brood, receiving and taking care of the ova, which throughout the period of incubation are firmly embedded by their bases in the soft abdominal membranes, until they are hatched. The ova of both Phyllopteryx and Gastrotokeus are of large size, of a delicate pale pink hue in the latter and a bright clear red in the first-named type. In many of the ordinary Pipe-fishes the eggs, carried by the male, are enclosed in membranous folds of the skin, while in the Sea-horses, Hippocampus, there is a true pouch or marsupium for their reception. As may be anticipated by the small aperture of the mouth in all of the members of this group here enumerated, their food is of the most minute description, consisting almost exclusively of Entomostracous crustacea and the larval forms of the larger species.

Some of the most outré-shaped and brilliantly coloured members of the Australian fish fauna are included in the very distinct group of the Plectognathi. This series is represented most conspicuously by what are known popularly in Australia as Leather Jackets and Cow-fishes, belonging respectively to the genera Monacanthus and Ostracion. Although the Plectognathi are usually represented as fish which are essentially characteristic of tropical or sub-tropical seas, the greater number and more brilliant of the Australian species pertain to the temperate waters of that Island-Continent, and are notably abundant on the extreme southern sea-board of Tasmania. As with the Syngnathidæ, a coloured plate, Chromo-Plate VII., has been set specially apart for the illustration of characteristic members of this group.

Of the Australian Cow-fishes, so-called with reference to the horn-like prominences which are usually developed from the frontal region of their indurated carapace-like integument, the Tasmanian species *Ostracion* or *Aracana ornata* is a brilliant example. The males and females of this species are so differently coloured that they were originally, as in the case of the British Cuckoo Wrass, *Labrus mixtus*, supposed to represent two distinct species, and had conferred upon them the separate titles of *Ostracion ornata* and *O. aurita*. The specific identity of the two species has been amply demonstrated in examples examined by the writer, one particularly interesting individual, an hermaphrodite, exhibiting the characteristic colours of the two presumptive species on opposite sides.

The Tasmanian Cow-fish rarely exceeds six inches in length, but is quite the gem of its tribe from an æsthetic standpoint. In the male fish, which is the more brilliant of the two, the ground colour of the body of the living fish is a bright grass-green above and on the sides, and pale lemon-yellow beneath; that of the tail or caudal fin orange-yellow; and the remaining pectoral, dorsal and anal fins a neutral transparent tint. Excepting the pale-yellow ventral area, the remaining surface of the of his body is traversed by broad, irregular, more or less interrupted, stripes of the most brilliant ultramarine blue, the edges of which are usually distinctly defined by narrow border-lines of dark chocolate brown. Two or three of the central blue body stripes are usually continued into the tail, forming upon that member loop-like patterns. The pale-yellow ventral region is further variegated by a broad reticulated pattern in pale blue.

It is especially remarkable of the broad blue stripes that ornament the general surface of the body that they are never of a precisely similar pattern in any two individuals. Some three or four, at the most, of the horizontal central stripes are continued uninterruptedly from the head or snout to the posterior region. Those belonging to the upper or dorsal area, more particularly, are usually broken up into short lengths or isolated spots, and this more rarely occurs also with relation to the central and lower stripes. An example of a characteristically marked male specimen of this Cow-fish is represented by Chromo-Plate VII., fig. A.

The female of the species, formerly known as *Ostracion aurita*, depicted opposite the letter B in the same Plate, possesses somewhat the same pattern markings, but is distinguished by altogether distinct tints. The ground colour of the body of the female individuals is chiefly a pale pinkish grey or dove-colour, with local flushes of a more decided pink, and the lower surface entirely pure yellow. The broad blue longitudinal stripes destinctive of the males are represented in the females by corresponding ones which are of a rich reddish-brown hue, and are rarely, if ever, broken up into short isolated lengths or spots. In place of this, however, the stripes commonly coalesce more or less extensively with one another in individually varying patterns, and very frequently in the neighbourhoods of the bases of the dorsal and pectoral fins assume an irregular helicoidal or spiral design.

The number of spinous processes developed on the carapace of this Tasmanian species are considerably in excess of those possessed by other familiar types, such as

Ostracion cornutum. In addition to the two horn-like frontal spines, two pairs are developed on the back, a single pair from the centre of the side, and two or three pairs on each side of the abdomen, making from twelve to fourteen in all. The supremely restless habits, bright colouration, and grotesque shape of the little Tasmanian Cow-fish, rendered this species a highly popular attraction at the writer's Hobart fisheries establishment, where a tank was temporarily devoted to the reception of a selected series of individually varying specimens. Among the more bizarre aspects exhibited by this Cow-fish reference may be more especially made to that presented in an end-on view as delineated in the two figures opposite the letter C in Chromo-Plate VII. In its young condition *Ostracion cornutum* is common in Summer in the shallow inshore waters of the innumerable coves and bays of the southern coast-line of Tasmania, while the larger, adult, individuals are frequently taken in the "grab-all" nets set by the fishermen in deeper waters for the capture of Silver Trumpeter.

The Leather Jackets or Trigger-fishes, referable to the genus Monacanthus, are among the most abundant of the representatives of the Plectognathi in Australian waters. They take the first of their popular titles with respect to the texture of their scaleless, finely granulated or hispid skins, which may be stripped from their bodies with the greatest ease in a form highly suggestive of a piece of fine kid leather. The second title of "Trigger-fishes" has been bestowed upon them with reference to the peculiar construction of the anterior dorsal fin, which in Monacanthus, as its technical name implies, consists of a single large sharply-pointed spine and its attached membrane. This spine is capable of elevation and depression at the will of its owner, and when erected is so rigidly held in its place that it will break before it can be pressed into its recumbent position by main force. There is, however, a small bone behind and at the base of this spine, which, on being moved mechanically or by the muscles of the animal, immediately causes the spine to drop down in a manner suggestive of the trigger and hammer arrangement of a gunlock. In the allied genus Balistes, with which the name of Trigger-fishes is more exclusively associated, this mechanical peculiarity is most conspicuously developed. Behind the large anterior spine there are two shorter and relatively slender ones, and the pressure of one of these when the fin is erected causes the immediate collapse of the larger spine. Several representatives of this genus frequent the coral reefs of the Northern tropical Australian sea-board.

Of the more cosmopolitan single-spined Leather Jackets, genus Monacanthus, no less than forty Australian species have been described. These may vary in size respectively from a few inches to about a foot and a half in length. Several of them inhabit the colder waters of the Southern Tasmanian coast-line, and one of these, *Monacanthus rudis*, has been selected as the central illustration in Chromo-Plate VII. It is worthy of remark that the individual here portrayed is a male fish, bedight in what might be denominated his wedding livery, his resplendent garb of blue and yellow being at other seasons of the year replaced by sober tints of brown and white. This fact was ascertained through the writer keeping this specimen for a considerable interval at the Hobart fishery establishment. A familiar instance of a British fish undergoing an almost equally remarkable colour metamorphosis is afforded by the male of the common Stickleback, *Gasterosteus aculeatus*, who, during the mating season, exchanges his everyday dress of silvery grey for a most gorgeous corselet of gold and scarlet and shining green.

In many of the common species of Australian Leather Jackets, notably *Monacanthus hippocrepis* and *M. Ayraudi*, lines, spots or suffused patches of rich cobalt blue are permanently conspicuous, rendering them among the gayest members of the Australian fish fauna. Added to this, many of the species are distinguished by an abnormal development of their fins or by body excrescences, which impart to them a most bizarre aspect. Thus, in one species, *M. megalurus*, a pouch-like membrane is developed on the chest; in many instances, horny, hook-like structures protrude in the male individuals on each side of the tail; while in one form, *M. Browni*, also taken in Tasmania, these horny hooks are replaced by a lozenge-shaped, tooth-brush-like patch of bristles.

Notwithstanding their somewhat abnormal appearance, and the fact that they represent an order which includes the highly poisonous Toad-fishes, Leather Jackets, though but rarely brought to the fish market, have been pronounced by connoisseurs to be most excellent eating, and, if skinned before cooking, as being equal to the much-esteemed Sole or Flounder. In such high estimation, in point of fact, were these fish regarded by one of the earlier Tasmanian settlers, an ancestor to a reigning judicial functionary, that these fish are still known to the local fisher-folk by the patronymic of their whilom appreciator. Leather Jackets are by no means regarded with favour by the ordinary line fishermen. The smaller species have a special knack of nibbling the bait off the hook with their trenchant chisel-like teeth, while the larger ones can with the same weapons easily sever a stout line or a steel hook of moderate thickness. When caught, they on the same account demand wary handling, a nip from the formidable teeth of a large specimen being liable to lead to a loss of flesh, blood, and—with any but evenly-balanced minds—serenity of language and temper.

One other interesting member of the order of the Plectognathi is included in Chromo-Plate VII. This is the Tasmanian Porcupine-fish, or prickly Globe-fish, as it is sometimes called, *Chilomycterus jaculiferus*, Fig. D. It very nearly resembles the common Porcupine-fish, *Diodon maculata*, of the tropical Australian coast-line but possesses more slender spines and other obscure points of distinction. When brought to the surface of the water it shares with the familiar tropical species the property of inflating itself with air into an almost perfect sphere, around which the spines stand rigidly erect. A fish thus floating and inflated was always observed to occupy a considerable interval, it might be half-an-hour, in getting rid of the injected air and thereby recovering the capacity to descend again into the profundities of its native element.

The Toad-fishes, belonging to the genus Tetrodon, which are very abundantly represented in Australian waters, possess the same power of inflating their bodies, and are on this account pre-eminently distinguished in Australia by the title of Blow-fishes. Several of the smaller varieties, and notably a handsome golden-green, black spotted form, *Tetrodon Hamiltoni*, are plentiful on the Tasmanian coast, but the larger species of a foot or more in length are chiefly limited to the tropics. It is worthy of remark that the members of this genus are notoriously poisonous, and that several fatalities have occurred in both Tasmania and on the Australian mainland through the injudicious participation in a meal of Toad-fish. A portrait of the above-named common and highly poisonous Tasmanian species is annexed.

TASMANIAN TOAD-FISH, *Tetrodon Hamiltoni*. TWO-THIRDS NATURAL SIZE.

A somewhat grotesque name has been bestowed on one member of the Plectognathi by the Fremantle fishermen, with whom the Toad-fishes are familiarly known as "Blow-fish." The species in question is one of the Trunk or Cow-fishes, *Ostracion lenticularis*, differing from the type previously described, by the entire absence of horn-like protuberances. In recognition of its obvious natural affinity with the more familiar Tetrodons the title improvised for it is the "Bony Blow-fish."

This list of Australian Phenomenal Fishes would be culpably incomplete without a brief reference to the so-called Port Jackson Shark, *Cestracion Phillipi*, which has been selected as the tail-piece illustration to this Chapter. The chief interest attachable to this fish is centred in its remote antiquity. It representing the only genus of a family group traceable back through the cretaceous and triassic to the more remote Devonian or carbouiferous formations. For a shark, the fish is of but small size, rarely exceeding four or five feet in length. A tolerable idea of its form and peculiarities may be obtained by a reference to the figure mentioned, which was taken literally from life. As there shown, the head is especially large and massive, and provided with conspicuous brow-ridges, while in front of each of the broad dorsal fins there is a strong spine, as is found in the Piked Dog-fish, Acanthias, of the British seas.

The armature of the mouth of Cestracion is highly characteristic. It consists in front of prehensile elements, but to the rear of these there is a pavement-like arrangement of crushing teeth, wherewith their owner breaks up and feeds upon various descriptions of hard-shelled Invertebrata. Oysters are a favourite food of this Shark, and, in consequence of its predilections for this bivalve, it has proved a formidable enemy to oyster growers in both Tasmania and on the mainland sea-board. At Spring Bay, in the former island Colony, the writer found it even necessary to fence round certain of the Government Oyster Reserves with closely wattled brush wood in order to protect the oyster stock laid down from this shark's depredations. In some localities Cestracion feeds almost exclusively upon Sea-Urchins or Echini, the sharp spines of which have apparently no other effect than the pleasant titillation of its palate. The proof of the extent to which this piquant food is favoured by this Shark is afforded by the fact that the entire pavement of teeth of captured specimens are not unfrequently permanently stained a deep purple, through constant indulgence in a dietary of the commoner purple Urchin. The name of Port Jackson Shark was conferred upon *Cestracion Phillipi* with reference to its having been first captured in that locality, to which it was then supposed to be restricted. It has, however, been since found pretty well throughout temperate Australian waters, and has been met with by the writer at such remote locations as Tasmania, Moreton Bay in Queensland, and Fremantle in Western Australia. The specimen figured on page 194 was taken at the last-named station. The Oyster-Crusher, Pig-fish, and Bull-Dog Shark are names by which it is locally known to Australian fishermen.

Phenomenal Sharks of the more ordinary type are by no means strangers to Australian waters. Examples of the Tiger Shark, *Galeocerdo Rayneri*, and the Blue Shark, *Carcharias glaucus*, have been authentically attested to up to twenty or twenty-five feet in length, and according to their measurement with relation to the boats alongside which these monsters have occasionally put in an appearance, may even attain to thirty feet. An original method of disposing of a large shark was recently brought before the notice of the writer. The fish, as commonly happens, was cruising around a steam-boat anchored in Sydney Harbour on the look out for flotsam and jetsam. The cook happened to have a pumpkin on the boil in the galley and, by a sudden inspiration, conceived the idea of throwing it to the shark. It was immediately seized and swallowed, followed by a commotion suggestive of a dynamite explosion. When the troubled waters sank to rest the shark lay lifeless on their surface! The experiment, though novel and successful in its object, was, it must be admitted, rather rough upon the shark.

The Elephant-fish, or Southern Chimæra, *Callorhynchus antarcticus*, undoubtedly claims a front place in the ranks of abnormal or phenomenal members of the Shark tribe. It has already received a brief share of attention at page 167, in conjunction with the reference to its effigy in Plaster of Paris, photographically reproduced in company with many other species in Plate XXVIII. The specimen No. 20, there portrayed, represented a full-grown female individual, which, after capture, deposited one of its singular, flattened, horny, egg cases. This structure, allowing for the absence of the horny tendrils at the four corners, somewhat resembles those of our British Dog-fishes, of the genus Scyllium. The edges of the case are, however, much more flattened and expanded, while, in order to secure the adherence of the organism to submarine objects, the whole under surface is coated with fine flocculent filaments, which are viscid and adhesive when the structure is first extruded from the oviduct. As shown by the replica of the egg case that was made in plaster in conjunction with the parent fish, its size is relatively considerable.

The chief interest attached to this type is its near relationship, and occurrence as a complementary form, to *Chimæra monstrosa*, the Chimæra or Rabbit-fish, and other species of the same genus, of the North Temperate and Arctic seas. The male individual of Callorhynchus shares with Chimæra the possession of a peculiar hinged spinous process, which is developed from the centre of the forehead. The true significance of this peculiar organ has never, as yet, been satisfactorily determined. The writer is disposed, however, from an examination of it in living fish, to believe that it is employed during the mating season as a grasping organ. The slender cartilaginous proboscis of the female of Callorhynchus, or the whip-like tail of that of Chimæra, furnish, in either instance, fulchra which can be tenaciously grasped by the organ in question.

PORT JACKSON SHARK. *Cestracion Phillipi*, p. 192.   ONE-TENTH NATURAL SIZE.

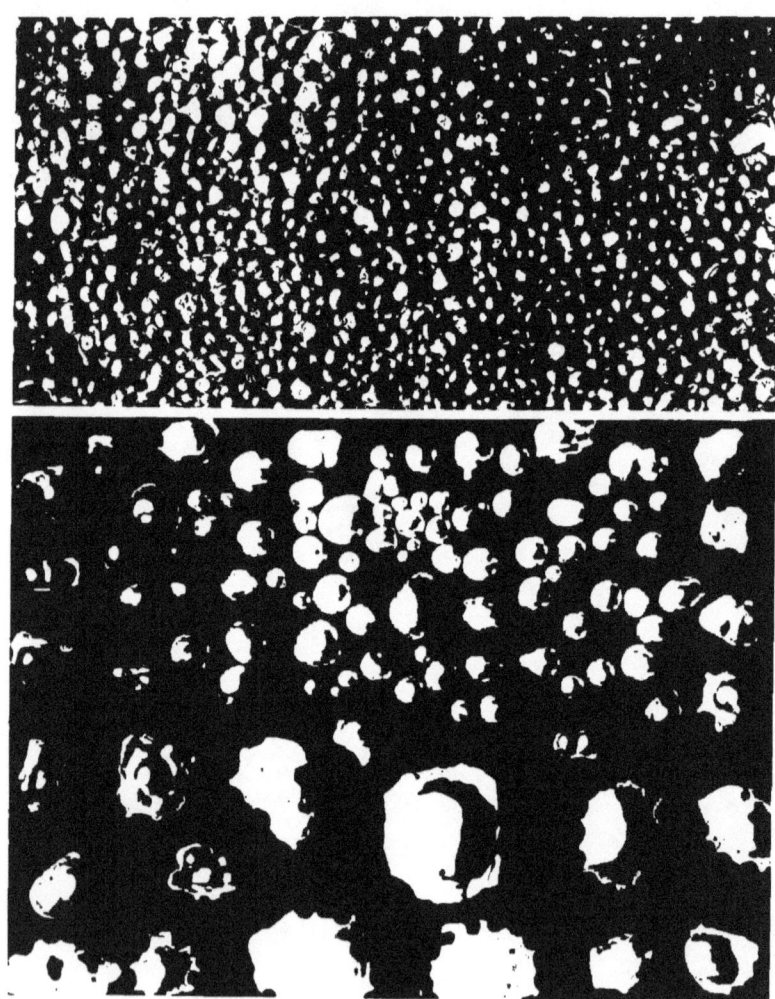

WESTERN AUSTRALIAN PEARLS

CHAPTER VII.

PEARLING LUGGERS IN BROOME CREEK, ROEBUCK BAY, WESTERN AUSTRALIA, p. 107.

PEARLS AND PEARL-OYSTERS.

CONSIDERABLE space was devoted in the author's recent work, "The Great Barrier Reef of Australia," to the subject of the Pearls and Pearl-shell fisheries of Queensland. It is not proposed in this Chapter to re-traverse that ground, but to chronicle a few data of interest relating to cognate subjects accumulated during the writer's later peregrinations in Western Australia.

As a matter of fact, a higher interest, from a naturalist's point of view, attaches itself to these fishing industries as pursued in this more recently exploited territory on account of the circumstance that they have here to deal with two very distinct varieties of pearl-oyster in place of being practically limited, as in Queensland, to a single type. In Torres Straits, the Gulf of Carpentaria, and along the northern confines of the Great Barrier Reef, where the Queensland fisheries are exclusively prosecuted, the shell sought after is the large essentially tropical species known technically by the name of *Meleagrina margaritifera*. This variety is obtained chiefly on account of the valuable nature of its shell, its substance representing the finest texture of Mother-of-Pearl, out of which knife handles, the largest descriptions of pearl buttons, and all other pearl articles of the best quality are manufactured. Although pearls of the most valuable description are also obtained from this shell, their acquisition is held to be of practically secondary import, while, through the inability of the boat-owners, in the majority of instances, to exert a strict control over the opening and cleaning of the shells, it is admitted that a very large percentage of the pearls obtained are surreptitiously disposed of. Notwithstanding this undesirable filtration, it has been estimated that the value of the pearls obtained ranges from about one-tenth to one-sixth of that of the shells, and as this in Queensland alone commonly averages £100,000, the item of pearls is by no means insignificant.

In Western Australia the large tropical Mother-of-Pearl shell is more or less abundant throughout the coast-line northwards of Exmouth Gulf. The main portion of the pearling operations is, however, carried out between there and King's Sound. North and east of this, as far as Cambridge Gulf, large areas of shell-producing grounds exist, but owing to the as yet unsettled condition of the adjacent mainland and the hostility of the natives towards boats' crews landing for wood and water, this section of the coast-line has so far remained almost undeveloped. Judging from the much indented reef and island-bestrewn character of this coast-line, and the observations made and impressions formed by several voyages between King's Sound and Cambridge Gulf, the writer is of the opinion that this section of the coast will in future years prove to be the most prolific of the Western Australian Pearl-shelling grounds. With it will probably be also associated Bêche-de-Mer fisheries, approximating in character and value to those of Torres Straits and the Great Barrier.

At the present time the fishery for Bêche-de-Mer on the Western Australian sea-board is scarcely worthy of notice, being limited to the takings of a few isolated fishermen, chiefly Chinamen, who ply their avocation in the neighbourhoods of King's Sound and Dampier's Archipelago. With the latter exception there are very few favourable reefs south of King's Sound of sufficient extent to yield a profitable fishery, while, as ascertained by the writer, the qualities produced are inferior to the Queensland types. As chronicled, however, in the preceding Chapter, the majority of the most esteemed Torres Straits and Great Barrier Reef species were discovered by the writer in Houtman's Abrolhos Islands, and though these do not appear to occur on the mainland coast-line up to and among the reefs of the Buccaneer Archipelago off King's Sound, it is by no means improbable that they may be met with again further north and east of that point.

The fishery for the larger tropical species of Mother-of-Pearl shell is conducted in Western Australian waters on precisely the same lines as in Queensland. In the earlier days, as in Queensland, the shell was obtained by simply wading on and collecting it from the reefs. Native or Malay naked divers were next employed to bring the shell up from relatively small depths, while finally, as the shell became exhausted in the shallower and was found to be obtainable in quantities only in the deeper water of from ten to twenty or more fathoms, recourse has been had to schooners and luggers equipped with the most perfected diving apparatus. From convenient ports, such as Broome and Cossack, luggers will run out to, and work independently on, the adjacent shelling grounds. Where, however, as mostly happens, these grounds are at a considerable distance from any port, a number of luggers belonging to a company or private firm are escorted to the grounds by a schooner which both acts as a tender for needful supplies and receives all the shell collected. Even under these conditions both schooners and luggers commonly return to port at spring tides when the currents are too strong to permit of the divers working. The photograph reproduced at the head of this Chapter represents one such occasion when a large number of the boats had come in for stores and shelter to Broome Creek.

The port of Broome in Roebuck Bay, may be said in respect to the Pearl-shelling industry of Western Australia, to occupy a position much akin to that of Thursday Island with relation to Queensland, it being the port furthest north to which the pearling fleet is accustomed to resort for supplies and repairs. As with Thursday Island (Port Kennedy), when first visited by the writer about a decade since, Broome is not very far advanced in the amenities and conveniences of modern civilisation. Thursday Island, however, revisited a few years later, had made such rapid strides as to possess a well-built jetty, obviating the necessity of landing in small boats. It could also boast of that indispensable anti-climax of British citizenship—a hansom cab. Broome revisited will, no doubt, a year or two hence, cap the precocity of Port Kennedy by the production of a motor car. In the interim, a substantial jetty at which vessels can land or embark freight and passengers at all tides is no doubt the *sine quâ non* of Broome, and possessed of it she should, with the enterprising guiding spirits now at the helm, speedily achieve distinction. In direct touch by cable station with the hub of the world, affording the greatest facilities for access to the Kimberley goldfields, and the last convenient port of call for steamers arriving at or departing for Singapore and the farther East, Broome possesses abundant claims for substantial recognition and support beyond those relating to its intimate connection with the Pearl-shelling industry.

Landing at Broome under conditions other than those of highest tide is, to the ordinary traveller, a gruesome undertaking. Even to the born naturalist a mud wade—and such mud—of close upon half-a-mile with, albeit, here and there a spotted whelk and here and there a winkle, all new to him, if not to science, to beguile his attention, becomes a somewhat toilsome pleasure. Such a wade, handicapped with a black cockatoo on his wrist, a cage of live lizards under his arm, boots slung round his neck, and a few unconsidered trifles of less account otherwhere disposed, represents the writer's last parting touch with Westralian soil. A new Broome, with a clean swept landing stage, is the devoutly-to-be-wished-for innovation looked forward to by the writer on the occasion of his next visit to Roebuck Bay. The foregoing experiences notwithstanding, mud-larking among the mangroves unencumbered with travelling impedimenta is undoubtedly, to the naturalist, a right royal treat that may be profitably reverted to later on.

In connection with his investigations of, and reports to the Government upon, the Mother-of-Pearl shell and other local fisheries, the writer spent some little time at separate intervals at the Port of Broome. He gladly avails himself of this opportunity of recording his grateful acknowledgments to Mr. Skelton Streeter and his worthy henchman, Mr. W. Male, for most liberal hospitality and accommodation, otherwise unattainable, accorded him while visiting the remote "nor-west." Mr. Streeter being, moreover, the owner of the largest fleet of Pearling boats that sailed from Roebuck Bay, and his station being the trade centre to which a large percentage of the shell and pearls derived from other local sources was continually flowing, the writer, during his sojourn there, enjoyed exceptional facilities for acquainting himself with the wealth of material derived from this district. A small corner from a couple of the many traysful of pearls that in this manner passed within the writer's purview has been photographically reproduced in Plate XXXII. The material embodied in this series is what might be denominated a mixed lot. It consists to a large extent of irregular-shaped, or what are known to the trade as "Barok" pearls; mixed with these, however, are a goodly number of first-class gems of both large and medium-sized dimensions. The group occupying the lower portion of the Plate is composed for the most part of pearl blisters, hollow excrescences of

THE WESTERN AUSTRALIAN "SOUTHERN CROSS" PEARL, p. 199.

REMARKABLE WESTERN AUSTRALIAN PEARL. "*L'Enfant Prodige,*" p. 202.

pearly substance, which, when symmetrically shaped and suitably mounted, are only distinguishable by the initiated from solid pearls. Pearls themselves often grow attached to the surface of the shell, and in course of time become enclosed by and buried within its matrix; these immersed pearls can be disengaged and trimmed or "faked" into perfect shape. Samples of such immersed pearls are conspicuous towards the left-hand corner of the Plate. In common with the "blisters" they are in the first place punched free from their shelly matrix, and in that rough form consigned to the hands of the expert.

A few choice pearls of more exceptional size and quality placed by Mr. Streeter at the writer's disposal for photographic delineation have been reproduced in the form of a tail-piece illustration to this Chapter.

The amount of pearls produced by any given weight of shell is an altogether problematic quantity, incapable of even approximate determination. There are, at the same time, grounds upon which the shell produces pearls in phenomenal abundance, such shell, however, being invariably of inferior quality. Pearls, in fact, as is tolerably well known, represent a diseased product of the mollusc. It might, indeed, be suggested that they partake somewhat of the nature of chalk-stones in the human subject. Were puns permissible, we might add the moral "Chacun à son goût"— not even excepting oysters. Although the best quality of Pearl-shell produces comparatively few pearls, those obtained from it are of the finest shape and lustre. As an illustration of phenomenal pearl production, it may be mentioned that when he was last travelling up the Western Australian coast about two years ago, the discovery was announced to the writer of a new bank of shell that had been made off the coast near Onslow in the Ashburton district. The shell proved to be of a very diseased and inferior description, but from a few tons' weight of it pearls to the value of no less than £10,000 were secured by one company.

Pearls may assume the most fantastic shapes. The two specimens photographed in Plates XXXIII. and XXXIV., obtained from Western Australian waters, are probably the most remarkable examples that have been hitherto discovered. The one portrayed in the former of these two Plates is the world-celebrated "Southern Cross." In place of representing, as might be reasonably supposed, an artificially arranged congeries of pearls of approximately equal form and size, it constitutes a natural cross-shaped group, firmly united to one another by their lateral surfaces, and in the precise relationship in which they were discovered on opening a shell collected during the fishery off the coast near Cossack in the year

1874. This remarkable pearl, or more correctly group of pearls, was brought to London and exhibited in the Western Australian Courts of the Colonial and Indian Exhibition of the year 1886, the value there placed upon it by the company to whom it then belonged being no less than £10,000. The opportunity of its advent to London was utilised for a most critical examination at the hands of experts to discover whether or not this product was a genuine *lusus naturæ*, or the outcome of a more or less assisted artificial combination. The gem, however, withstood the severest tests applied, and it emerged from the ordeal with the added lustre of high scientific testimony as to its *bonâ fides*.

A purchaser of the "Southern Cross" at the above price was not forthcoming, and it still remains in the hands of a Western Australian Syndicate, from whom it may yet be bought at a very material reduction upon the fancy sum originally placed upon it. While in Western Australia the author enjoyed the privilege of the possession of this wondrous Cross for two whole days and nights, in order that he might examine and immortalise it with his camera. The responsibility of sleeping with a gem under one's pillow for the loss of which damages to the extent of £10,000 might be claimed was somewhat nerve-stirring, and would not have been as lightly undertaken, and probably not as promptly conceded, in London as in the as yet unsophisticated capital of Western Australia. The author's investigation of the "Southern Cross" has led him in no way to dissent from the verdict previously pronounced. Through an accidental fall since its original discovery, the two adherent pearl elements at the foot of the combination have become slightly loosened from the preceding five, but the manner in which the whole series fit into one another by reciprocal convex and concave surfaces, or, so to say, shallow cup and ball articulations, leaves little or no room for scepticism.

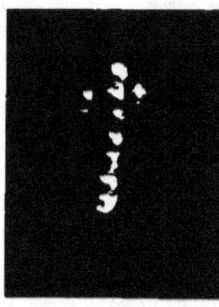

THE "SOUTHERN CROSS" PEARL. NATURAL SIZE.

The Cross, in its entirety, consists of what may be defined as nine amalgamated pearls, seven of which constitute the main shaft, and the remaining two, disposed laterally and almost, but not quite, symmetrically on opposite sides of the second pearl in the central row, represent the arms. The superficial likeness in size and contour of the central shaft, regarded separately, to a row of plump, closely-adpressed marrow-fat peas, as they lie in a newly-opened pod, is very suggestive. In order to preserve it from accident, and to facilitate its handling, this pearl cross has been mounted in a light open gold setting that in no way interferes with the full display of its unique character. It was under these conditions that it was laid upon a pearl shell corresponding in dimensions with that from which it was originally taken, and photographed as portrayed in the accompanying plate of its precise natural size. For the second photograph of this gem, reproduced on the preceding page, and representing it in its original unmounted condition, the writer is indebted to Mr. Maitland Brown of Geraldton, Western Australia, one of the Members of the Syndicate to whom this celebrated Cross belongs.

Much speculation has been indulged in respecting the probable nature of the foreign matter or nucleus, if any, that has formed the foundation of this remarkable pearl combination. It is suggested by Mr. Edwin Streeter, in his interesting treatise "Pearling and Pearling Life": London, 1886, that such a nucleus is possibly represented by an adventitiously intruded fragment of serrated seaweed. There is, as a matter of fact, a species of seaweed, *Hormosira Banksii*, abundant in Australian waters, that takes the form of concatenations of spheroidal nodes which might be likened to strings of pearls. Pearl is, unfortunately, one of the substances impermeable to the recently discovered Rontgen rays; otherwise, the solution of this mystery of the "Southern Cross" might be easily achieved. Maybe, however, in the near future, a new X.X., X.X.X., or other occult luminant will be evolved which shall possess the property of laying bare and naked the nuclei of Pearls, the marrow within our bones, and even the quantity and quality of the packing of our brain-cases. The physician's diagnosis of the eligibility or otherwise of candidates for Hanwell, or of paterfamilias' determination of the most appropriate career for the training of his verdant olive-branches, will, under such conditions, be a lightsome task.

As an illustration of the *rôle* that adventitious substances may take in the fashioning of the form of pearl concretions, it may be mentioned that there is on view in the Conchological Gallery of the British, Natural History, Museum a Western Australian pearl shell into which a small fish, Fierasfer, which not unfrequently occurs as a commensal of the Pearl-oyster, having intruded, died, and, in lieu of ejectment, became imprisoned and completely enswathed in a nacreous cerement. Through the kind courtesy of the Director, Sir William Flower, the writer has been provided with facilities for taking a photograph of this remarkable specimen, which is reproduced overleaf. It sometimes happens that a little correspondingly commensal crab,

cc

referable to the genus Pinnotheres, dies and becomes similarly embedded within a pearly matrix.

The second exceedingly remarkable example of pearl formation referred to on a previous page, which has been selected for special illustration, is the one portrayed in Plate XXXIV. This photograph is a life-sized replica of a Western Australian Mother-of-Pearl shell, with its attached pearly excrescence, contained in the author's collection. This most extraordinary pearl is remarkable not only for

PEARL-OYSTER, WITH PEARL-EMBEDDED FISH, CONTAINED IN THE BRITISH MUSEUM COLLECTION. TWO-THIRDS NATURAL SIZE. p. 201.

its size, for it measures just two inches in its largest diameter, but for the fact that it presents a most singular resemblance to a human head and torso. Some experts

PEARL-OYSTER LAID OPEN, SHOWING AT A DOUBLE PEARL, AND AT B SMALL COMMENSAL CRAB. HALF NATURAL SIZE.

have associated with this pearl an essentially infantile cast of countenance—child-like and bland, like that of the heathen Chinee—which to some minds has proved a direct stimulus to a fuller faith in the late Charles Kingsley's enunciated doctrine of the existence of Water Babies. A German acquaintance, on the other hand, to whom it was submitted, recognised in it a most distinct likeness to his illustrious countryman, Prince Bismarck, and suggested that the specimen only required the addition of a gold-spiked helmet to perfect the portrait. This example, being for the present on public view, in company with a considerable series of pearl and pearl-shells and other indigenous products lent by the writer to the Western Australian Court of the Imperial Institute, those specially interested in such *lusi naturæ* are afforded an opportunity of examining and deciding for themselves as to the most manifest resemblances of this infant prodigy.

Included in the collection above referred to are several pearls of fantastic shape, some of them, notably, resembling the bodies of spiders and various insects. One of these practically double pearls is photographically reproduced to a scale of a little less than one half of its natural size in the lower figure on page 202, which represents it as it was originally exposed to view, embedded in the living tissues of the pearl-oyster. This specimen is the more interesting since it likewise reveals the presence of one of the little commensal Pea-crabs, Pinnotheres, already referred to as not unfrequently occurring in this bivalve. Another little crustacean " commensal "— so-called to distinguish it from a predaceous parasite; it being dependent upon its host only for comfortable lodgings—takes the form of a tiny lobster of a transparent hue, sprinkled with red. This species, which is known to science by the title of *Alpheus avarus*, is more plentifully met with in the Mother-of-Pearl shells on the Queensland coast and is figured in Plate XIV. of the writer's book descriptive of the fish and fisheries of the Great Barrier Reef.

As attested to on a previous page, not many years since—Pearl-shelling on the Western Australian coast, having commenced in the year 1868—shell was to be obtained in quantities by simply wading and gathering it from the inshore reefs. These were the happy times when the Pearl-sheller could rapidly make his pile, effecting a grand coup perhaps in a single day by the purchase of a pickle bottle full of pearls from the unsophisticated natives for no more substantial a consideration than a pound of bad tobacco. Times have changed since then, the former inshore reefs have been stripped clean, and it is with much toil, trouble and a frequent loss of human life that profitable returns are filched from the deeper water from which the shell is now alone to be obtained in abundance.

Within a measurable distance of time it will probably come about that the banks at present workable will become exhausted, and the goose with its golden eggs, like the oyster industry in some of the South Australian colonies, will be well-nigh, if not absolutely, done to death. The one remedial measure for this untoward condition of affairs is undoubtedly that of artificial cultivation. This expedient may not perhaps recommend itself to the generality of the reapers of to-day, to whom the outlook for the harvesters of the morrow is a matter of supreme indifference. The waste involved to a considerable extent through the destruction of immature shell has been checked in Queensland waters by the enactment of an Act of Parliament framed on the recommendation of the writer, prohibiting the taking of pearl-oysters for the market of less than certain judiciously prescribed dimensions. Greater hope for the future, however, undoubtedly lies in cultivation, and towards the practical demonstration that such operations were, though hitherto unaccomplished, perfectly feasible, the writer devoted some time and attention in both Queensland and Western Australia.

Full particulars regarding the *modus operandi* pursued, and the results arrived at in Queensland, being recorded in the author's "Great Barrier Reef" volume, those data will not be recapitulated in detail. It will suffice in the present connection to relate that the shell was brought in from the outer grounds and laid down in natural lagoons in the Coral reefs of Thursday Island, and that it both throve and multiplied under such conditions. The greatest difficulty connected with these operations was found to be the transport of the living pearl shells, which, unlike ordinary oysters, were found to be exceedingly impatient of removal from their native element. It is absolutely necessary, in fact, that they shall be immersed in pure and frequently changed sea water throughout the voyage. As the result of the success attending the Thursday Island experiments, certain of the Torres Straits fleet owners have carried out similar operations on a practical scale, one of these, due facilities being now provided, leasing a large area from the Government between Friday and Prince of Wales Islands for the purpose. A photograph of a portion of the area of the Coral Reef at Thursday Island, which was the scene of the writer's first successful experiments, is reproduced in the upper portion of Plate XXXV. The four corner posts of one of the cultivation frames in which the shells were for safety's sake deposited are clearly seen projecting above the surface of the nearer coral pool. At high tide the entire area of this reef is covered by from two to three fathoms of water, over which there then rushes a deep current of mill-stream-like strength and velocity.

The Archipelago of Islands in Torres Straits undoubtedly presents exceptional facilities for the inauguration of Mother-of-Pearl shell cultivation. Not only on account of the innumerable sheltered reefs and bays that are there eligible for the purpose, but with regard to the fact that this extreme Northern district lies outside the range of the devastating hurricanes which during the North-West monsoon are liable to visit the more southern tropical zone, and which might cause incalculable damage to Pearl-shell beds laid down in shallow waters.

PLATE XXXV.

EXPERIMENTAL PEARL-SHELL CULTIVATION SITES: A. CORAL REEF, THURSDAY ISLAND. B. MANGROVE THICKET, ROEBUCK BAY.

South of King's Sound on the Western Australian sea-board, the coast-line is for the most part far too open and exposed, and, within the tropics, subject to the risks indicated in the preceding paragraph, for the encouragement of prospects of successful Pearl-shell cultivation. From King's Sound northward, however, which is outside the hurricane zone, there are almost limitless areas that might be eligible for such an industry when the country is more settled. Making the most of the few facilities that were available, the writer was enabled to demonstrate the possibility of cultivating the large Mother-of-Pearl shell in Western Australian waters under seemingly altogether unfavourable conditions. The site elected in this instance was no other than a mangrove swamp, close to Broome, in Roebuck Bay, which had a firmer bottom than the general run of these areas, and in which natural ponds of a foot or two in depth were left by the retreating tide. At the writer's suggestion, and on his supplying him with cultivation frames as used at Thursday Island, Mr. G. S. Streeter obtained a suitable selection of living pearl-oysters and laid them down in one of the indicated ponds. As the result, within a year of the initiation of the experiment, the shells had not only increased in size but commenced to propagate, several young ones being attached to the parent shells and to the substance of the wood and wire frames.

A photograph of this site in the mangrove thicket forms the lower of the two illustrations in Plate XXXV. This was necessarily taken at extreme low tide; two of the cultivation frames have been lifted out of the pool on to the surrounding bank, and one of them is propped open to display its contents. Among these a matured shell, which has been purposely tilted against the lid of the frame, distinctly shows one of the locally grown young shells attached to its surface. Except for the presence of these oyster-frames it might be supposed that this picture represented a piece of inland sylvan scenery, a happy interblending of wood and stream, in the depths, maybe, of Epping Forest. Appearances are, however, in this case eminently deceptive. At high water of spring tides these trees are more than half submerged by the sea, which not unfrequently comes rolling through them and breaks with considerable force upon the beach on the shoreward side of this forest-like mangrove band. The trees composing this woodland scene consist almost exclusively of what are known as the White Mangrove, *Avicennia officinalis*. A very characteristic feature of this type are the somewhat asparagus-like vertical sprouts, known locally as "Cobbler's pegs," which are developed in more or less thickly aggregated bunches from their horizontally extending roots. Numbers of these structures are visible close to the bases of the isolated larger trees, and form a thick carpet on the farther side of the water in the illustration quoted.

Although it was thus proved that the large Mother-of-Pearl shell would live and even multiply amid such apparently uncongenial surroundings as a mangrove swamp, it is not suggested that such locations should be selected in preference to their natural reefs for cultivation purposes. When, however, it happens, as at Broome, that there are no contiguous suitable reef areas, mangrove ground, having a firm shell and gravel bottom, is capable of being turned to substantial practical account. Such areas can be utilised not only as convenient or necessary store ponds for the temporary accumulation of supplies of shell to be subsequently transported to more remote cultivation beds, but also for the conduct of operations and experiments bearing upon the growth and even possible artificial production of pearls. The question, at all events, of the feasibility of cultivating the large species of Mother-of-Pearl under the most varying conditions within its indigenous tropical area of distribution was now fully solved, and the steps next initiated were with the view of ascertaining to what, if any, extent such cultivation might be practical outside those limits. An oyster whose shells have a market value of from £100 to £200 per ton, and a matured individual pair of which may weigh as much as ten, twelve, or even fourteen pounds, is, from a commercial standpoint, far too valuable an asset among Nature's products to be abandoned to ruthless decimation, if not ultimate extinction, at the hands of reckless and irresponsible fishermen.

There is no reason, indeed, why the bulk of the easily accessible, and consequently denuded, shallow water coral reefs, from which the shell was originally gathered, should not be again restocked and leased by the Australian Governments to enterprising and responsible companies or individuals on lines nearly corresponding with those that are at present applied to Oyster banks in Queensland, where, at the author's suggestion, a start has been already made. Western Australia, and the Northern Territories of South Australia, alike possess facilities for the establishment of this highly profitable industry.

The site selected by the writer for the first extra-tropical experimental cultivation of *Meleagrina margaritifera* was in Shark's Bay, Western Australia. This Bay, lying roughly between the parallels of 25° and 26½° south latitude, is noteworthy for the production of a smaller and much less valuable species of Pearl-shell, which receives full attention later on. In the neighbourhood of Dirk Hartog Island, towards the southern extremity of this Bay, there are extensive shallow water coral reefs, composed almost exclusively of the encrusting or frondose expansions of Madrepores, belonging to the genus Turbinaria. This coral group, it has been observed by the writer on various parts of the Coast of Australia, represents the one that is most abundantly developed in extra-tropical waters, it being capable of surviving lower temperatures than any of the more ordinary tropical Madreporidæ. Nevertheless, the presence and exceedingly luxuriant growth of this Coral, combined with the circumstance that many shell fish and other organisms, indigenous to the waters in which the larger Pearl-shell grow, inhabited Shark's Bay, encouraged the anticipation that this species also might be artificially induced to become established there.

The Government approving of the suggested experiment, it was forthwith carried into practice. The chief difficulty that presented itself was the transportation of the living shell from the nor'-west to Shark's Bay. During the writer's cruise of investigation in the Government schooner "Meda" to the Pearling grounds, some two dozen healthy living pairs of shells were procured from the fleet then working near the Lacepede Islands, and, being kept in tubs of sea water that was constantly changed, were ultimately brought in safety, with only a single loss, to their appointed destination. Here they were placed in frames, similar to those used

YOUNG TROPICAL PEARL-OYSTERS, ATTACHED TO A PORTION OF THE PARENT SHELL, GROWN IN SHARK'S BAY, WESTERN AUSTRALIA. NATURAL SIZE. p. 208.

at Thursday Island and Broome Creek, and lowered into suitable situations on the coral reef near Dirk Hartog Island. Examined at intervals, the shells were found to be progressing favourably and increasing in size. Finally, on the frames being taken up and opened in the writer's presence, in November, 1894, precisely twelve months from the date upon which they were first put down, a number of young ones were, as happened at Broome, discovered growing on the parent shells. While on the boat's deck, and before being again lowered to their coral bed, the opportunity was utilised of securing several photographs of these interesting specimens. A portion of one of these, which shows two young examples of the imported species attached to a corner of the parent shell, is reproduced in the illustration on the preceding page.

The possibility of growing this large tropical species of Mother-of-Pearl shell in Shark's Bay in place of, or in conjunction with, the small indigenous type was thus practically demonstrated. What has been shown to be capable of accomplishment in that colder inlet of the sea could undoubtedly be repeated with even better prospects of success among the coral reefs and lagoons of Houtman's Abrolhos. These islands, as pointed out in a preceding Chapter, though lying considerably to the south of Shark's Bay are, owing apparently to a warmer current from the Indian Ocean impinging upon them, tenanted by a far more essentially tropical marine fauna, including more especially types of Bêche-de-Mer that are most abundant in Torres Straits. The character of the reefs and lagoons at the Abrolhos, combined with their short distance and easy access from the commercial port of Geraldton, render them, in the writer's mind, particularly eligible for the inauguration of the cultivation of the larger species of Mother-of-Pearl shell on a substantial commercial basis. The chief obstacle to be surmounted in the event of this suggestion being turned to practical account will undoubtedly be the transportation to the Abrolhos of a sufficient quantity of living shell to lay down as breeding stock. To accomplish this on a commensurate scale one or more large well-boats, as used in Tasmania for the transport of living fish, might be most advantageously employed. The wells in this case, however, should be modified in such a fashion that a more abundant circulation of water should be admitted, and at the same time horizontal screens or partitions should be added for the purpose of keeping the shells in separated layers, through which the water could freely percolate.

In the carrying out of the above suggested acclimatisation of *Meleagrina margaritifera* at the Abrolhos Islands, Shark's Bay would undoubtedly form a valuable entrepôt for the temporary storing of the shell, pending its delivery at its destination. From the nearest tropical pearling grounds, off Onslow or Exmouth Gulf, much by degrees could be accomplished through the medium of small but systematically recurrent consignments in tubs of constantly changed sea-water placed on board the coasting steamers, which make both Onslow and Shark's Bay ports of call. Through the adoption

PLATE XXXVI.

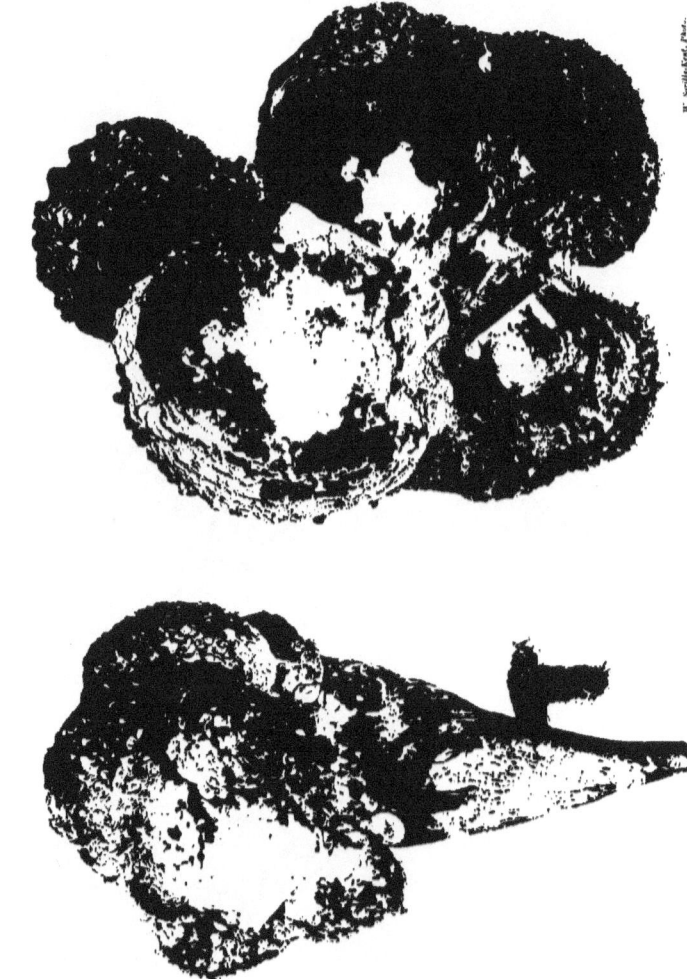

of this plan, additional instalments for the Shark's Bay grounds have, as a matter of fact, been recently transported to the Dirk Hartog cultivation grounds, under the able management of Mr. John Brockman, the Shark's Bay Pearl Shell Fishery inspector.

Some attention may now be given to that second and smaller species of Mother-of-Pearl shell which is indigenous to Shark's Bay and which in former years constituted a by no means unimportant fishery. This is the *Meleagrina imbricata* of conchologists, known to the trade as Shark's Bay shell, since, though occurring in more or less abundance throughout almost the entire northern moiety of the Australian coast-line, it is only in Shark's Bay, Western Australia, that it grows in such abundance as to constitute a systematic fishery. Considerable difficulty has been experienced in precisely fixing the technical title of this shell. The above name was originally awarded to it by Reeve, but he at the same time conferred upon other examples from the Australian Coast, which are evidently young or local variations only of the same species, many additional titles, including notably those of *irradians*, *lacunata* and *fimbriata*, any one of which is also applicable. *Meleagrina imbricata*, in the British Museum collection, however, representing the type most nearly coincident with the shell's normally matured form, the name has been adopted in these pages.

The growth habit of this Shark's Bay shell is very distinct from that of the large tropical species. While the latter type usually occurs as solitary individuals, which in their matured condition lie loosely on the reef or sea-bottom, *Meleagrina imbricata* grows in dense clusters attached to one another or to other objects by a permanently present "byssus," or bundle of thread-like filaments. A similar anchoring byssus is also possessed by *Meleagrina margaritifera* in its younger and half-grown stages, but is dispensed with on its arriving at maturity. Two characteristic groups of Shark's Bay shell, in the one instance forming a social cluster and in the other growing on the apex of a species of Pinna, another notable byssus-secreting bivalve, are illustrated in Plate XXXVI. This species of Pinna or Razor shell, as it is locally known, with its attached clusters of Pearl shells, often occurs as crowded colonies, which cover acres in extent. Not unfrequently as many as forty or fifty Pearl-oysters are clustered together on a single Pinna shell, so that the number yielded by even a single acre of them is very considerable. In the larger of the two clusters represented in the Plate quoted, the shells are depicted at about two-thirds of their natural size, the two central ones only being matured specimens. This index to their dimensions will suffice to show how small this species is compared with its tropical congener, *Meleagrina margaritifera*.

DD

In the earlier days the Shark's Bay shell was collected exclusively on account of the pearls that were obtained from them, the shells having no market value. More recently the shells, though relatively small and thin, have been extensively utilised for the manufacture of the smaller and poorer descriptions of pearl buttons and other ornaments, and if of the best quality have realised from £40 or £50 to £65 per ton.

The fishery for this shell is conducted on an entirely distinct system from that applied to the large tropical type. As it grows thickly, much after the manner of ordinary oysters, on shallow banks or over large areas of the sea bottom, it is either gathered by the hand at low spring tides from what are designated the "Pick-up banks" or is taken with sailing craft and oyster dredges from the deeper water, one cutter often working as many as from four to six dredges simultaneously. As the result of reckless dredging and the indiscriminate destruction of every description of shell, young and old, in former times for the acquisition of the pearls they might contain, the Shark's Bay fishery was about three years since depleted almost to the verge of annihilation.

One of the principal missions delegated to the writer by the Western Australian Government about that time was the investigation of this fishery and the possible prescription of some remedial measures. These were submitted and adopted in the form of the restriction of the fishery to the taking for sale of mature shell only, the closing of overworked banks until such time as they had recovered sufficiently to permit of moderate but not exhaustive fishing, and more particularly the encouragement of the establishment of systematically cultivated private fisheries, as represented by the leasing by the Government of banks of varying dimensions to private individuals. Notwithstanding the circumstance that some years must necessarily elapse before this fishery recovers its former prosperity, the measures adopted have already begun to yield highly satisfactory results. Many of the more important banks are fast becoming restocked with shell; the facilities for leasing areas for cultivation have been extensively utilised; and, as a consequence of the restrictions placed upon the age and condition of the shell taken, a marked improvement in the quality of the exported article has ensued.

A fair idea of the general aspect and appointments of a Shark's Bay Pearl-shelling Station is afforded by Plate XXXVII., which contains two views taken by the writer at the principal settlement in the Bay known as Fresh Water Camp. A most delusive title! Galvanised iron, drifting sand-dunes, a torrid sun, and odours of

PEARL-SHELLING STATIONS, FRESH WATER CAMP, SHARK'S BAY, WESTERN AUSTRALIA

decaying shellfish reign supreme. Little cause for marvel that amid these surroundings long residence there without change generates a most decided tendency towards the development of hepatic and splenetic disorders.

The view given in the upper illustration of the Plate just quoted embodies everything that is essential in the equipment of a Shark's Bay Pearling Station. Pearling cutters anchored in the Bay. Heaps of shell brought in by the boats and awaiting attention at the hands of the native boys and "belles" who are extensively engaged here to perform the savoury task of detaching the half-putrid gem-bearing fish from their shelly tenements. To the left, on the beach, may be seen a number of barrels set upright, into which are consigned the abstracted shellfish, of which more anon. A half-windlass, half-barrel-organ sort of apparatus in the mid-ground is the happy inspiration of a local genius for the reception and rapid revolution of the roughly-cleaned shells, whereby all the brittle outer and useless margins are rubbed off and fall through the coarse wire-netting framework to the ground beneath. In the yet more immediate foreground the cleaned and bagged-up shells are ready for dispatch to the London market. The lower of the two pictures gives a nearer view of the same station from a different standpoint, in which the shell supplies and much of the paraphernalia above described are still more clearly defined.

The pearls yielded by the Shark's Bay shell are of a somewhat unique character. While a large portion of them are of the ordinary milk-white or opalescent tint, a not inconsiderable percentage are a brilliant straw or golden hue. "Golden Pearls" from the "Golden West" represent a happy and altogether appropriate conjunction. Although not ranking at present in the trade so high as their colourless compeers, there can be no doubt that, from an æsthetic standpoint, these golden pearls possess a richness and warmth of tint that, to many minds, is incomparably finer. In the upper half of Plate XXXVIII., a photographic representation is given of the separated valves of a Shark's Bay shell, and in the hollow of the one to the left a small series of these golden pearls were temporarily deposited. Both the shell and pearls are in this instance represented at about three-quarters of their natural size. A yet finer series of this particular description of pearls derived from Shark's Bay has been recently placed on view among the writer's loan collection to the Western Australian Court of the Imperial Institute.

The *modus operandi* of abstracting the pearls from the animal substance of the Shark's Bay shells differs in a marked direction from that practised with relation to the large tropical species. With this last-named type the pearls are almost invariably obtained during the initial process of cutting open the valves and detaching the fish. Not unfrequently, when opened fresh, the firmer muscular part of the fish is saved, strung up in the rigging to dry, and, in this form, supplies a very palatable material for soups, curries, or stews. Its flavour as an article of food may be said to coincide more nearly with that of the Scallop than the ordinary oyster. In the case of the smaller Shark's Bay species, the greater portion of at least the larger pearls are secured during the like operation. The flesh of each individual fish is, however, not minutely examined, but consigned wholesale to the tubs previously mentioned. In these tubs the accumulated mass, locally known as "pogee," is allowed to stand and putrify, maybe for a year or more, occasional stirrings being given to it, until at length it assumes a purely liquid condition. Arrived at this state, the liquor is poured off, and a greater or less number of pearls, which apparently suffer in no way from their prolonged putrescent surroundings, are picked out from among the sediment. The accumulation of pogee tubs in and around Fresh Water Camp, invests that settlement, as might be imagined, with an odour of un-sanctity which is most peculiarly and distinctly its own, and which for penetrating pungency might give points to the celebrated two-and-seventy Stinks of Cologne. The pogee tub is, moreover, a power in the land for the adjustment of social differences. If the wind be in the right direction, an aggrieved party can inflict the most condign punishment on an offending next-door neighbour. He must at the same time give due heed to the fact that the weather-cock is proverbially shifty, and that the original "stirrer up of wrath" runs the risk of having his own measure meted unto him again.

In addition to *Meleagrina imbricata*, which is the typical Shark's Bay commercial Pearl-shell, it occasionally happens that solitary examples of the similar-sized but more cup-shaped *M. fucata* are taken in the dredge. This species is identical with the so-called "lingah" shell of the trade lists, which is obtained abundantly in the Indian seas and in the Persian Gulf. It has been observed by the writer to occur sparingly in Torres Straits and at many other stations on the Australian Coast. Lingah shell, though of inferior quality, is the most formidable foreign competitor with the Shark's Bay species in the European market. The small black-edged Pearl-shell, *Meleagrina Cumingii*—as distinguished from the larger Polynesian black-edged variety—is also occasionally collected from the reefs near the South Passage in Shark's Bay, but in insufficient quantities to be of commercial value.

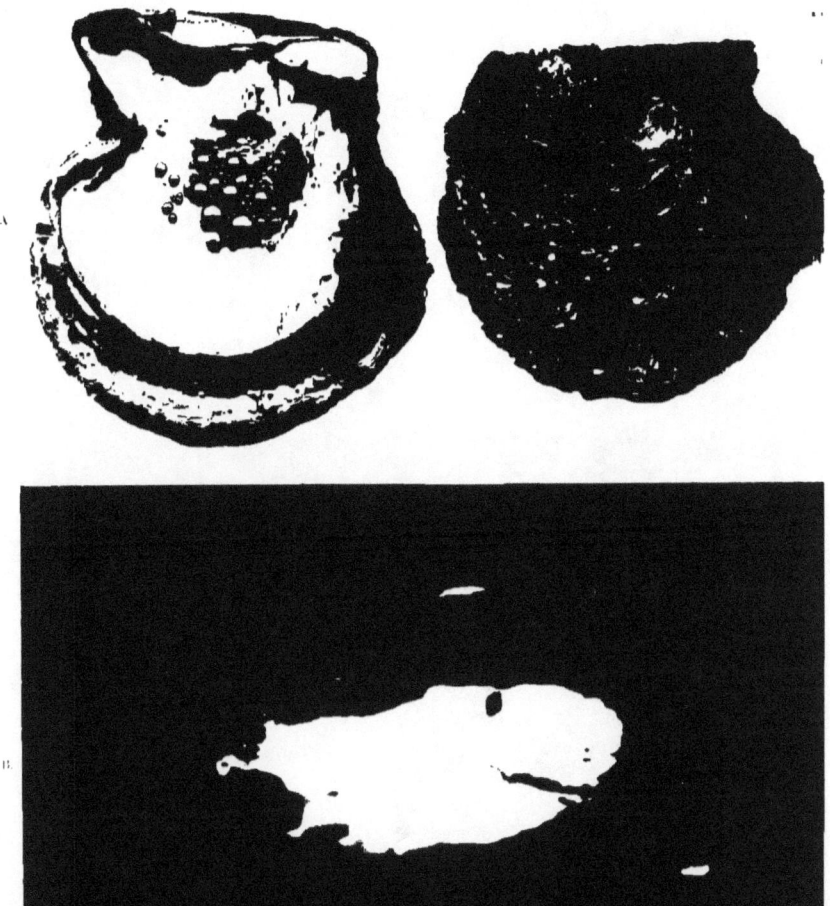

A. SHARK'S BAY GOLDEN PEARLS.   B. ARTIFICIALLY-PRODUCED PEARL.

On the coast of Queensland, where it is in some districts tolerably abundant, occasional consignments to the home market realize from £60 to £70 per ton. By way of illustrating the up-to-date market values of the various descriptions of Mother-of-Pearl shells, we reproduce here an excerpt from the latest Report upon a recent sale held by the well-known Shell Brokers, Messrs. Henry Kiver and Co., and to whom the writer is indebted for this abstract.

HENRY KIVER & CO.'S
REPORT ON THE MOTHER-OF-PEARL SHELL, &c., SALES, 14TH OCTOBER, 1896.

MOTHER-OF-PEARL SHELLS, &c.

WESTERN AUSTRALIA.—378 packages offered and all sold, good clean bold and medium £8 7s. 6d. to £8 15s., slightly yellow ditto £7 15s to £7 17s. 6d., medium £8 10s. to £8 17s. 6d., medium and chicken £8 15s. to £9 5s., chicken £9 2s. 6d., to £9 10s., good slightly defective £5 15s. to £6, fair ditto 92s. 6d. to 107s. 6d., pieces £6 10s. to £6 12s. 6d., and dead and stale 45s. to 77s. 6d. per cwt.

PORT DARWIN.—227 packages offered and all sold, good to fine bold medium £7 12s. 6d. to £8 2s. 6d., yellow Manila character £6 10s., good medium £8 15s. to £9 5s., yellow ditto £6 to £6 5s., chicken £9 10s., yellow ditto £6 to £6 2s. 6d., defective wormy 80s. to 92s. 6d., good pieces £6 5s. to £6 15s., and yellow ditto 90s. to 97s. 6d. per cwt.

SYDNEY* AND QUEENSLAND.—2,204 packages offered and all found buyers, fine selected bold £9 to £9 12s. 6d., fair to good bold and medium £7 10s. to £8 7s. 6d., ditto rather grubby and yellow £7 5s. to £7 7s. 6d., medium £8 to £8 12s. 6d., pieces £6 5s. to £6 10s., defective ordinary to good 80s. to 117s. 6d. per cwt. Country sorted sold at £8 to £8 15s. for medium, £7 7s. 6d. to £7 17s. 6d. for stoutish bold and medium, £7 2s. 6d. to £7 12s. 6d. for bold slightly defective; 85s. to 95s. for defective; and 45s. to 47s. 6d. for common ditto.

MERGUI.—Only 21 packages offered and 16 sold, bold slightly yellow £6 10s. to £7, medium £6 2s. 6d., pieces 95s., good defective £6, and ordinary ditto 87s. 6d. to 90s. per cwt.

NEW GUINEA.—72 packages offered and sold, good to fine bold £8 5s. to £9, yellow ditto £6 10s. to £7, medium and chicken £8 5s., defective £5 to £6 17s. 6d., and pieces £5 17s. 6d.

MANILA.—710 packages offered and 373 sold, fine stout bold £9, fair to good bold slightly yellow £6 7s. 6d. to £6 15s., yellowish medium £6 to £6 2s. 6d., good defective £5 to £6 7s. 6d., ordinary ditto 90s. to 97s. 6d., and pieces 92s. 6d. to 97s. 6d. per cwt.

MACASSAR.—39 packages offered and 11 sold, fair bold £7 12s. 6d., ordinary yellow ditto £6, yellow medium £6, good ditto £8 12s. 6d., and wormy 90s. to 115s. per cwt.

BOMBAY.—389 packages offered and partly sold, fair stout bold 90s. to 92s. 6d., medium 90s., small 90s. to 92s. 6d., and small and oyster 85s. per cwt.

EGYPTIAN.—191 packages offered and sold, bold 82s. 6d. to 87s. 6d., medium 85s. to 87s. 6d., medium and small 85s. to 87s. 6d., small 85s., and oyster 77s. 6d. per cwt.

PANAMA.—132 packages offered and partly sold at 46s. to 50s. for fair, and 30s. per cwt. for ordinary.

SHARK'S BAY.—39 packages offered and all bought in. (Previous sale prices throughout the year 1896 have varied from 27s. to 50s. per cwt.)

LINGAHS.—3,844 packages offered and only a small portion sold, good to fine stout 22s. 6d. to 30s., and low 5s. to 6s. per cwt.

---

* This title of "SYDNEY" refers to Queensland shell gathered by Sydney boats. Pearl shell is not a product of New South Wales.—S-K.

The lower of the two illustrations incorporated in Plate XXXVIII. has already appeared, but with less clear effect, in the author's book "The Great Barrier Reef." It represents a photographic fac-simile of the outcome of one of various tentative experiments made by the writer to induce the production of pearls by artificial means. The result obtained is obviously to such an extent encouraging that the prosecution of kindred operations on a larger and systematic scale may be commended to the attention of those who may carry into practice that artificial cultivation of the larger tropical Mother-of-Pearl shell which the writer has now demonstrated to be distinctly feasible.

The subjoined tail-piece illustration portrays a little group of fine Western Australian Pearls photographed by the writer at Mr. G. S. Streeter's Pearling Station, Roebuck Bay.

"RHINOCEROS ROCK," ROEBUCK BAY, WESTERN AUSTRALIA.

CHAPTER VIII.

MARINE MISCELLANEA.

OUR above somewhat sensational illustration will, we fear, encourage anticipations of the record, in this Chapter, of the fossil remains of some monstrous marine or terrestrial Colossus—such a Titan, in fact, that, compared with it, the huge Dinosaurs of Professor Marsh, for all their accredited eighty or ninety feet of length, would have shrunk into insignificance. To correct, as speedily as possible, any misconception that may have arisen, it must be explained that the seemingly cyclopean skull of palpably rhinocerotic contour, represents but one out of a number of the fantastic wave and weather-carved sedimentary rocks that fringe the coast-line of Roebuck Bay, Western Australia. Did space permit it, a voluminously-illustrated chapter might be devoted to this special subject.

ASCIDIAN-COVERED ROCKS, ROEBUCK BAY, WESTERN AUSTRALIA.

W. Saville-Kent, Photo.

As matters stand, this rock may conveniently serve as a guide-post to that particular spot in the Bay, whence, when the tide was low, a considerable number of the marine forms illustrated or referred to in these pages were collected or observed.

Although so high up within the tropics, the reefs of Roebuck Bay, which render navigation to the port of Broome extremely intricate, are not of coral, but are stratified replicas of the much-interrupted friable red sand-stone cliffs that line the shore. Coral grows here, but very sparingly compared with a little farther to the north and east. In place of it, certain other marine organisms are abundant to an extent that has not been observed by the writer on any other part of the Australian coast-line. The zoological group most in evidence among the reefs at low tide, at what is known as Entrance Point, close to the "Rhinoceros Rock" of our Chapter heading, is that of the Ascidians. One particular type, referable to the genus Colella, literally festoons every ledge and crevice of the much-eroded rocks in such a manner that their aspect, as seen from a little distance, presents a remarkable resemblance to stalactitic formations. The luxuriant growth of these Colellæ, together with the peculiar petrological resemblance suggested, will be recognised on a reference to the accompanying corner illustration, which is composed of the joined-up replicas of several photographs that were taken of contiguous areas. In the lowermost section the Ascidians are taken from a nearer point of view, and being more considerably distended with water, distinctly reveal their organic character. Otherwise, the upper portion of these illustrations might pass muster as depicting scenery from the stalactite caves or grottoes of Derbyshire. Seen at close quarters, these Ascidians are of gelatinous consistence and of a transparent grey hue, sprinkled throughout their lower inflated areas with minute bright blue spots. These spots, examined with the aid of the microscope, are found to represent the separate bodies of the many hundred zooids or individuals that are colonially associated in each of the hanging ovate masses or capitula.

The exceedingly roughened character of the upper surface of the rocks in the views under notice is also attributable to the accumulation of organic entities. Each minute excrescence that contributes towards the general roughness is, in point of fact, the conically-pointed shell of a species of Barnacle or Acorn Shell, and they are, in many instances, piled one upon another in serried aggregations. Another animal group which, while unobserved on the Queensland coast, is very conspicuous in Roebuck Bay, is that tribe of the Alcyonaria, or soft flexible corals, that includes the genus *Spoggodes*, and its allies. Examples of the type most abundant in the vicinity of the rock scenes here illustrated were observed in their contracted condition, when the tide was down, to present a symmetrically spheroidal contour, with all their characteristic elongate, sharply-pointed, calcareous spicules projecting from the periphery of the sphere, like the spines of an Echinus or Sea Urchin. They are, doubtless, under these conditions effectively protected against the attacks of wading birds, crabs, or any other ordinary enemies to which they might otherwise be exposed when left high and dry or covered by but a thin sheet of water. The transformation of these spheroidal spinous masses of Spoggodes, when kept in sea-water, into erect, tree-like growths, in which the most delicate transparent tints of pink, lilac and pale yellow predominated or were variously combined, was a marvellously fascinating spectacle.

A very striking feature of some of the reef pools, at Entrance Point, in Roebuck Bay, was the brilliant pink and yellow hue of certain organisms that bestrewed the rock surfaces. These were in places so closely crowded together as to constitute continuous patches of several square yards in extent. On near examination it was found that the component units of these colour patches consisted of a species of Holothuria or Bêche-de-Mer, technically named *Colochirus anceps*. This variety, which was obtained sparingly by the writer with the use of the dredge in Torres Straits, North Queensland, would appear to find the conditions on the tropical Western Australian sea-board especially favourable to its growth and multiplication. A coloured representation of this species is included in that Plate of the writer's book on the "Great Barrier Reef" illustrative of Queensland Bêche-de-Mer, of which group it is a non-commercial type. A photograph of a cluster of over a dozen of these Roebuck Bay individuals, taken vertically through the water, with their tentacular crowns in several instances partly extended, is reproduced to a scale of one-third of their life size in the accompanying illustration. A remarkably fine and possibly new representative of apparently the same genus Colochirus was obtained by the writer in King's Sound, further north. Its extended length was as much as nine inches, its body colour pale lilac with bright vermilion acetabular ridges, and the expanded tentacles orange scarlet with yellow tips.

Another somewhat abnormal example of colour development which affects the coast scenery of the foreshore of Entrance Point, Roebuck Bay, when visible at low spring tide, invites brief notice. The above tide conditions subsisting on the occasion of the writer's first visit to the Port of Broome, his attention was arrested by the presence on the foreshore of what, as viewed with

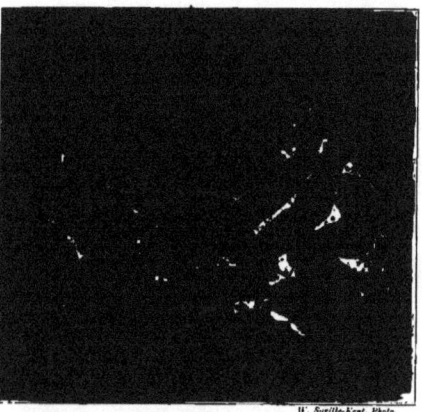

SOCIAL HOLOTHURIANS, *Colochirus anceps*. ROEBUCK BAY, WESTERN AUSTRALIA.
ONE-THIRD NATURAL SIZE.

glasses from the steamer's deck, appeared to be masses of some solid form of coral of a bright scarlet hue. Among the innumerable species of Madreporidae observed and collected by the writer on the reefs of the Northern and Eastern Australian sea-boards, no coral of such a tint had been met with. The earliest opportunity was consequently seized of repairing to this reef, in the anticipation of securing a notable scientific novelty. The goal arrived at, these great expectations were to some extent disappointed by the discovery that the masses were corals indeed, but that the conspicuous colouring was entirely adventitious, being derived from the presence of a brilliant scarlet parasitic sponge, which had apparently killed the coral and permeated itself throughout its skeletal tissues. In this manner, this sponge species possessed much in common with the orange-coloured British parasitic type, *Cliona celata*, which, as is well known, bores into and in time will completely infiltrate and destroy the shells of oysters and other molluscs. Several of the solid forms of Madreporidæ, representing such genera as Favia, Cæloria and Goniastræa, were thus found to be completely infiltrated with this sponge. Specimens were secured, and have been contributed to the British Museum, but it is to be regretted that the scarlet pigment tissues of the parasitic organism lose all their colour soon after removal from their native element.

The Sea Anemones observed by the writer on the Roebuck Bay reefs, and indeed on the Western Australian coast generally, were, in almost all instances, identical with types met with in Torres Straits and on the Great Barrier, and have for the most part been already figured or described in the author's volume bearing the above-named title. At the same time additional observations were made, and satisfactory photographic pictures secured of certain of these species, which invite a brief notice or reproduction in this Chapter. Among the more noteworthy of the species recorded from the Queensland coast were two giant forms, referable to the genus Discosoma, whose expanded discs might measure as much as, or more than, eighteen inches in diameter. The author's friend, Professor Haddon, a specialist in this branch of zoology, having collected one of these types in Torres Straits, paid the author the compliment of associating it with his name, and he has in reciprocation conferred on the second species the title of *Discosoma Haddoni*. The most interesting feature concerning these giant Anemones, recorded by the writer in his "Barrier" book, is the circumstance that both of these two species fulfil the *rôle* of hosts to fish belonging to the genus Amphiprion, a distinct specific form being consorted as a so-called "commensal" or lodger with each variety of anemone. Coloured drawings of these two species, with their respective commensal fish, are included in the author's previous volume, as also a photograph, taken vertically through the water, of *Discosoma Haddoni*.

In Western Australian waters, and more particularly in the vicinity of the Lacepede Islands, the writer found the species upon which Professor Haddon has conferred the name of *Discosoma Kenti*, in more abundance. It was here also accompanied by a fish commensal—in fact, two distinct varieties—but neither of these, though belonging to the same genus, was specifically identical with the species found by the writer consorting with it under corresponding conditions on the Queensland coast. Both of them were referable to the genus Amphiprion of the family of the Pomacentridæ. These fish are characterised by their brilliantly contrasting tints, broad alternating bands of scarlet and white, or black and white being usually predominant. In the two Queensland types, *Amphiprion bicinctus* and *A. percula*, both fish are notable for their orange-vermilion hue, this ground colour being diversified in the first-named species by two, and in the second one by three transverse white bands. Mention was also made of a species that had been met with by the writer at Port Darwin, apparently identical with *Amphiprion melanopus*. In this type a single white band is present which, traversing the fish's head and passing under its chin, communicates to it the most ludicrous appearance of having its face tied up for the toothache. A somewhat similar single white-banded species of Amphiprion was found sparingly associated with *Discosoma Kenti* at the Lacepede Islands. The more common form, however, found consorting sociably with this anemone was a species having, like *Amphiprion percula*, three white bands, but the body in place of being scarlet was almost black, and the fins were scarlet without the black and white borders distinctive of the last-named form. In many respects the species appears to coincide with the *Amphiprion Clarkii* of Cuvier and Valenciennes, the distribution of which has been recorded as extending from Mozambique to Ceylon, Singapore, and China.

In addition to the two above-named fish, a small species of crab, corresponding in shape and size with the little English flat crab, *Platysoma platycheles*, and apparently referable to the same genus, was found amicably sharing with the fish the free lodging and sheltering protection afforded by the huge anemone's convoluted disc and spreading tentacles. From among a number of plates exposed a negative was secured, reproduced in the lower half of Plate XXXIX., in which an example of this anemone with three attendant fish, and also one of the commensal crabs above referred to, is distinctly shown. One of these fish is photographed on the point of making a swimming excursion around the anemone's disc, and another one is, except for its projecting black and white head and shoulders, buried among the anemone's tentacles. This picture, taken vertically through the surface of the water with a whole plate camera, represents the anemone with its attendant fish on a scale of almost precisely one-half of its natural size. It is at the same time worthy of note that the anemone is by no means fully extended, and that its disc when completely opened out would cover an area having a diameter equal to close upon four times

PLATE XXXIX.

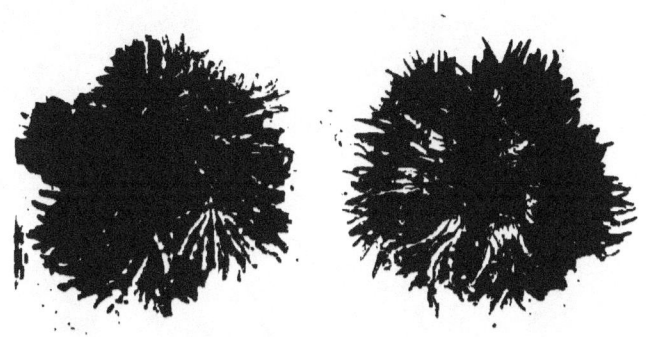

A. CYGNET BAY SEA ANEMONES, *Condylactis* sp. One-half nat. size, p. 222.

B. GIANT SEA ANEMONE, *Discosoma Haddoni*, WITH COMMENSAL FISH AND CRAB. One-half nat. size, p. 229.

that of its photographic portrait. The tentacles would also be of a much more elongate contour.

Discosoma Haddoni, the second species of giant Australian anemone discovered by the writer on the Queensland coast, is chiefly distinguished from the preceding type by the circumstance that its tentacles are spherical, except for their attachment by a short cylindrical footstalk. This feature communicates to the expanded disc the appearance of being, as it were, bestrewn with beads. Examples of this bead-tentacled species of Discosoma were met with as far South as Shark's Bay, on the Western Australian coast, and were there also accompanied by a commensal species of Amphiprion altogether distinct from the orange-red fish with two white bands previously observed consorting with this anemone in Queensland. The Shark's Bay fish had three white bands, the centre of which was continued over the upper margin of the second dorsal fin. The caudal and pectoral fins, and also the lips and irids, were lemon-yellow; otherwise the body and remainder of the fins were almost black. It will probably prove to be a new species. The colours of these two giant Discosomæ were found to vary considerably. A dominant tint of the tentacles of the species last referred to was a bright apple-green, or this tint mixed more or less with lilac grey on a fawn-coloured disc. These hues were observed to prevail in examples collected in both Shark's Bay and Torres Straits. In no instance, however, was the subulate-tentacled species, *Discosoma Kenti*, met with in Western Australian waters having bright blue tentacles, as not infrequently occurs on the Great Barrier Reef and in Torres Straits. An example of this strikingly beautiful variety is figured in the writer's previously cited volume. The representatives of this anemone observed at the Lacepede Islands and elsewhere on the Western Australian coast were usually of a light grey or lilac tint, with sometimes brighter lilac or crimson tentacle tips.

The two species of fish, Amphiprion, referred to as found associated as commensals with the Lacepede Islands Discosomæ, were also met with by the writer consorting with a smaller and altogether distinct species of sea anemone. The largest examples of this anemone when expanded rarely exceeded, or even equalled, six inches in diameter, and when growing as isolated individuals were never found accompanied by commensal fish. Their more customary growth habit was, however, to congregate in clusters in the fissures of the coral rocks, their adherent bases and columns being thrust deep within the rock crevices, leaving their densely united masses of extended tentacles alone visible. The Opelet Anemone, *Anthea cereus*, of the British seas, commonly exhibits a similar aggregated growth plan. It was among these thickly clustered tentacles that the commensal fish nestled for shelter, as within the more voluminous tentacular folds of the isolated Discosomæ.

This socially consorting sea anemone was found on closer examination to correspond very nearly with a new species observed in Torres Straits, and figured and described by the writer in his "Great Barrier" volume under the title of

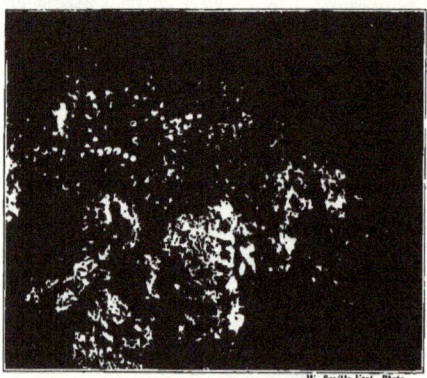

BLADDER-TENTACLED ANEMONES, *Phyosobrachia sp.* ONE-THIRD NATURAL SIZE.
W. Saville-Kent, Photo.

Physobrachia Douglasi. The most marked peculiarity of this type was the contour of the tentacles, which in their condition of full extension were inflated in a bladder-like manner at their distal extremities. The shafts of the tentacles of the Western Australian examples were usually either a transparent dark myrtle green or a clear brown, and the inflated extremities pure white or palest lilac with a minute crimson apical tip. A fairly successful photograph of a small area of a reef crevice thickly populated with this particular anemone, necessarily taken vertically through the surface of the water, is reproduced in the accompanying illustration. This anemone group represents one of many that were observed on the reefs at Gantheaume Point, Roebuck Bay, but with which no fish commensals, as at the Lacepede Islands, were found consorted.

A characteristic Sea Anemone that was obtained by the writer in Beagle Bay, Western Australia, midway between King's Sound and Roebuck Bay, is illustrated by the photographs reproduced in the upper moiety of

Condylactis sp.

Plate XXXIX. Unlike the preceding forms, it is not a rock or reef dwelling species, but takes up its abode on the sandy foreshores, having a long cylindrical column, which extends five or six inches through the sand to a stone or shell fulcrum to which it is affixed, while the tentacular disc is expanded on the sand surface. The colour of the normally concealed column is usually a bright scarlet, while the tentacles are mottled with short alternating bars of greys, browns, or olive green. A noteworthy peculiarity in this species is the circumstance that the individual tentacles often present a distinctly nodular or moniliform contour, each node coinciding with the alternating colour bands. The anemone, moreover, in its ordinary condition of expansion shows a marked tendency to pucker its oral disc and radiating tentacles into six symmetrically even folds. Both of these conspicuous characteristics are distinctly shown in the photographs from life of this species previously cited, as also in the smaller illustration representing two closely approximated individuals reproduced on the preceding page. This fine anemone, which is apparently referable to the genus Condylactis, has been met with by the writer at Port Darwin and also on the Queensland coast. The expanded disc of the largest individual observed measured seven inches in diameter, and the height of the elevated column but little short of the same measurement.

Another sand-frequenting Anemone that occurs under conditions substantially identical with those recorded of the foregoing type is photographically depicted in the figures on the next page by two examples in varying conditions of extension. This Anemone is remarkable for its stinging properties, which equal those of a nettle, and is apparently identical with the type originally described by Quoy and Gaimard, under the title of *Actinodendron alcyonidium*. The individuals here figured were obtained at the Lacepede Islands. They were of a much more brilliant tint than those met with by the writer on the Queensland coast, of which an illustration and description are given in his volume on the "Great Barrier Reef." The previously observed Queensland specimens were in all instances of mixed light brown and whitish tints, and in such respect closely coincided with the hue of their sandy surroundings. The Western Australian examples bore in their native element a much nearer resemblance to growing tufts of seaweeds, in which tints of green, yellow, and light orange red were variously interblended. These colours, in fact, as exhibited by the animal in a condition of semi-extension as depicted in figure A. of the two photographic reproductions on page 224, were wonderfully suggestive of a bunch of mignonette. The shafts of the tentacles, in pursuance of this simile, were bright green, the secondary subdivisions a pale yellowish-green, and the terminal, finely separated tentacular filaments, of that peculiar orange-red hue characteristic of the ordinary growth of the flower with which comparisons have been instituted. The smooth

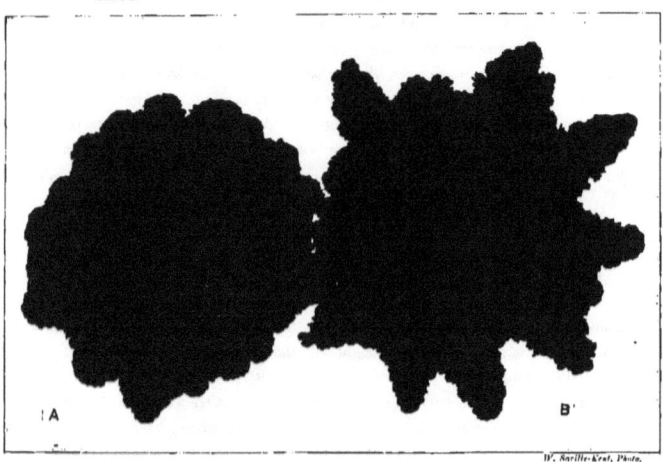

STINGING ANEMONES, *Actinodendron alcyonidium*. TWO-THIRDS OF NATURAL SIZE. *p.* 223.

central area or disc which surrounds the mouth was in the Lacepede Islands specimens of the same pale yellowish-green as the greater mass of the tentacles, but diversified with spots of brown. This character is distinctly shown in the example, B, to the right of the two individuals here figured.

Among the more remarkable representatives of the Sea Anemone or Cœlenterate tribe figured and described in the author's volume relating to the Queensland Great Barrier system, was the species upon which the writer provisionally conferred the title of *Acrozoanthus australiæ*. It belonged to that particular group of the Sea Anemones which is technically known as the Zoantharia, and whose members are recognisable from the circumstances that the zooids or individuals form more or less extensive socially united colonies in place of separating asunder and maintaining an independent existence, as happens with the ordinary Sea Anemones or Actinaria. As a rule, these social Zoantharia form encrusting colony-stocks on stones, shells, or other submarine objects. One noteworthy form, Palythoa, is remarkable for its constant association with the Glass Rope Sponge, *Hyalonema Sieboldii*, on the erect rope-like spicular stalk of which organism it develops its encrusting cœnosarc. The question as to whether the erect fascicle of elongate siliceous spicula encrusted by the polyp was the product of this organism, or an integral part of the subspherical sponge body, usually, but not invariably—it first falling to decay—found seated at its apex, was the subject of a prolonged and most heated controversy in the earlier days of biological investigation. The names of Dr. J. E. Gray and Dr. Bowerbank will be remembered by all contemporary zoologists as the doughty champions of these respective interpretations.

The Zoantharian discovered and described by the author also possesses the somewhat unusual distinction of an erect supporting fulcrum. The fulcrum in this instance is, however, altogether different in character to that of the commensal Palythoa. In place of a relatively solid axis, *Acrozoanthus australiæ* is built up on a hollow or tubular support of parchment-like consistence, which, when denuded of the encrusting polyps, exhibits a singularly symmetrical zigzag growth plan. Its angular projections are, moreover, usually developed on the same plane. The polyps in the living organism invest this structure completely with their united flesh or cœnosarc, but project in the most conspicuous clusters from the alternating prominences of the tubular support. As figured and described in the author's original account of this organism, the individual polyps are attractively coloured, the semi-contracted zooids more particularly presenting a rounded button-like contour, in which the tints of bright emerald green and red-brown specklings are predominant. When fully expanded, the brighter green hues of the column and sphinctral regions are concealed by the radiating, red-brown tentacles.

Acrozoanthus was first collected by the writer on the foreshore at Port Darwin, subsequently at Cambridge Gulf, Torres Straits, along the Great Barrier Reef, and finally at Roebuck Bay in Western Australia. It is thus shown to be indigenous to all quarters of the tropical Australian sea-board, and is in that respect a fitting claimant to its allocated specific title. With the exception of the gatherings made in Western Australia, the examples observed or obtained on the north and eastern Australian coasts, and from which the original description was drawn up, were growing in isolated colony-stocks only, or, at the most, in groups of two or three contiguous polyparies on the foreshore or in the reef pools of the districts named. The abundance, however, in which the polyp-denuded tubes were found among the flotsam and jetsam cast upon the beaches testified to its growing in considerable abundance in some less accessible and probably deeper area of the sea bottom.

The writer's later quest for this interesting type on the coasts of Western Australia was rewarded by its discovery in the neighbourhood of Gantheaume Point,

Roebuck Bay, in much more abundant colonies than had been before observed, and also under conditions that yielded a satisfactory solution of the difficulties that had previously prevented the determination of its precise zoological position and affinities. The occurrence of a Zoantharian or polype community that built for itself a hollow tubular axis was hitherto unknown, and the interpretation that first presented itself to the writer's mind with relation to the earliest observed examples obtained at Port Darwin in the year 1888 ("Great Barrier Reef," p. 154) was that the organism was not improbably a worm tube, parasitically encrusted by a typical Zoanthus. This first impression, which was also shared, and has since been more emphatically advocated, by another cœlenterate specialist, Professor R. C. Haddon, has eventually proved to be correct. At the same time, the fact that in none of the examples previously obtained could any trace of a worm be found, added to the essentially unique character and growth contour of the tubular structure, necessitated the provisional treatment of the investing polyps and their nodulated hollow support as integral elements of the same organism.

From the abundant material obtained at Roebuck Bay, the full exposition of the nature and habits of this interesting type is rendered practical, and more especially with the aid of the photograph reproduced in Plate XL. Fig. 1 of this Plate depicts the terminal half of a typical basal core denuded of its investing polyps, as washed up on the coral beach. The regularity of the alternating growth-nodes and their projection on the same plane are particularly distinct in this specimen. Fig. 2 portrays an abnormal, bifurcating, tube. Fig. 3 in the same Plate represents a complete, living colony-stock reproduced to a scale of one-third of its natural size, which, having been photographed in its native element, shows many of the polyps in a state of semi-expansion. This specimen, with others, had to be transported across country and, being for many hours isolated from its native element, the polyps refused under such circumstances to display their tentacular crowns to greater advantage. A portion of the terminal half of this compound polypary is reproduced more nearly the natural size in Fig. 4, in which the grouping of the polyps on the projecting angles of the axial core is very distinctly shown.

The most interesting figure in this Plate is, probably, the bottom one, Fig. 5, which illustrates the precise conditions under which this organism grows and is left for a while uncovered at low spring tides. As many as twenty-five separate polyparies or colony-stocks of Acrozoanthus are here discernible on one side of an isolated rock, which is, throughout the interspaces, coated with a matted growth of seaweeds,

PLATE XL.

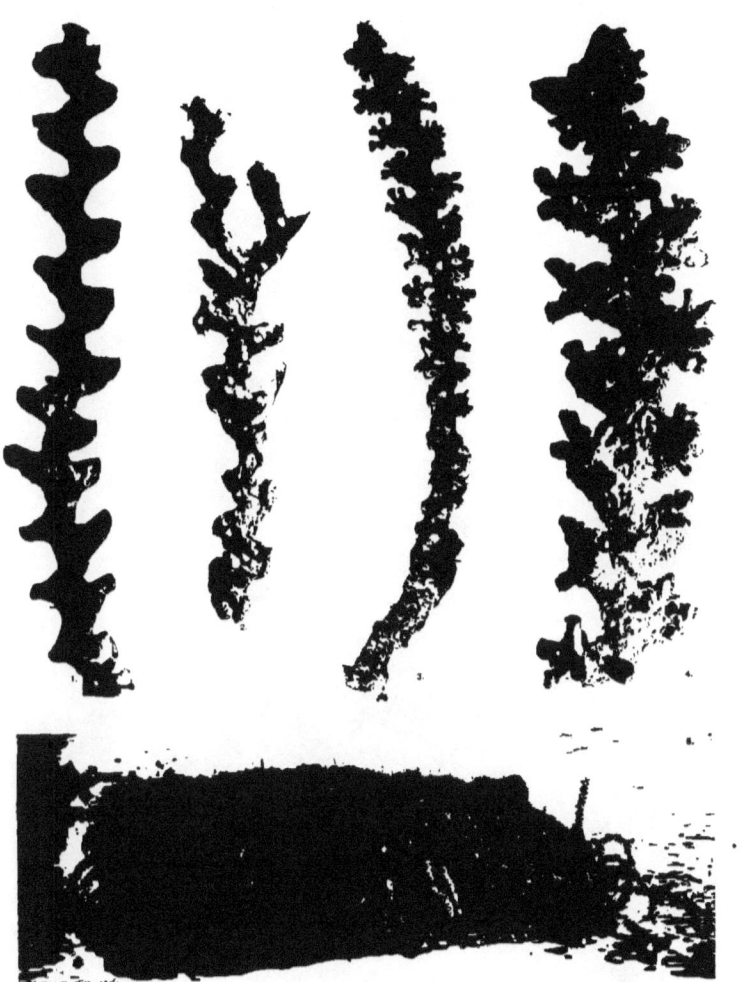

sponges, hydroid zoophytes and other organisms. As shown in the photograph, the weight of the clustered polyps on their nodulated axes is such as, in most instances, to bend the organisms prostrate when uncovered by the tide, though as soon as the water rises again they assume the vertical position indicated by one of the colony-stocks to the extreme right.

By carefully detaching some of the polyparies from this rock and examining them on the spot, the mystery of the hitherto seemingly abnormal constitution of their axial cores was revealed by the discovery, within the central hollow of several of them, of an elongate centipede-like worm. This was in some instances as much as a foot in length. The worm belonged to the family of the Nereidæ, and in that group to the genus Nereis, of which there are many representatives in British seas. It is a common property of these worms to exude from their body surface a mucilaginous secretion, which hardens in the water to a silk-like consistence. Frequently, layer upon layer is added until the tube assumes a felted texture, or it may have incorporated within its substance particles of sand, shell, or any other contiguous foreign substances. The hollow, zigzag, felted tube-core of Acrozoanthus was the direct product of the enclosed worm, but it owed its erect form and anomalous zigzag shape exclusively to the directing influences of the commensal zoophyte. Lengths of this felted worm tube, but without any attendant polyps, were found adherent to the rock amongst the tangle of seaweeds and other growths that covered it. In this phase they resembled the simple prostrate, cylindrical tubes of the ordinary Nereids.

As soon, however, as the polyp attaches itself to the growing end of the tube, it appears to impart to the worm a tendency to construct its domicile at a tangent from the basal structure, and thenceforward a sort of zigzag race ensues between the worm and the polyp to get the upper hand. The polype, in fact, by rapid sub-division and spreading of its connecting flesh, or "cœnosarc," soon threatens to, and presently does, overgrow the terminal aperture from which the Nereis is wont to thrust its head and tentacular cirrhi, and to fish in the surrounding water. Finding itself baulked in that direction, the worm makes a new start in the same plane, and at a divergent angle of about 60 degrees. Before, however, it has added another inch to its dwelling tube, the polyp has once more overtaken it, and converted its front entrance into a *cul-de-sac*. This process is repeated again and again with almost mathematical uniformity, and so at length, by the accumulation of a score or so of obliquely ascending gradations, the organism in its complete form, as here photographed, is finally produced. Arriving at this stage, the worm apparently grows tired of being continually outflanked, and either retires sulkily to its basement apartments or altogether abandons the field of its repeated discomfiture to the triumphant polyps. This interpretation affords a logical explanation of the fact that the majority of the adult growths are untenanted by worms, and for the circumstance that it was not until an extensive colony, as here recorded, was available for investigation that the precise nature of this anomalous compound organisation, however shrewdly guessed at, could be absolutely determined.

The Coral fauna of the Australian seas, and more especially with relation to its redundant development on the Queensland coast, has been so fully dealt with in the author's previous volume, "The Great Barrier Reef of Australia," that the extensive reference to this subject that might have otherwise been appropriately allocated to this Chapter would be scarcely justified. Considerable space has, moreover, been devoted to this very prominent marine zoological group in Chapter V., dealing with Houtman's Abrolhos Islands. A conspicuous feature in that section was the attempted portrayal, in something of its natural tints, of one of the very characteristic coral growths in Pelsart Island Lagoon. As a counterfoil to that picture, the portraiture of an entirely distinct type of coral reef formation, with the corresponding reproduction of the life colours and environments of its component units, has been reproduced as the frontispiece to this volume. The particular scene depicted in this instance is an area of the tidally-exposed fringing reef in the vicinity of the Palm Islands, North Queensland. Several photographic views of corresponding and neighbouring reef areas are represented by Plates VI. and X. of the author's "Barrier Reef" volume, but, necessarily under such circumstances, in simple monochrome. The attempt here made to portray such a reef scene with an approximation to its natural colouring and as sketched on the spot will suffice, if crude, to convey to those unfamiliar with such scenery a more realistic and natural aspect than can be imparted by an ordinary photograph.

This individual reef scene having, as a matter of fact, been taken from a point closely adjacent to that of the photograph reproduced in Plate VI., No. 2, of the author's above quoted volume, the context descriptive of that illustration is in the main applicable to the present one. The basis of the exposed reef-area in the foreground is in either case composed of a solid mass of a coral species referable to the genus Porites, which is remarkable for both the minute size of its component cells or calicles, and at the same time for the colossal dimensions to which the compound masses or coralla may ultimately attain. The example here figured measures over thirty feet in diameter, and ten or twelve feet deep, but originated, it may be anticipated, many centuries ago, in a single minute, anemone-like polyp, scarcely one-eighth of an inch across. It is this Madreporic genus that commonly forms the bases of the outer or deep-water edge of growing coral reefs, and is especially constituted by its hardness and density to oppose a successful resistance to the impact of the breakers. As indicated in our coloured illustration, the tint of the living Corallum of this species of Porites, *Porites astreoides*, was in this instance of a delicate lilac or lavender hue. It is, however, subject to considerable variations, and in other localities was observed by the writer to assume diverse shades of light ochre, brown, golden yellow, pale lemon, and also a delicate pink.

The circumstance that the solid Porites masses constitute the basis upon which an infinite variety of species of lesser dimensions become established is amply demonstrated in both the accompanying coloured plate and many of the photographic reef-views reproduced in the writer's previous volume. The coral species most conspicuously represented in the coloured plate now under discussion are referable chiefly to the more or less solid Astraeaceae, as typified by such genera as Goniastraea, Symphyllia, Caeloria, Maeandrina and Mussa, with here and there a few projecting branches of Madrepore or Stag's-horn corals. The variety of tints that may be associated with such an assemblage of distinct types is almost indefinite, and, in conjunction with the marvellous hues of the overlying or surrounding waters in these tropic latitudes, produces a picture that must be actually witnessed to be believed in. While it has been found possible to do but scant justice to this topic in these pages, the present opportunity may be advantageously utilised in indicating to the artistic world what an entirely original and practically inexhaustible field for high achievements lies fallow and awaiting development at their hands in the shape of the reef scenery of the Australian and other coral seas. To the marine artist more especially it opens up a vista of hitherto undreamt-of possibilities.

A group of the Madreporaria or Stony Corals, which received a brief share of attention among the many others in the author's work on "The Great Barrier Reef," was the genus Turbinaria. As previously met with on the northern and eastern Australian coasts, nothing especially abnormal was placed on record concerning the species collected. With relation, however, to the development of this generic group recently found by the writer to obtain in Western Australian waters, some more noteworthy data may be chronicled.

The genus Turbinaria will be familiar to many Australian readers by reason of the fact that the cup or vase-shaped coralla of several of its varieties, popularly known as Cup Corals, grow attached to the large Mother-of-Pearl shells in deep water, and, obtained in their naturally united state from the divers, are in considerable favour as card-baskets or other table ornaments. In addition to these cup-shaped varieties there are various other growth-modifications of the genus Turbinaria which may take the form of flattened discs, encrusting or mound-like masses, or of an innumerable host of leaf-like or diversely convoluted laminæ.

By a fortunate coincidence, Mr. H. M. Bernard, M.A., who is continuing the cataloguing of the Madreporaria in the British Museum Collections commenced by the late Mr. George Brook, had selected the genus Turbinaria for his first attention, and was consequently prepared to deal with the extensive series of Australian forms collected and presented to the National Collection by the writer, almost immediately on their arrival. How substantial an accession this series proved may be gathered from the following facts. The catalogue of the genus, now published, indicates that fifty-seven species of Turbinaria are contained in the British Museum Collection. Of these, twenty-eight specific forms were collected and contributed by the writer from the Australian coasts, and out of them no less than twenty were determined by Mr. Bernard to be new to science. Of some species, e.g., *Turbinaria peltata*, and *T. conspicua*, over twenty individual coralla were included in order to illustrate the growth phases and remarkable number of modifications of which one specific type was susceptible. The writer's entire Australian collection numbers, in fact, no less than one hundred and twenty-six specimens out of the total of two hundred and sixty examples from all sources representing the genus Turbinaria described in the British Museum Catalogue.

Apart from the satisfactory score won for the Australian members of this generic group from a numerical standpoint, the author has, with relation to the specimens most recently obtained from Western Australia, enjoyed the gratification of enriching the National Collection with larger examples of Madreporaria or Stony Corals than had been hitherto possessed by either the British or any other zoological museum. The most remarkable of these examples have been correlated by Mr. Bernard with the titles of *Turbinaria conspicua* and *T. peltata*, and were obtained in both instances from Shark's Bay. The photographs taken by the writer of two magnificent specimens of these two corals, and showing them as now occupying positions of honour in the British Museum Coral Galleries, are reproduced on pages 232 and 233. In the matter of greatest mass and weight the palm must undoubtedly be awarded to the example of *Turbinaria peltata*, in which the hemispherical aggregation of superimposed or coalescing lamellae measures a yard in height and from four to five feet in diameter. Greater elegance, with little inferior bulk, undoubtedly attaches itself to the finer specimens of *Turbinaria conspicua* depicted on page 233. The largest examples secured of this Turbinarian also measure as much as three feet in height and about four and a half feet in diameter, but, in place of forming more or less solid hemispheres, take the shape of bush-like masses of erect convoluted plates or folia which may be isolated or united with one another in every conceivable fashion. The series of examples of this species contributed by the writer to the British Museum Collection numbered no less than twenty-four, and illustrate every phase of growth, from tiny cups, less than one inch across, to large bushes many feet in circumference. Several of the less advanced developmental phases are, as shown in the photograph reproduced, exhibited in the same table case. A series of small cups, which represent the initial growth phases of the two specific forms, *Turbinaria conspicua* and *T. peltata*, natural size, as gathered by the writer growing in close contiguity to one another on the reefs at Port Denison, Queensland, are portrayed in the top illustration of the same page.

In conjunction with the matured coralla of the two species of Turbinaria depicted on the same and opposite pages, these photographic replicas might be suggestively labelled "small beginnings and big endings." Four out of five of the nascent coralla contained in the top illustration of page 233, represent the life-sized initial growth of the relatively huge foliaceous coralla portrayed to a scale of about one-twentieth immediately beneath it. There is, moreover, on the farther side of the group of three, towards the left, a yet smaller corallum of the same species that measures only a quarter of an inch across and contains but four polyp cells or calicles. *Turbinaria peltata*, the massive mound-forming species protrayed in its matured condition on the next page, is represented in the group of "small beginnings" by the single, almost circular, disc with about twelve notably large cells or corallites. Some idea of the aspect of a young corallum of *Turbinaria peltata* as it appears seen through the water with its polyps expanded may be gained by a reference to the photograph from life of an allied coral, a cespitose Dendrophyllia, *D. axifuga*, reproduced overleaf. The corallum is necessarily under these conditions completely hid, the organism presenting the aspect of a group of daisy-like sea anemones. As previously recorded and figured in the author's volume on the "Great Barrier Reef," the corallum and polypes of *Turbinaria peltata* are more usually of a light cream or whitey-brown hue, but vary from this to a most delicate rose pink.

The trouble and labour involved in detaching the larger coral specimens here figured from their ocean-bed, together with their subsequent packing and dispatch, were, as may be anticipated, somewhat considerable. A heavy flat-bottomed lighter, with blocks and tackle, was requisitioned in the first instance to raise the more ponderous masses after they had been loosened from their attachment with crowbars and carefully secured in lashings by an experi-

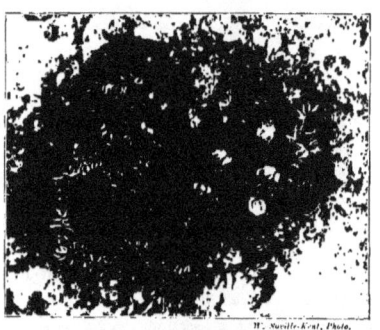

LIVING CORAL, *Dendrophyllia axifuga*, WITH POLYPS EXPANDED.
ONE HALF NATURAL SIZE. p. 231.

enced diver. They were then found to be too heavy to be raised on deck and were towed in and deposited as high up on the beach as the tide would permit. At low ebb they were then uncovered for the space of an hour or two. This circumstance permitted of the operation of their bi-section with a cross-cut saw in order to reduce them to a calibre that would allow of their being brought on shore and packed for

TURBINARIA PELTATA.
Shark's Bay, W. Australia.
Saville-Kent Coll.

MARINE MISCELLANEA.

YOUNG CUP CORALS, *Turbinaria conspicua* et *T. peltata*. PORT DENISON, QUEENSLAND. NATURAL SIZE.

transport. Even then troubles were not at an end. The huge cases, in the absence of jetty, landing stage, or any suitable machinery, were as much as a dozen men could handle and place on board the steamer's tender, and it was with the utmost difficulty that even so large a company could be mustered in this sparsely populated district.

The large, foliaceous forms, *Turbinaria conspicua*, were packed intact, being, while almost equally bulky, of lighter structure. This species, it may be observed, covered extensive areas in Egg Island Bay, near Dirk Hartog Island, Shark's Bay, representing the site referred to in a previous Chapter, which was selected by the writer for pearl-shell cultivation. Seen in their growing condition, they presented a striking contrast to the dried skeletons exhibited in the museum. In the living state, the coralla of this species mostly vary from grass to a light glaucous green, while the entire

FOLIACEOUS CUP CORALS, *Turbinaria conspicua*. SHARK'S BAY, WESTERN AUSTRALIA.
SAVILLE-KENT COLLECTION, BRITISH MUSEUM.

GG

area of their growing surfaces is thickly studded with the expanded polyps. These, when fully extended, are about three quarters of an inch in diameter, possess twenty-four almost uniformly even subulate tentacles, and are coloured a still lighter yellowish green.

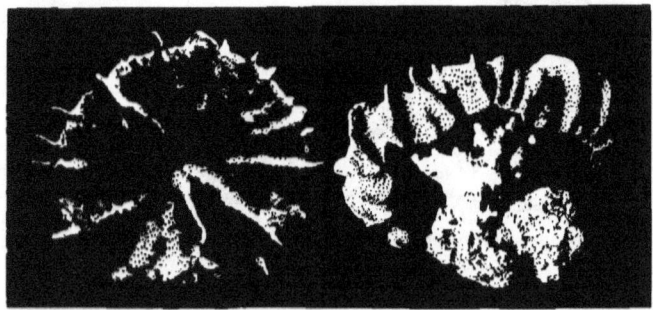

W. Saville-Kent, Photo.
REVOLUTE CUP CORAL, *Turbinaria revoluta*, S.-E. SHARK'S BAY, WESTERN AUSTRALIA. ONE-THIRD NATURAL SIZE.

An extensive reef area of *Turbinaria conspicua*, viewed through the clear water, while drifting over it in a boat, presents, on account both of the colour and contour of its component coralla, a most remarkable resemblance to subaqueous plantations of Brobdingnagian, crinkled-leaved, savoy cabbages. As has been previously remarked in Chapter V., the genus Turbinaria is a coral group that would appear to attain to the zenith of its development in the somewhat cooler waters on the outskirts of the tropics, and has been observed by the writer to flourish under such conditions, to the exclusion of other species, in the Gulf of Carpentaria and in South Queensland waters. Nowhere, however, has it been found by him to attain to such a plenitude of development, with regard both to the number of varieties and the magnitude of their individual coralla, as in Shark's Bay, Western Australia.

A final example of this very characteristic generic group is photographically portrayed on this page. It represents a specimen also gathered by the writer at Shark's Bay, which has been retained in his collection as a sort of "ewe lamb" out of the flock of its comrades donated to the National Museum, which it will probably rejoin later on. This specimen takes the form of a shallow cup, between eight and nine inches in diameter, mounted on a short, stout, central supporting stalk or pedicle. In addition to presenting the familiar shallow cup-shaped contour with raised central ridges common to many species, this Turbinaria possesses the notable peculiarity of having the entire outer margin of its rim, to a tolerably uniform depth of two inches, developed downwards and inwards again towards its central axis, in the form of an ornamental frill or border. This revolute edge, which represents the growing margin, is, moreover, decorated with a considerably larger number of vertical ridges than are visible on the upper surface of the corallum. Overlooking the existence of its revolute border, this specimen most nearly resembles the species described in Mr. Bernard's Catalogue under the title of *Turbinaria bifrons*; it differs essentially from that type, however, in the relatively minute size and sunken character of the polyps cells or calicles and in other details having a purely technical importance. This specimen, evidently representing a new and hitherto undescribed species, is, pending a full description elsewhere, provisionally associated in this volume with the title of *Turbinaria revoluta*.

Shark's Bay can boast of some notable sponges as well as corals. One of the most remarkable of these is figured in the illustration overleaf. By a singular coincidence, the type of this sponge was originally described by the author in the "Proceedings of the Zoological Society for the year 1871," under the name of *Caulospongia verticillata*. The specimen, while vaguely labelled "Australia," was from an unknown locality, and was placed in the author's hands for description by Dr. J. E. Gray, F.R.S., then Keeper of the Zoological Department of the British Museum, in which Institution the writer was, at that time, an assistant. That originally described type measured but little over a foot in length, and comprised but a single apical cone. The specimen of the same sponge with an authenticated habitat recently added by the writer to the National Collections is upwards of three feet in height, and, in addition to the main central cone, has two smaller ones, each about the size of the original type specimen, symmetrically developed on either side of the central one. This sponge, it may be added, belongs to the so-called "Horny" or "Keratose" group, in which the sponge skeleton is composed mainly of horny matter, after the manner of that of the sponge of commerce, *Euspongia officinalis*. The striking feature in the present form is that it consists of a central more solid stalk-like axis, around which are developed closely growing flattened whorls or verticils of the finer tissues. The author was more especially indebted for the acquisition of this fine sponge to Mr. Ludwig Stross, a former resident at Fresh Water Camp, Shark's Bay.

The sponge figured in juxtaposition to the type last described hails from the hotter and more remote waters of Torres Straits, and is remarkable as being an almost unique variety of the so-called "Neptune's Goblet" or "Neptune's Cup" sponge, *Poterion patera*. In its normal form, this sponge takes the shape of a huge cup or goblet, with a supporting stalk, and, as a whole, is commonly three or four feet in height and a couple of feet in diameter. In this instance, the sponge possesses no basal stalk, and in place of a deep, cup-shaped body, is spread flat out like a tray or salver of over three feet in diameter, having its margin slightly turned up, after the manner of an ordinary tray rim. From the centre of this basal tray, two other lesser, irregular cup-like developments of the sponge mass are raised one above the other to the height of little over one foot. Mounted flat, this sponge formerly did duty in the author's collection as an appropriate show-table for a choice assortment of Torres Straits shells and is figured under these conditions on Plate XLII. of his volume on the "Great Barrier Reef." This fine specimen has likewise fallen a victim to the insatiable appetite of the Sponge and Coral Galleries of the British Museum, where it is now on view.

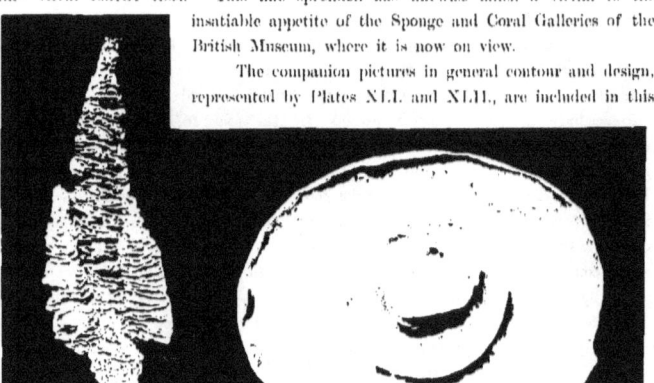

The companion pictures in general contour and design, represented by Plates XLI. and XLII., are included in this

PLATE XLI.

Collotype, W. & S. Ltd. W. Saville-Kent, Photo.

YOUNG TURTLES, *Chelone mydas*, PHOTOGRAPHED WHILE SWIMMING. One-third nat. size, p. 237.

PLATE XLII.

Collotype, W. & S. Ltd.   W. Saville-Kent, Photo.

JELLY FISH, *Discomedusæ*, PHOTOGRAPHED IN WATER FROM LIFE. One-third nat. size, p. 237.

Chapter with the more express purpose of demonstrating the potentialities of the camera for rendering marine zoological subjects in a state of active locomotion in their native element. In each instance the camera has been arranged vertically, in conjunction with the apparatus figured in Plate XXVII. of the author's book on the "Great Barrier Reef," and described at length in that volume. Bright sunshine falling uninterruptedly upon the subjects and an instantaneous shutter were necessarily indispensable accessories. In the first picture we have a group of eight young turtles, *Chelone mydas*, collected at the Lacepede Islands just as they had tumbled out of their egg-shells on an adjacent sand bank, and were liberated for their first swim in a large basin of sea-water. A few sprays of sargassum weed were added to the water with the double intent of garnishing the picture, and of in some measure restraining the almost too exuberant gyrations of the infant Chelonians. A few of the more characteristic attitudes of the limbs during consecutive phases of natation is well exemplified in this illustration.

The picture opposite to the turtles, Plate XLII, portrays a score of a remarkably beautiful jelly fish collected among the reef-pools in King's Sound, which were in a like manner photographed as they swam freely in a basin of sea-water. These jelly-fish are represented to a scale of about one-third of their natural size, and in life were variously tinted with soft shades of olive and bottle-green, scarcely two specimens being precisely alike. While referable to the group of the Discomedusæ, and among these to the sub-order of the Rhizostomæ, it has not been found possible as yet to identify them with any previously described species, and they may possibly prove to be new to science. A marked peculiarity in their habits was their tendency to float with their mouths and tentacles uppermost and to lie in that position in the reef shallows, presenting under those conditions a considerable resemblance to certain of the branching-armed sea anemones, such as Phymanthus or Heterodactyla.

Combining, as is befitting, the *utile cum dulci*, we might suggest that charming designs for an original dinner-service might be evolved from the foregoing and cognate zoological subjects. Could, for instance, a pattern be more appropriate than Plate XLI, for the serving up of Calapash and Calipee at my Lord Mayor's banquet? The jelly-fish, again, would pass muster with the arrival of the *glacées*. Should, in fact, any enterprising firm be smitten with the "notion," we have in our mind ideas galore wherewith to decorate and adorn the Minton or other delf for the most gargantuan feast.

This Chapter of Marine Miscellanea would be scarcely complete without some brief reference to an example or two of the more remarkable Australian Crustacea. There are Crabs on the Southern Coast-line, to wit, *Pseudocarcinus gigas*, of Brobdingnagian proportions, and excellent eating, beside which our homely *Cancer pagurus* is a dwarf. A specimen of this crab, taken from Tasmanian waters and presented by the writer, will be found in the South Kensington Crustacean Gallery. Its shell or carapace is recorded as not unfrequently attaining to a diameter of as much as two feet. In its vicinity is another representative of the same class, also hailing from the Island-Colony, which probably represents the largest freshwater Crustacean in the world. This is *Astacopsis Franklinii*, a near ally of our small river crayfish *Astacus vulgaris*, but it attains to no less goodly a weight than 8 or even 10 lbs. avoirdupois. The species is much esteemed for the table; it is, as proved by the writer, easily kept and fed, and is eminently worthy the attention of practical pisciculturists. The distribution of this magnificent Crustacean is very limited, it being indigenous, like the Tasmanian Black-fish, *Gadopsis marmoratus*, to the rivers only of the northern district of the Colony. The Murray and other rivers of the Australian Continent likewise produce a very fine crayfish, *Astacopsis serratus*, but it is much inferior in size, if not in flavour, to the Tasmanian type. The specimen of this Crustacean given by the writer to the National Collection was somewhat remarkable as being an albino or cream-coloured variation of the customary reddish-brown or greenish-tinged race. This specimen, however, by no means represents the full size to which the adults will attain, being indeed a cast shell or ecdysis of a living specimen which was for a considerable while in the writer's keeping. This, as also the Murray river crayfish, is popularly known in the Colonies as the "Fresh-water Lobster." Of the true lobster, genus Homarus, there is no representative in Australian waters, but an abundance of marine crawfish referable to the same genus as the British species, *Palinurus vulgaris*.

The crab which is the prime favourite in the central sea-ports from Sydney, on the east, to Perth and Fremantle, in Western Australia, is the so-called Blue Crab, *Neptunus pelagicus*. It belongs to the group known as swimming crabs, on account of the paddle-like modification of its hinder claws. Though not much larger than our English shore crab, *Carcinus mœnas*, it is excellent eating, and, being taken in abundance with the most simple tackle, comprising merely some string, raw meat and a small landing net, provides an attractive stimulus to fishing excursions innumerable, which are patronised by angling enthusiasts of every age and status. Immediately on, and even before, entering the tropics, the large Mangrove Crab, *Scylla serrata*, almost equal in dimensions to the British "Parton," becomes fairly plentiful, and is held in high esteem. As its popular name implies, the favourite haunt of this crab is the dense mangrove thickets, wherein it excavates deep burrows, from which the experienced fisherman is wont to dislodge it with a long hooked wire.

Apart from the economical varieties, Australian waters produce an infinity of bizarre Crustacean types. Not the least notable among these are the relatively small mangrove-frequenting crabs belonging to the genus Gelasimus. As many as seven species have been recorded as inhabiting Australia. They all share the same eminently gregarious habits, forming regular warrens with their associated burrows on the sea-shore, or more commonly among the mangroves. The special structural distinction of the members of this genus is the huge relative size of one of the larger fore-claws, or "chelæ," of the male individual. Sometimes it is the right and sometimes the left claw which exhibits this abnormal development; though, as shown in the photographs of one of the most conspicuous North Australian species, reproduced on the following page, in as many as five out of six of the crabs figured, this larger chela is the left one. The particular species, *Gelasimus coarctata*, here portrayed, is remarkable for its brilliant colouring, as well as for its grotesque shape. Except the back of the carapace, which is glossy black, the huge claw, the under surface, and all the other limbs are a most brilliant scarlet vermilion, inclining to crimson. The aspect, in fact, of a number of these crabs seen from a little distance, as they lie stationary at the mouths of their burrows, against the dark background of mud and mangrove, is that of hot, glowing coals. Presently one of these living coals will move out towards another one. It is the literal carrying of a firebrand into the enemy's camp. The two presently become engaged in a deadly combat, there is a rapid retreat of the vanquished warrior, and he disappears into his hole like an extinguished spark. Advance a step nearer, and the whole community as suddenly and simultaneously vanish, leaving the spectator in possession of the seemingly utterly barren and untenanted mud-flat. If he stands still for a brief interval, they one by one re-appear at the mouths of their burrows and illumine the landscape with their glowing tints. In the third figure, from the top of the group here portrayed, a male crab of this species is represented very nearly the natural size. The female is somewhat smaller, and possesses two very tiny pincers, as homologues of the large and little claw of the male. Two examples of this relatively unarmed sex may be readily identified in the lower portion of the accompanying illustration, one of them occupying a position immediately above the two champions fighting in the left-hand corner. With reference to their highly pugnacious propensities and scarlet colouring, the species is commonly distinguished by the local titles of the "Fighting" or "Soldier Crab." The eyes in the members of the genus Gelasimus are, as will be recognised by their portraits, remarkable organs. Being set at the extremities of long moveable footstalks, their owners command an extensive all-round vision, a fact which accounts for the readiness with which these crabs take alarm and disappear when approached. These elongated eyestalks are, moreover, possessed of completely independent powers of motion, and when, as often happens, a single "optic" only is lowered into the elongated groove that is provided for its reception, the action is ludicrously suggestive of a sagacious wink.

Broome Creek, of which an illustration is given and to which reference is made in Chapter V., is conspicuously rich in Gelasimi. In addition to this commonest scarlet type, *G. coarctata*, the writer met with no less than four other notably distinct species as the result of a few hours' "mud-larking." As a rule, all of these several varieties were found to frequent independent zones, in relation to the tide level. One form, equal in size to the scarlet species, and, in this

MANGROVE FIGHTING CRABS, *Gelasimus coarctata*. TWO-THIRDS NATURAL SIZE.

instance, living in its vicinity, had the large, but somewhat shorter and thicker, fighting chæla in the male cream colour, while the body and all the other limbs were a slaty black. The smallest species of all, having a carapace, or shell, scarcely half an inch in diameter, makes its burrows in sandy situations, far up on the beach near high spring level. Its strikingly delicate colouring included a pale lilac carapace, rose pink legs, and a large lemon-yellow fighting claw. Out on the mud-flats below the mangrove zone there was a species nearly as large as the scarlet one, but much less abundant, having a purple carapace, lilac and ochre yellow ambulatory limbs, and a portion only of the large fighting chæla orange red. This species is possibly identical with *Gelasimus curiatus*. The fifth variety observed at Broome was smaller than the preceding one, had the carapace in the male bright blue centrally, with a brown anterior border. The ambulatory limbs were pale yellow, and the fighting and rudimentary chæla a most delicate rose pink. It was most abundant in a zone between that of the scarlet type and the form last referred to. Probably one or more of these Broome Gelasimi will be found on nearer investigation to be new to science.

An essentially characteristic Australian group of the Crustacean class is that represented by the genus Mycteris, and including what are familiarly known on the Australian littoral as " Army Crabs." Two species have been described, *Mycteris longicarpus* and *M. platycheles*, which differ from one another chiefly in the relative length of their limbs. Tasmania in the South, and Port Jackson and Botany Bay, New South Wales, on the Eastern Australian sea-board, represent their range of distribution, as recorded in Mr. Haswell's Monograph. The larger crab, *M. longicarpus*, has, however, been observed by the writer in abundance in Moreton and Wide Bays, on the Queensland coast, and the second or an allied form at Carnarvon and as far north as Roebuck Bay, in Western Australia. Both species agree with one another in the remarkable habits which have gained for them their above named popular title. Sandy-flats, in more or less sheltered bays that are extensively exposed at low-water, represent the conditions under which they are most abundantly met with. They are eminently gregarious, and in the situations indicated associate together in numbers that may often be more correctly expressed in terms of thousands rather than of hundreds. When the tide is high nothing is seen of them, nor, probably, are they visible till some time after the tide is down. Then, as though by magic, armies of them will rise up from beneath the previously barren sand, and, assembling in battalions, march in open order, the separate companies it may be in different directions, as though amenable to strict military discipline. It commonly happens, moreover, that several larger male individuals lead the van of these moving battalions, and seemingly fulfil the *rôle* of commanding officers. If followed at a little distance, they will continue their march at an accelerated pace, but if overtaken or pressed hard, they

ARMY CRABS, *Mycteris longicarpus*. NATURAL SIZE.

will disappear beneath the surface of the sand almost as suddenly as they emerged above it. The body of these crabs being usually pale blue or lilac, and all the limbs light flesh colour, they constitute, when seen in masses under their characteristic auspices, a conspicuous feature in the landscape.

The structural peculiarities of Mycteris distinguish it from all other crab genera, and indicate and explain the singular celerity with which its members burrow into or underneath the sand, together with their general adaptation to their sandy habitat. The body in these crabs is almost globose, but more dilated above. The so-called foot-jaws or external maxillipedes differ from the usual formula in being set nearly vertical, are much the wider superiorly, and form in their conjoined condition an inverted cone. When descending into or burrowing through the sand, the apex of this cone is driven foremost; a spiral motion is communicated to the body

W. Saville-Kent, Photo.

ARMY CRABS, *Mycteris platycheles*. "A DESPERATE MÊLÉE."

by the action of the ambulatory limbs and claws and the complete organism thus constitutes itself a sort of animated centre-bit. While the writer had not the good fortune to secure a satisfactory photograph of the Army Crabs on the march, the accompanying snapshot taken of about eighty living individuals engaged in a desperate mêlée, added to the adjacent life-sized drawing, will convey a tolerable idea of the aspect and contour of these Crustacean oddities. In this photographic replica, two contending crab armies are ostensibly striving for the mastery. A near examination will reveal several couples among the general scrimmage engaged in single combat, and one or more encuirassed warriors lying *hors de combat* on the field of battle.

In the author's volume on the "Great Barrier Reef," some space and a coloured plate have been devoted to the description and delineation of a few of the more remarkable members of the Starfish tribe characteristic of that region. The handsome cerulean blue, *Linckia lævigata*, the jewel-bestudded Cushion Star, *Culcita grex*, and the massive Nodose Cushion Star, *Oreaster nodosus*, are among the more noteworthy types included in that record. As is the case with the fish, however, the colder waters of the Tasmanian seas produce, if not so abundant, yet a goodly number of forms that vie with the tropical species in the brilliancy of their tints. Even in our British seas the gorgeous crimson and scarlet livery of the familiar Sun Star, *Solaster papposa*, is almost aggressively irradiant with the reflected glow of, as it were, tropic climes.

The Tasmanian Starfish types that have been chosen for reproduction in Chromo-Plate VIII., have been selected with the object more particularly of illustrating a few of the almost kaleidoscopic series of colour variations to which one particular species is susceptible. The most prominent form in this group, represented by Figures 1 to 6 in the Plate quoted, belongs to the same genus as the little *Asterina gibbosa*, or "Starlet" of the British coast, but in this type, *Asterina calcar*, attains to a much more considerable size, and is notable among starfish for its octagonal fundamental structure. With the great majority of starfishes and all other members of the same class, that of the Echinodermata, the number five, or a pentagonal formula, is dominant with reference to both the numbers of arms, angles, and internal structural details. The examples of *Asterina calcar* here figured were all obtained within a few yards from one another on the rocky foreshore of Spring Bay, on the East Coast of Tasmania. Several plates might have been filled with as conspicuously divergent tinted individuals of the same species. No two examples, indeed, are precisely alike. The latent possibilities possessed by these many-tinted Starfish for utilisation for decorative purposes will possibly occur to the æsthetic mind. Surely in the no-distant future some departure will be made from the monotonous repetition of impossible floral inanities that are year after year foisted upon the public as the latest triumphs of inventive genius in the matter of wall papers, draperies, and cognate subjects!

Dame Nature teems with new suggestions in both form and colour that appeal most urgently for recognition at the hands of the decorative artist. Not the least noteworthy among these is her wealth of treasures yielded by the sea. As an initial notion in that direction, what a vista of original distinction and success is open to the artist who, turning his back upon the egregious conventionalities and bastard banalities of every flower that blooms, shall strike out a new path! Taking, say, a wall paper for his theme, a well-thought-out design might have the body of the subject represented by coral branches. There are a thousand or more varieties to choose from of every form and tint, and these might again be indefinitely diversified by the inclusion or otherwise of their living flowers. For the dado of such a paper what is more capable of lending itself to elegant and artistic uses than the Pentacrini or stalked Feather-stars, many of them with their tall, graceful and finely divided articulations singularly suggestive of peacock's feathers? The frieze, again, by way of harmonious compliment to the dado, might be suitably composed of a cordon of Sea-stars. Here, in form as well as in colour, there is, both literally and metaphorically, a perfect galaxy of beauty to select from. Asteroids and Sun-stars of the first magnitude and flame-like lambency, down to Starlets, Asterinæ, scarcely half-an-inch in diameter, with every intermediate gradation, are, in point of fact, awaiting in the Sea-star firmament the epoch of their artistic recognition.

Between the lines of the larger *Asterina calcar* figured in Chromo Plate VIII., three or four smaller forms have been intercallated which are conspicuous for their brilliant colouration and shape. The violet-tinted example, Figure 9, occupying the centre of the plate quoted, is an antipodeal representative of the familiar "Bird's foot" Sea-star, *Palmipes membranaceus* of the British seas. This specimen, with many others, was dredged up by the writer in the estuary of the Tamar river, North Tasmania. The majority of these individuals were coloured a dark dull crimson, but this and one or two other examples, evidently the aristocracy of their race, were clad, as delineated, in royal purple. Their rows of bright yellow acetabula, or sucking feet, which decorate and adorn their under surfaces, suggest the further simile of chains of gold. The specimen here figured represents a half-grown individual, the adult ones measuring from three to four inches

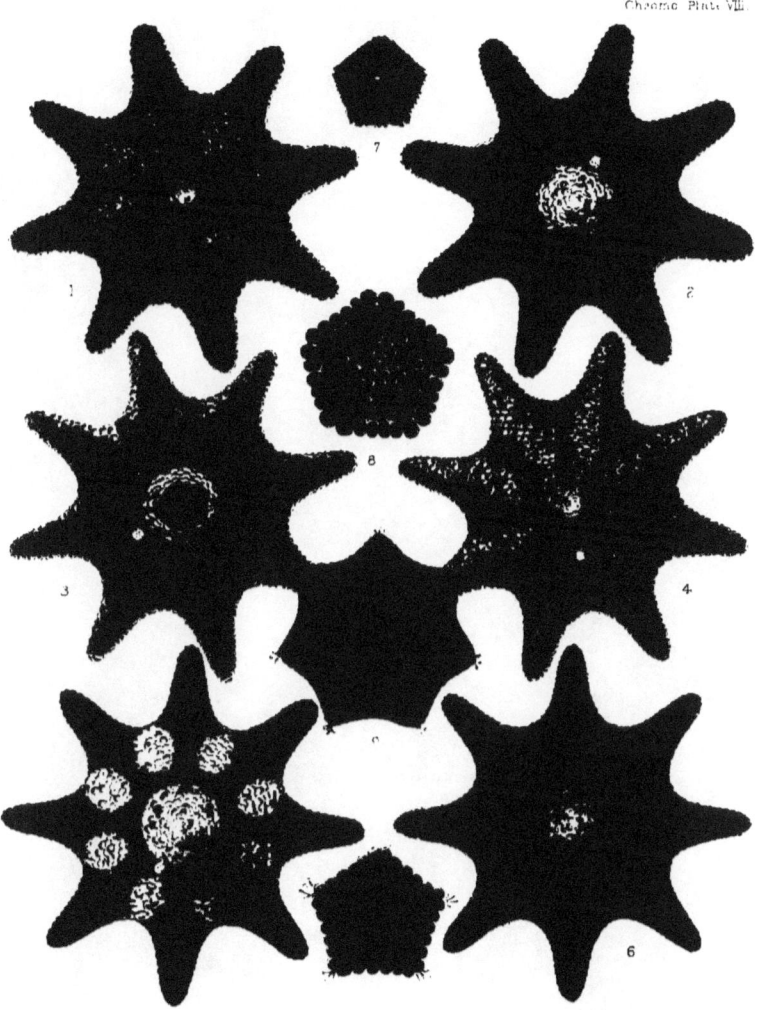

in diameter. It is noteworthy that this Sea-star, in common with the Asterinæ previously referred to, departs from the pentagonal structural formula. The majority of the specimens dredged were six-rayed or hexagons, while the one figured is heptagonal. It would seem that Sea as well as Society-stars at the Antipodes have decided to abate somewhat of the hard and fast lines of the Old World conventionalities.

A perfect little gem—small, but of the first water—is portrayed by the diminutive, irreproachably pentagonal type bedight in crimson and scarlet situated near the top of the Plate. It also belongs to the genus Asterina, but of an as yet undetermined species, and was dredged by the writer in Western Port Bay, Victoria. An essentially characteristic Australian species concludes this series. It is represented by the two individuals respectively numbered 8 and 10, notable for their perfectly flat pentagonal shape and the tesselated pattern of their component structure and ornamentation. The species, *Pentagonaster australis*, is abundant in Tasmania and throughout the Australian sea-board. More usually the tint of these Sea-stars is a light ochreous yellow, a colour which, in conjunction with their flattened symmetrical shape, imparts to them a remarkable resemblance to some sort of fancy biscuit, for which, if exposed for sale in a confectioner's window, they would undoubtedly pass muster. The two colour-variations figured in Chromo-Plate VIII. were obtained by the writer in Tasmania.

The marine zoological organism that finds most favour in the public estimation at the close of the day's doings is that luscious bivalve, the Oyster. We have consequently reserved it by way of a valedictory *bonne bouche* to this present Chapter. The Common Oyster, *Ostræa edulis*, flourishes, or, more correctly, used to flourish, in all of the Southern Australian Colonies. The demand, however, has overtaken and nearly annihilated the supply of this particular bivalve in both Victoria and Tasmania, its former headquarters, and the oyster catch most extensively now throughout the length and breadth of Australia is the so-called Sydney Rock Oyster, *Ostræa glomerata*. Although formerly obtained in the greatest abundance in New South Wales, over-exhaustion has greatly diminished the output from that colony, and the largest supplies are now derived, and exported to all of the other colonies, from Queensland.

So much has been said regarding the Oysters and Oyster fisheries of Queensland in a chapter extensively devoted to this topic in the author's recent work on the "Great Barrier Reef," that it is not proposed to deal with this subject at any length on the present occasion. The same species of Rock Oyster is found, though not so abundantly, up the Western Australian coast, and a considerable quantity are systematically consigned from Shark's Bay and stations higher up the coast to Perth and Fremantle. The species attains to the zenith of its development a little south of the tropics, and on this account, other conditions being favourable, it grows to the greatest perfection in Moreton and Wide Bays in Queensland, and their corresponding parallels on the Western sea-board. The experiment has been initiated by the writer of laying Rock Oysters down in the Swan River estuary at Fremantle. During the writer's stay in the colony these oysters had increased in size and commenced to propagate, though whether the species can be permanently acclimatised in a station so far south of its natural habitat remains to be demonstrated.

In prehistoric years the Swan River estuary was the site of enormous banks of the ordinary cold water or Common Oyster, *Ostræa edulis*, which, in the Australian colonies, is most commonly known in contradistinction to the Rock species as the "Mud" variety. Portions of the river bed are at the present date solid masses of this oyster's shells, and similar accumulations on either side mark the former much more considerable area of permanently salt water. With the process of time the river channel and its connecting reaches have become more or less extensively silted up, and so it has at length been brought about that where, formerly, water sufficiently salt for the growth of oysters was permanently present, it has been, as now during flood seasons, so long replaced by that which is perfectly fresh that the oysters have been destroyed. A somewhat similar process of oyster extermination by natural causes was in course of actual operation in the Tamar estuary, Tasmania, some years since, when the writer was investigating and reporting to the Government of that Colony on the causes of the decadence of the local oyster fisheries and the prospects of rehabilitating them. On this occasion, it being winter and the floods out, a crust of ice had to be broken to obtain access to the few surviving bivalves.

In Western Australia the most promising area for the establishment or re-establishment of prolific oyster fisheries is, beyond doubt, in the vicinity of Albany and King George's Sound. It formerly produced the so-called "mud" variety, *Ostræa edulis*, in great abundance, and there are even yet a few surviving from the reckless depletion of the beds that was practised in years gone by. On the writer's advice, Government oyster-breeding reserves are being established in the most suitable locations in this vicinity, from which, in course of time, it is anticipated that the former prolific beds may be again restocked. Denudation has, however,

PLATE XLIII.

A.—CORAL ROCK OYSTERS, *Ostrea mordax*, GREAT BARRIER REEF, p. 247.

B.—ROCK OYSTERS, *Ostrea glomerata*, KEPPEL BAY, QUEENSLAND, p. 247.

unfortunately proceeded so far that considerable difficulties are experienced in the matter of obtaining a sufficient stock for the insurance of substantial results within a reasonable period of time. Port Lincoln, on the South Australian coast, as a matter of fact, represents the nearest and about the only spot where this most estimable oyster still abounds in sufficient quantities to be available for importation and laying down.

For the resuscitation of the exhausted, and the maintenance of the existing, oyster fisheries throughout Australia, there can be no doubt that methods of artificial cultivation will have to be ultimately resorted to on a much more extensive scale, and on more scientific principles, than have been hitherto pursued. Much has been written upon this subject by the writer in his previously quoted volume and in Government Reports to various of the Australian Colonies, and it will suffice here to refer only to the illustrations of the apparatus introduced by him with marked success at the Antipodes for the catchment of the free swimming oyster spat. This apparatus, as shown in the top figure of Plate XLIV., consists merely of lengths of board, the Australian "split pailing," having a brick attached at each end, and a wire loop handle. The under surface of the board, immediately before being placed in the water, is thinly coated with Portland cement. Three such "collectors," with their crops of attached Rock oysters, which were laid down and taken up by the writer in Moreton Bay, Queensland, after an immersion of about three months, are portrayed in the lower photograph reproduced in the same Plate. This description of spat collector was found to be equally efficacious when previously, and for the first time, employed by the writer on the Tasmanian Government Reserves for the artificial collection of the so-called "Mud" Oyster, *Ostræa edulis*, which is held to be only an Antipodeal variety of the British "native."

A couple of hitherto unpublished photographs representing remarkable growths of the ordinary Australian Rock Oyster, *Ostræa glomerata*, and also of an allied species, *Ostræa mordax*, are reproduced in Plate XLIII. The growth of the first-named species, occupying the lower half of the Plate, was photographed by the writer in the Narrows, Keppel Bay, near Rockhampton, Queensland. It portrays what may be denominated a virgin oyster reef as exposed to view at extreme low tide. The representation of *Ostræa mordax* in the upper of the two prints is a typical illustration of the tendency of this species to form definite zones of growth around restricted areas of the rocks to which it is attached. This zone of most luxuriant development coincides, as a matter of fact, with half-tide mark; scarcely an oyster is to be seen above it, and they are very thinly scattered beneath its abruptly defined lower edge. The depth of this veritable oyster girdle is about eighteen inches, and it projects from the surface of the rock to a thickness of over one foot.

This and other photographs illustrative of this remarkable growth condition were taken by the writer in the vicinity of Rocky Island, off Cooktown, Queensland, one of the many granitic island formations that are hereabouts intercallated among the Great Barrier system.

The popular title of the Coral Rock Oyster has been given by the writer to *Ostrœa mordax* with reference to the fact that it occurs in great abundance and to the exclusion of the ordinary Rock oyster, *O. glomerata*, on the outlying rocks and reefs of purely coral origin throughout the tropical Australian coast-line. A very singular modification of this oyster species is represented in the adjoining process-prints. Through growing in crowded juxtaposition much after the manner shown in Plate XLIII., the lower or attached shell elongates indefinitely, while the outer or opercular one retains its normal dimensions. The originals of these illustrations, which are reproduced life-size, were obtained in the neighbourhood of Keppel Bay, Queensland. A very closely corresponding variety of the ordinary Rock Oyster, *O. glomerata*, produced under parallel conditions, has been more recently obtained by the writer at Shark's Bay in Western Australia.

Australia is, *par excellence*, to the unsophisticated Britisher, the land of incongruities and topsy-turvydom. Christmastide is a mid-summer festival; the swans are black;

CORNUCOPIA-LIKE VARIETY OF OSTRÆA MORDAX. KEPPEL BAY, QUEENSLAND. NATURAL SIZE.

THE AUTHOR'S OYSTER CULTURE APPARATUS, p. 217.

cherry-stones grow outside the fruit; flies eat the spider, and oysters grow on trees, *cum multis aliis*. The last-named quotation alone invites present attention. The circumstance of oysters growing upon trees is both illustrated and described at some length in the writer's previous volume. The two descriptions of Mangroves, *Rhizophora mucronata* and *Avicennia officinalis*, popularly known as the Red and White Mangroves, are there shown to be the trees to which the Australian Rock Oysters most systematically adhere. In the case of the Red Mangrove, the oysters are usually attached to the arched many-branching aerial shoots characteristic of that species which are left high and dry with every fall of the tide. When growing on the White Mangrove, the oysters adhere chiefly to the innumerable vertical respiratory shoots or so-called "cobbler's pegs" that are developed from the subterranean, widely radiating, ordinary roots. Commencing on these slender shoots, the oysters frequently increase to such an extent as to constitute solid masses or banks of oysters that may be several feet in thickness. Examples

DWARF OYSTERS, *Ostrea ordensis*, ORD RIVER, WESTERN AUSTRALIA. NAT. SIZE.

W. Saville-Kent, Photo.

of both these forms of oyster growth are characteristically represented in Plate XXXIX. of the author's "Barrier" book.

The most notable example hitherto recorded of oysters growing upon trees is probably afforded by that delineated in the illustrations on page 249. These portray a new and remarkably minute species of oyster discovered by the writer in the estuary of the Ord River, Cambridge Gulf, Western Australia, when accompanying, as a guest, the surveying cruise of H.M.S. "Myrmidon," in the year 1888. With reference to its recorded habitat, the name of *Ostrea ordensis* was conferred upon this species in a paper entitled "Oysters and Oyster Culture in Australasia," contributed by the writer to the Auckland, New Zealand, meeting, 1891, of the Australian Association for the Advancement of Science. *Ostrea ordensis* grows not only on the roots, stems, and respiratory shoots, or "cobbler's pegs," of the White Mangrove, *Avicennia officinalis*, but also on its leaves. As shown, in fact, in the photographic figures reproducing the specimens their exact natural size, as many as forty or fifty individual oysters may be crowded together on a single leaf, measuring about two inches in length. That these oysters had attained to a state of maturity was established by the fact that, on being opened, they were found to be crowded with well-developed embryos. The growth zone of these oysters was, moreover, considerably nearer high-tide mark than that affected by the only other type, a stunted race of *Ostrea glomerata*, which grew sparingly on the rocks at a lower level in the near vicinity. It happens, as a consequence of their inhabiting this high-level growth area, that these oysters are covered by the tide for a few hours only during the course of the day. During neaps, indeed, they may be left uncovered for several consecutive days. There is a famous oyster, celebrated in song, accredited with a penchant for walking upstairs. The particular variety here figured has, at any rate, manifested the predilection to climb higher up a tree than any oyster species previously described.

A companion picture to the "Rhinoceros Rock" heading to this Chapter has been selected for its tail-piece. The rock scene in this instance portrays a somewhat remarkable sandstone formation on the foreshore of Sweer's Island, a member of the Wellesley group, in the Gulf of Carpentaria. The anomalous nature of this rock formation is recorded in Dr. Fitton's Geological Appendix to Captain King's "Survey of the Coasts of Australia," Vol. II., 1826, and is briefly referred to on page 570, in the following words:—"In Sweer's Island, a hill of about fifty or sixty feet in height was covered with a sandy calcareous stone, having the appearance of concretions, rising irregularly about a foot above the general surface without any distinct ramifications. The specimens from this place have evidently the structure of stalactites, which seem to have been formed in the sand." The tubular character of some of the examples of these and similar stalactite-like concretions collected, and their attribution to the passage of calcareous or ferruginous solutions through the sand masses of which they are fundamentally composed is further referred to at pages 621 and 622 of the same treatise.

The photograph here reproduced depicts this peculiar strata under conditions more favourable for the illustration of its characteristic features than obtained at the epoch of its earlier observation. Sweer's Island, when visited by the writer in 1891, had a few years previously been the scene of a violent hurricane. The low sandstone cliff that forms the subject of the accompanying photograph, then taken, had been completely submerged and undermined by the abnormal waves, and was broken up into disrupted fragments that bore a by no means remote resemblance to masses of a Cyclopean growth of the Organ-pipe coral, *Tubipora musica*.

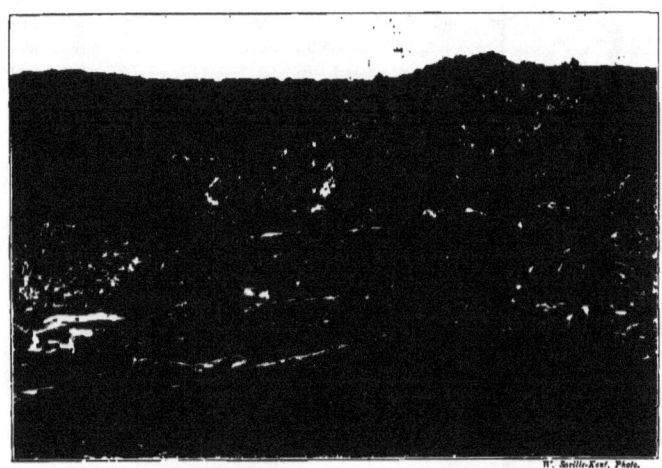

STALACTITE-LIKE ROCK CONCRETIONS, SWEER'S ISLAND, GULF OF CARPENTARIA, NORTH QUEENSLAND.

CHAPTER IX.
INSECT ODDITIES.

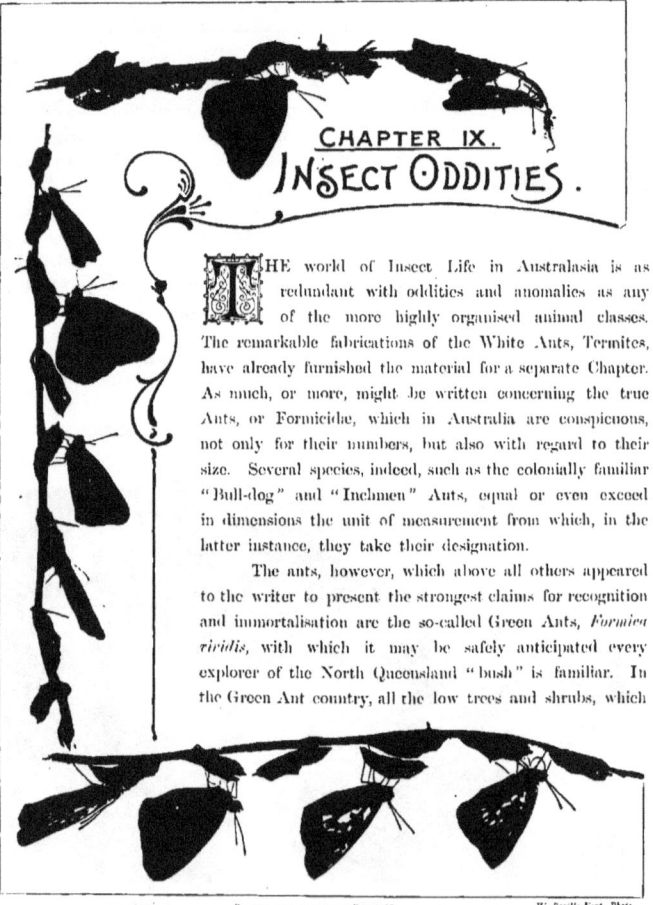

"BUTTERFLY BIRTHDAYS," p. 287.   W. Saville-Kent, Photo.

THE world of Insect Life in Australasia is as redundant with oddities and anomalies as any of the more highly organised animal classes. The remarkable fabrications of the White Ants, Termites, have already furnished the material for a separate Chapter. As much, or more, might be written concerning the true Ants, or Formicidae, which in Australia are conspicuous, not only for their numbers, but also with regard to their size. Several species, indeed, such as the colonially familiar "Bull-dog" and "Inchmen" Ants, equal or even exceed in dimensions the unit of measurement from which, in the latter instance, they take their designation.

The ants, however, which above all others appeared to the writer to present the strongest claims for recognition and immortalisation are the so-called Green Ants, *Formica viridis*, with which it may be safely anticipated every explorer of the North Queensland "bush" is familiar. In the Green Ant country, all the low trees and shrubs, which enter mainly into the composition of the "bush" or "scrub," appear to be suffering from a plague of caterpillars. Not that the leaves of the trees are themselves devoured, but they are all spun together with webbing, after the manner practised by many Lepidopterous larvæ, and in bunches which vary in calibre and dimensions from two or three leaves only to masses five or six inches or more in diameter. Pushing one's way through the plantations infested by these ants, one soon becomes unpleasantly conscious of their presence. They fall down in showers from their shaken nests and are prone to fix themselves with their sharp, powerful jaws to whatever exposed area of epidermis may present itself. Fortunately, they do not, like many of their congeners, sting as well as bite, though, all the same, two or three that have, after the manner of their kind, insinuated themselves beneath your shirt collar, or probably further extended their peregrinations before selecting an anchorage, can evolve from their human victim a highly creditable display of gymnastic capacities, together with a by no means unusual accompaniment of fluent rhetoric. In aspect the Green Ant is undoubtedly one of the most elegant of its tribe. The green pellucidity of its body rivals that of the beryl, while the colour and texture of its limbs may be most aptly compared to amber. Beauty, in the case of the Green Ant, is more than skin-deep. Their attractive, almost sweetmeat-like translucency possibly invited the first essays at their consumption by the human species. Mashed up in water, after the manner of lemon squash, these ants form a pleasant acid drink which is held in high favour by the natives of North Queensland, and is even appreciated by many European palates.

That these Green Ants should be capable of spinning silk seemed such an anomaly that the elucidation of their *modus operandi* attracted the writer's attention on more than one of the occasions of his visits to the north. It was, finally, when examining the nests of these ants and their ways in the bush a little way out of Cooktown in July, 1890, that the enigma was solved. It was then found that the ants in their matured state took no distinct part in the weaving, though they were at the same time instrumental in requisitioning their immature grubs or larvæ to fulfil this task.

The way in which these helpless weaklings were impressed for active service and trained to labour for the public weal may be more clearly understood by a reference to Figs. 1 and 2 in Chromo-Plate IX. Fig. 1 in this Plate portrays a relatively small nest of this ant species so far composed of but a few leaves, which, as sketched, were in process of being united together by a silken tissue. To accomplish this undertaking a number of the Green Ants were assembled on the outside and were engaged in rows in gripping and drawing to their closest proximity the edges of the adjacent leaves. This strained position, with every muscle at the fullest tension, was maintained, it might be, for hours together without its being possible to detect the rationale of the ants' manœuvres. The weaving and binding of the leaf edges to one another was probably progressing throughout this interval, but on the inner side only of the approximated edges.

Presently an ant was seen to emerge from the interior of the nest, carrying with it a whitish object, which on nearer examination proved to be one of the matured grubs or larvæ. This grub was carried to the scene of one of the labouring parties occupied in holding the leaf edges together. Grasping the grub by the centre of its body, the ant then held it in such a position that its extended head could just come in contact with one of the gripped edges, to which the grub sought to attach itself by exuding a glutinous silk fibre from its mouth. No sooner, however, had the larva effected a silken attachment to one leaf edge than it was immediately transported to the opposite one, to which, in like manner, it endeavoured to effect a secure anchorage. The silken cord was drawn out and carried across the intervening space, and, being fastened again to this opposite edge, made the first complete stitch to unite the respective leaf edges. This simple process was repeated over and over again until the silk supply of the larva was exhausted, when it was carried back to the nursery and another grub substituted in its place. The larvæ thus selected and utilised as weaving shuttles were in all cases matured individuals that were on the point of entering upon the pupæform condition and had consequently their silk glands fully developed. So soon as a sufficient amount of silk fibre had been woven between and over the opposing leaf edges to overcome their natural tendency to spring apart, the labouring ants relaxed their grip, and either commenced operations on other leaf edges or retired into their nest. The rare occasions on which it was found possible to detect these insects engaged upon their remarkable weaving operations favour the belief that they are usually accomplished at night.

The leading data, here recorded, concerning "The weaving properties of the Australian Green Ant" were communicated by the writer in a paper bearing the foregoing title, contributed to the Queensland Royal Society's Meeting of May 15th, 1891. Nest-weaving phenomena of a similar character were subsequently observed by Mr. H. N. Ridley of an allied Malaccan species, *Formica smaragdina*, and

Chrome Plate IX

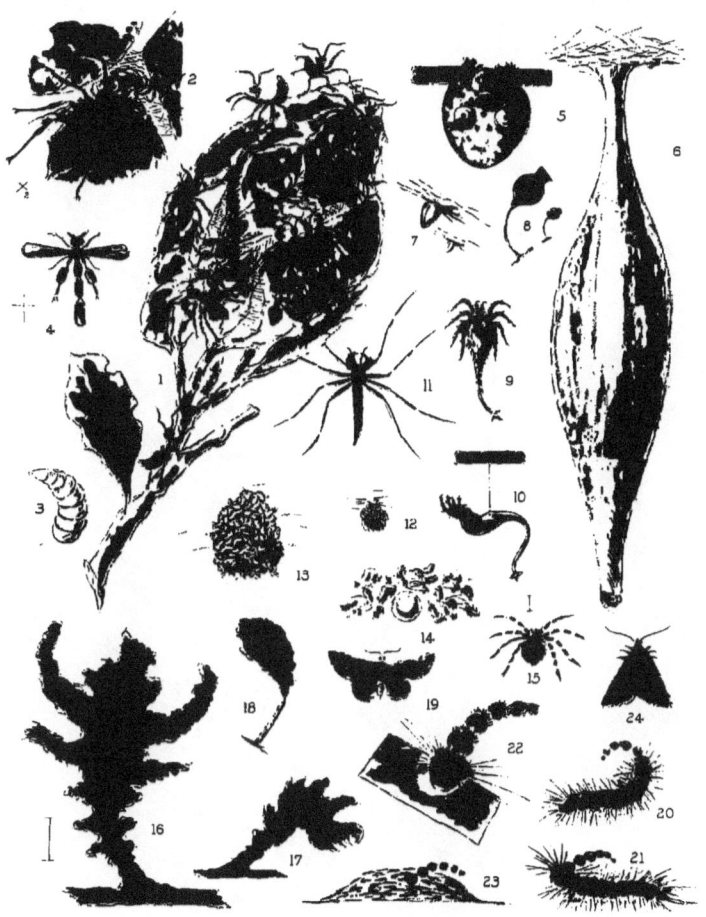

are briefly chronicled in the Proceedings of the Entomological Society for the year 1894.

The strength and tenacity of jaw manifested by the North Queensland Green Ants in the architecture of their nests is also displayed by them in the capture of living prey. The writer frequently observed instances in which an unfortunate beetle, or other insect, had fallen a victim to their hunting expeditions. It was in such case pinioned fast to the ground by a radiating circle of its merciless captors, every limb being dragged out to its fullest extension. Secured in this helpless position the poor insect was then speedily dismembered and transported piece-meal to the Green Ants' larder.

A singular little winged insect, apparently of the Hymenopterous order, was observed hovering over and occasionally lighting upon one of the Green Ants' nests under observation. As shown in the enlarged drawing of it, given in fig. 4 of Chromo-Plate IX.—the natural size being indicated by the cross lines in its vicinity—the most remarkable features of this fly were the gouty-like enlargement of the central joints of the hind limbs and the somewhat coincident abnormal dimensions of the last abdominal segment. While hovering, both these weighted limbs and the abdomen hung perpendicularly to the otherwise horizontal plane of the little creature's body. From the special attention that this fly concentrated upon the Green Ants' nest, it would appear probable that it was in some way related either as a commensal or a parasite with the hymenopterous community.

An exceedingly interesting account of the practical utility to which the gripping powers possessed by certain of the larger species of Ants may be applied was contained in a paper contributed by Mr. R. Middleton, to the Meeting of the Linnæan Society for February 6th, 1896. According to a communication received by that gentleman from Mr. Miltiades Issigonis, of Smyrna, extensive use is made by the Greek Barber Surgeons of the Levant of a large species of ant for the purpose of holding together the edges of incised wounds. The ant, held with a pair of forceps, opens its mandibles wide and is then permitted to seize the edges of the cut which are held together for this purpose. As soon as a firm grip is established the ant's head is severed from the body and remains tenaciously holding the cut edges of the wound in the position most favourable for its speedy healing. Mr. Issigonis had seen several natives with wounds in course of healing to which seven or eight such decapitated ants' heads were attached. There can be no doubt that Australia from among her larger " Bull-dog " and " Inchmen " races could furnish a phenomenally redoubtable contingent for the exercise of this most ingeniously improvised application of the healing art.

The remaining insects proper included in the Plate now under discussion belong to the order of the Lepidoptera, or Butterfly and Moth tribe, and are more especially noteworthy with reference to the bizarre aspect of their larvæ or caterpillars. In neither case does the perfect insect present features of note. Figs. 16 to 19, depicting the first of these two forms, represent the transformation phases of a small moth, referable to the genus Phorodesma. The larva is of the Semi-looper or Geometer tribe, and feeds upon a green lichen abundant on the Eucalyptus trunks in the neighbourhood of Brisbane. As delineated in the enlarged back view, Fig. 16, it is, while quiescent, quite impossible to distinguish it from a small lichen frond.

To accomplish this disguise the caterpillar bites off small fragments of the lichen on which it feeds, and, apparently with the aid of a glutinous secretion from its mouth, causes them to adhere, not only to the entire body surface, but also to the long bristles that are symmetrically developed on either side. It is this latter circumstance more particularly which imparts to the insect such a frond-like aspect. The profile view given in Fig. 17 represents a caterpillar which has somewhat outgrown its coat, the head and pro-legs being distinctly visible. A somewhat analogous method of disguising itself for protective purposes is resorted to by certain of the Spider Crabs of the British seas, referable to the genus Stenorhynchus, which are in the habit of decorating their carapaces and limbs with detached fragments of seaweeds to such an extent that they may be readily mistaken for growing plants. The pupal stage of this lichen-feeding Phorodesma is delineated in Fig. 18. As will be recognised, the cocoon is mounted on a slender pedicle, and lichen fragments are extensively incorporated into its substance. The perfect moth, enlarged to twice its natural size, is delineated in Fig. 19.

The remaining Lepidopterous type that invites brief notice is represented by Figs. 20 and 24 of Chromo-Plate IX., and has been identified for the writer by Sir George Hampson, of the British Museum Entomological Department, with *Dimona porrigens*. The larva in this instance feeds on various species of the Australian gum trees or Eucalypti, and has been observed by the writer in both Tasmania and the Botanic Gardens of Adelaide, South Australia. The notable peculiarity in this instance is the circumstance that with each skin-casting, or "ecdysis," the caterpillar retains the entire head-skin attached to the short hairs at the back of its new one.

It thus comes about that, on arriving at its adult state, this caterpillar carries with it an accumulation of five or six of its discarded head-masks; the bizarre appearance thus presented is grotesquely suggestive of the old drawings of the Jew pedlar, with his stock-in-trade pile of battered hats. The most characteristic illustration of the singular aspect of this much be-hatted caterpillar is afforded by Fig. 22, where it is drawn as seen eating its way through a leaf. On preparing for the chrysalis stage and the renouncement of the pomps and vanities of its hitherto earth-bound vegetative existence, the caterpillar weaves an ovate silken cocoon, on the outer wall of which, as shown in Fig. 23, it jauntily plants its ultimately discarded hat pile. In the cocoon here figured, the fabricator, being of an evidently æsthetic turn of mind, has gnawed off and interwoven with its substance fragments of the coloured lining of the box in which it was confined.

The somewhat silhouette-patterned border illustration reproduced on the first page of this Chapter, while introduced primarily for decorative purposes, claims brief explanation. It represents the twig-attached pupæ or chrysalides and newly-emerged imagos of one of the many butterflies belonging to the Pieridæ family common to Australia. The particular species is apparently identical with, or near to, *Belenois Clytie*, and notable for its gregarious habits. The three chrysalis-covered sprays here photographically reproduced from life, represent but a very small fraction of the mass from which they were originally gathered in the Adelaide Botanic Gardens.

Availing ourselves of the licence sanctioned by the Entomological Society, which admits Spiders provisionally and by courtesy to the rank of Insects, the introduction of some notable "oddities" is permitted. Figs. 5 to 15, in Chromo-Plate IX., previously quoted, portrays a few of these spider types, remarkable either for their individual form or colour, or for the singularity of their architectural products. Some of these can, indeed, claim distinction on both counts. Fig. 5 presents, in its external contour, but little to distinguish it from the ordinary garden spiders of the genus Epeira, but, at the same time, may be described as being of a somewhat flattened, obovate shape. Colour, however, here comes in, and plays a part that, in conjunction with its environments, invests this Arachnid with unique interest. The ground tint of this spider, as shown in the illustration, is a delicate lilac, with individually variable shadings. Superimposed on this, near the centre of the body, are two smooth, slightly elevated, pale yellow, circular, eye-like spots.

Taken in conjunction with the flattened obovate body, the entire organism, with the relatively small cephalothorax and limbs turned in an opposite direction,

KK

presents the most singular resemblance to the head of a small snake, or of a lid-less, goggle-eyed, Gecko Lizard. When seen under its customary natural conditions, at the end of the cylindrical silken tunnel among the foliage that it weaves for its habitation, this suggested simile is the more remarkably obvious. The useful purpose of this mimetic adaptation, if such it be, to this spider as a protective element is evident. Among the several popularly-quoted anomalies distinctive of the Australian region, the circumstance that the spider is the victim of the, or, more correctly, of a certain species of, fly has been already noted. The particular flies that thus turn the table at the Antipodes upon the spider are solitary members of the Wasp or Hornet tribe, genus Sphex, and its allies. These Hymenoptera construct clay nests of conspicuous size, a sheltered corner of the verandah or dwelling room of one's bungalow being a favourite location. Before laying their eggs within the nest, however, provision has to be made for the emerging grubs. This invariably consists of a holocaust of spiders, which are stung, not to death, but simply to a state of coma. They are then carried off and hermetically walled up with the eggs in such a number that they provide a continual feast of fresh meat for the Hornet larvæ from the time they are hatched until they enter upon the pupa stage.

The fat soft-bodied Epeiræ are special favourites with these Assassin Wasps, and the species now under notice doubtless owes much of its immunity from wasp persecution to the circumstance of the superficial resemblance of its body to a reptile's head. The apparent pair of baleful phosphorescent eyes gleaming at the extremity of the dimly-lighted recess occupied by the spider, would, it may be anticipated, immediately deter its winged enemy from rashly penetrating within the portals of a darksome chamber, where it would, itself seemingly, run the greater chance of being devoured.

The cocoon constructed by this Epeira for the enclosure and protection of its eggs, represented in its natural size in Fig. 6, is a conspicuous object and is fabricated with considerable ingenuity. Four or five of these egg cocoons are commonly constructed by an individual spider, and are hung freely under a leafy shelter, from an irregularly webby basement in the position indicated. They may vary among one another in being of stouter or more slender relative dimensions, but at the same time collectively agree in their essentially pyriform shape, and more particularly in the exceedingly narrow tubular terminal orifice. This contracted contour of the exit passage is no doubt a special provision against unwelcome intrusion of insect enemies from without.

The writer availed himself of the opportunity of submitting these and other spider sketches to Messrs. R. J. Pocock and F. O. Pickard Cambridge, the Arachnid Authorities at the British Museum. While it was not found possible to establish the identity of the spider just described and figured, it being probably new to science, some interesting data were elicited concerning its characteristic cocoons. Examples of these fabrications had, it appeared, been acquired by the Museum Collection between fifty and sixty years ago. Up to the present time, however, no data were available indicating what description of spider or other insect constructed them. In another form included in the author's sketch book, the egg cocoons were perfectly spherical, half an inch in diameter, smooth on their surface, and of a light-brown tint with darker striations. The contour and size of these egg cocoons so nearly coincided in contour and dimensions with the seed capsules of the indigenous convolvulus, Ipomea, that they might be easily mistaken for such a vegetable structure. The spider, moreover, when at rest with legs doubled close to its body, so closely resembles the egg cocoons in shape and markings, though of lighter colour, that it too might be taken for an older, bleached and battered example of the same seed vessel. Samples of these egg cocoons were also contained in the British Museum Collection, but without any record as to their relationship. The writer's included sketch of the spider sufficed for its identification with Koch's figure of *Celænia excavata*, so that the organism and its products are now satisfactorily correlated with one another.

The Spider form represented by Figs. 7 and 8 of the Plate now under notice, is referable to the genus Argyrodes, and is remarkable for the circumstance that it does not spin and abide in an independent web, but takes up its residence on the snare woven by some larger and stronger species, preying there on the smaller midget-sized flies and other insects that are beneath the notice of its accommodating host. The form here figured was most abundantly observed by the writer commensally associated on the wide-spreading geometrical webs of the relatively gigantic spider *Nephila fuscipes*, in Cape York Peninsula, North Queensland, of which a figure is given in Plate XLV. An identical or very closely allied species was also abundant on the webs of *Argiope regalis*, in the neighbourhood of Brisbane. One of the most remarkable features of this type is its burnished silver-like sheen. Scattered in some numbers over the surface of the web as they frequently are, and of all ages and sizes, from almost invisible pin-points to the adult dimensions of about one-eighth of an inch in diameter, they glisten in the sun like minute globules of mercury. An alternative simile, and the one which is probably a more correct interpretation of their adaptive significance, is their conspicuous resemblance to web-entangled dewdrops. Masquerading under this delusive cloak, they would no doubt escape the attacks of many enemies such as birds or predatory wasps, to which they would be otherwise subject in the exposed positions they are wont to occupy on their adopted webs. The egg cocoon constructed by this little commensal spider is a very elegant structure, shaped like a Greek amphora, and mounted on a slender pedicle, by which it is either attached to the web or to a convenient point on the adjacent foliage. The contour of these egg cocoons, as illustrated in Fig. 8 of Chromo-Plate IX., is also remarkably suggestive of that of the stalked sheaths or loricæ of various of the minute Flagellate Animalcules figured in the third volume of the author's earlier published work "A Manual of the Infusoria."

Several other species of Argyrodes were observed by the writer, more especially in North Queensland. All of them possessed the same habit of sharing on friendly terms the webs of larger spiders, but lacked the silvery lustre of the type above described. The dimensions of two of these Argyrodes were more considerable than those of the silver species, and they were both characterised by their more elongate gibbous contours. Stalked egg cocoons of a somewhat narrower ovate shape than that of the variety figured were observed attached in a hanging position to the horizontal guy-suspended snare of a species of Linyphia that one of these types had selected for its abode in Cape York Peninsula. The genus Argyrodes is widely distributed, and many of its members are credited with contenting themselves with anything but a peaceful and subordinate share of the web to which they attach themselves. One species in particular, which is referred to in Dr. McCook's Monograph of American Spiders, and upon which he has conferred the appropriate title of *Argyrodes piraticum*, habitually kills and devours the constructor of the web it elects to occupy and takes sole possession of it on its own account.

The Spider, *Arachnura Higgensii*, represented by Figs. 9 and 10 of the above-quoted Plate, is remarkable for the singular elongation of the hinder moiety of its body, which, to a considerable extent, recalls the structural modification distinctive of the Scorpions or higher Arachnida. This elongate extremity is not, however, as in that group, composed of distinctly differentiated articulations, but is uniform in character with the remainder of the body, and both flexible and also retractile to about one-half of the length indicated in the illustration. As shown in the profile view of the species suspended by its web, portrayed in Fig. 10, the spinnaret occupies a position on the under surface of the body that is precisely coincident with the creature's centre of gravity. The long-tailed Spider here figured was obtained by the writer in the Botanical Gardens of Ballarat, Victoria. A closely-allied representative of the same genus was also found by him in some abundance, spinning an irregularly meshed snare among the Ti-tree bushes, Melaleuca, on the banks of the Prosser's river, Tasmania. The bright-green, crimson-striped Spider, apparently a species of Tetragnatha, represented in Fig. 11 of the Insect Plate, is more notable for its abnormally elongate shape and for its conspicuous resemblance, when it rests with extended legs, to the many red-veined foliaged plants among which it takes up its abode. This single example noted was collected by the writer in the Botanic Gardens at Bowen, Central Queensland.

A remaining spider form included in Chromo-Plate IX. invites brief notice. It is represented by Figs. 12 to 15. This type is apparently referable to the genus Theridium, and is remarkable more especially with relation to its habits and the singular environments of its egg cocoon. It was observed by the writer in the neighbourhood of Derby, at the head of King's Sound, Western Australia, taking up its abode in the fissures of the gnarled trunks of the older Baobab or Bottle-trees, *Adansonia rupestris*. The spider, a small brown one, presents no special features of interest, and neither does the web, which is of the irregularly meshed order. Suspended in the snare, however, there is generally present a little cupola-shaped mass, which, on near examination, is found to be composed superficially of the emptied skins and disjointed limbs of a small species of black ant upon which this spider habitually feeds. The interior of this ant aggregation is hollow, and is found to contain in its upper confines the spherical silken egg cocoon of its fabricator, which it has most effectively and ingeniously concealed from view. It sometimes happens that two or three of these egg domes are suspended within one web, and while the bee-hive or cupola shape depicted in Plate IX. represents the most ordinary form, they are occasionally of a much more slender and elongate contour. An allied American form, *Theridium riparium*, is recorded in Dr. McCook's treatise as forming somewhat similar elongate conical nests, the external surface of which is strengthened and rendered opaque by the addition of a thickly entangled coating of minute pellets of clay.

A common, but at the same time notable, spider in the Queensland "scrubs," which also frequently takes up its abode in the Brisbane suburban gardens, is portrayed on the following page. It is apparently identical with the *Argiope regalis* of Ludwig Koch. The photograph here reproduced represents the spider one-quarter only of its natural size, and it is consequently a tolerably large species. It is also a handsome one, being of a warm chestnut brown hue with bright yellow bands across the dorsal surface of the body, which gives to it a somewhat wasp-like aspect. The example in the illustration having had its ventral surface turned towards the camera, this detail is not apparent. The feature most distinctly indicated in this picture is the singular supplemental additions to its ordinary orb-shaped web; these take the form of four zig-zag, thick, ribbon-like bands or rays of the web-substance disposed obliquely and at nearly right angles, and in such manner as to form the letter X. Argiope, in fact, as here shown, stole a march on Professor Röntgen centuries since in the invention and practical utilisation of mythical X-rays. As may be discerned in the accompanying photograph, the spider, when at rest in the centre of its web, has its eight legs disposed in four pairs, which are precisely parallel with, and are respectively anchored for support to, each of the four radiating ribbons. Dr. Henry McCook, in

QUEENSLAND "X-RAY" SPIDER, *Argiope regalis*, WITH RIBBON-STRENGTHENED WEB.
ONE-FOURTH NATURAL SIZE.

his "American Spiders and their Spinning Work," includes descriptions of several United States species of the genus Argiope, with especially copious details of the largest and commonest form, *Argiope cophinaria*. This type is also in the habit of strengthening its snare with ribbon-like zig-zag cords. The ribbon-webbing in this form is, however, limited to a single band descending downwards from the web's centre or "hub," or there may be a second one in the same straight line ascending vertically above it. The immediate centre of the web is also in this type usually covered in by a densely woven sheet of webbing. In certain other American species, notably *Argiope argyraspis* and *Uloborus mammeatus*, somewhat similar ribbon-like accessory web-bands are added more abundantly, and in concentric lines around the centre of the orbicular snare.

Reference has been made on a preceding page to the circumstance that a small spider form referable to the genus Argyrodes commonly occurs as a commensal or fellow lodger in the orb webs of both the huge tropical *Nephila fuscipes* and the Argiope last described. The Nephila is itself deserving of brief notice. It is among, if not altogether, the largest of the Australian snare-weaving Arachnids. The body alone is commonly as much as two inches long, while the spread of the expanded limbs may measure as much as seven or eight. The colours in this form are not so varied as in many of the smaller species. The abdomen is light brown, the cephalothorax and the greater portion of the limbs a dark glossy chestnut brown, almost black; there are dull orange patches at the joints of the limbs that show most conspicuously underneath, and the palpi are yellow with black tips. A covering of fine golden down-like hairs clothes the upper surface of the thorax. The snare constructed by this fine spider consists of a central circular area which may measure three or more feet in diameter, while the guy- or foundation lines may bridge over interspaces of many yards betwixt the forest trees and undergrowths that it frequents. A rough drawing of a fine example of one of these Nephilæ, occupying its normal position at the centre or "hub" of its snare, together with other details bearing upon the life-habits and surroundings of the species, are embodied in the lithograph, Plate XLV. overleaf. It was originally made by the writer of a specimen that was under his notice several consecutive days in the tropical scrub adjacent to Mr. Frank Jardine's homestead at Somerset, Cape York Peninsula, close to the Albany Pass. The large orb-weaving spider here represented is, as in all corresponding examples, the female and the superlatively larger of the sexual individuals. The male is, in fact, a most insignificant epitome of his colossal spouse, and, like the little Argyrodes before mentioned, is a mere privileged hanger on to his mistress's estate.

In the illustration here reproduced, two of the relatively minute male Nephilæ are in attendance, the one who has evidently made a favourable impression being admitted to such terms of intimacy that he is reposing peacefully on his partner's body. A second, would-be, suitor may be seen occupying a position towards the lower, right-hand corner of the web. Courtship among Spiders is a proverbially hazardous undertaking, the female, where she is the larger or even where no disparity of size subsists, frequently killing and eating her enamoured and too rashly approaching swain. Among the Nephilæ and other types in which the male is of such diminutive relative size, these cannibalistic propensities are apparently held in abeyance. The husband is too unsubstantial a morsel for even the tickling of his partner's palate, and his safety is consequently secured by his very insignificance. On several occasions the writer observed a male individual approaching the female here delineated without eliciting from her any signs of hostility, with the exception that on one occasion, when he slowly walked up to within reach of her terrible fangs, she simply pushed him away with her palpi. Repulsed in front, he then made a strategic detour to the rear, and was soon to be seen gaily disporting himself on the plump outlines of his tolerant consort. Between the bloated body and chitinous head or cephalothorax of the female, there was necessarily on the dorsal side a conspicuous notch or cleft. To the Liliputian male this was a formidable gulf, and he accordingly wove across it a silken carpet, which allowed him to run with unimpeded ease from stem to stern of his veritable mountain of delight.

The method by which these huge female Nephilæ dispose of their captured prey differs substantially from that pursued by the smaller members of the tribe. The Spider class is, as a rule, represented as being of entirely suctorial habits, feeding simply on the abstracted juices of their winged victims. In the species now under consideration, large flies and even beetles taken in the web were, as observed by the writer on several occasions, literally chewed up and swallowed in their entirety. The irony of fate inseparable from greatness is aptly illustrated in the specimen here figured. This Spider is itself a victim to a minute blood-sucking sand-fly, apparently belonging to the genus Simulium. One of these parasites, with immersed rostrum and distended body, may be observed adherent to the right-hand margin of the Nephila's abdomen as presented to the reader. Several of the little burnished-silver Argyrodes are, as were seen in life, stationed at various points of the Nephila's web here delineated. As previously mentioned, these diminutive commensal spiders add their slender lines, and devour those smaller flies taken in their host's snares which are beneath his, or more correctly her, august notice. In this manner they also probably fulfil a useful function, and repay their host's hospitality by preying upon the minute parasitic sand-flies by which the larger Arachnids are prone to being persecuted.

Much more might be written concerning the Australian Arachnida. The foregoing notes will, however, suffice to indicate how wide and interesting a field

Plate XLV.

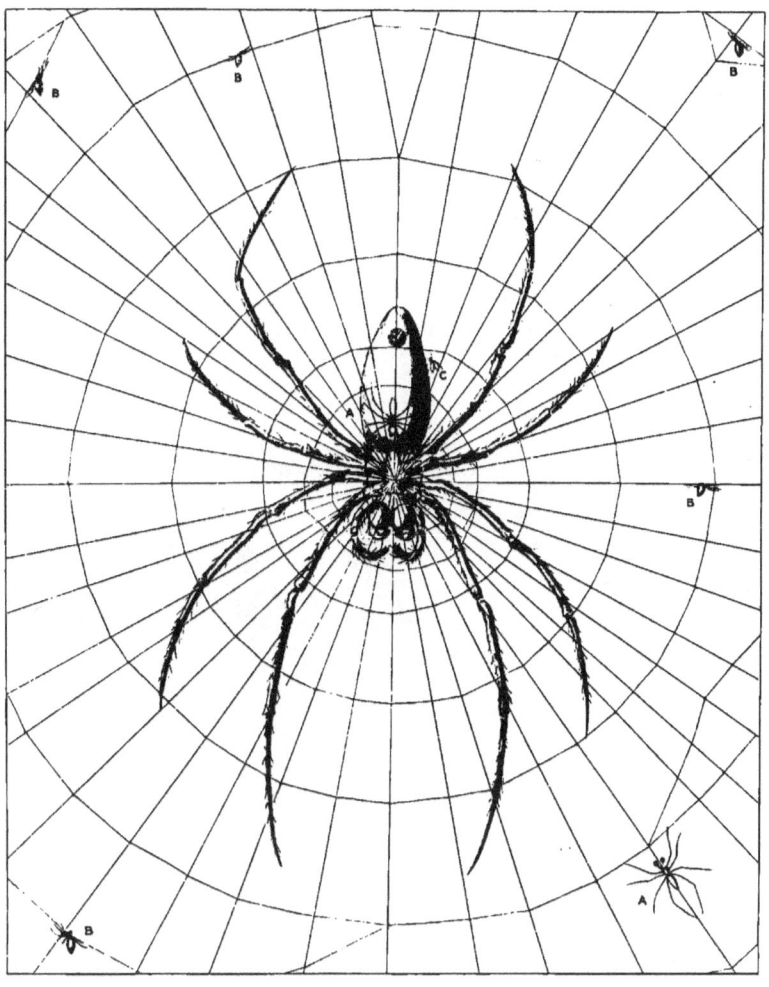

QUEENSLAND SPIDER, Nephila fuscipes, Natural size, p 263.
A A DIMINUTIVE MALES. B.B. COMMENSAL SPIDERS, Genus Argyrodes. C. PARASITIC SANDFLY.

INSECT ODDITIES. 265

this group alone provides for original investigation. The drier details of the morphological structure and technical nomenclature and affinities of the more commonly occurring species have to a considerable extent been already chronicled, notably through the labours of Dr. Ludwig Koch. There is, however, still a preponderating number of undescribed species, while of but few of those already known have any data been placed on record concerning their most interesting life-habits and idiosyncrasies, their individual peculiarities of snare construction, or special and often most singularly shaped receptacles woven for the protection or concealment of their eggs. Following out the lines adopted in Dr. H. C. McCook's magnificent three-volume Monograph of the "Orb-weaving Spiders of the United States," there is undoubtedly open to the talent and energies of an enthusiastic Australian arachnologist an opportunity of producing an equivalent if not an even more voluminous and fascinating treatise.

The spider figured below from life belongs to the non-orb-weaving family of the Heteropodidæ ; a timber and house-frequenting species familiarly known to Australian settlers by the title of "Tarantula" or, to the less illiterate, as a "Triantelope."

WESTERN AUSTRALIAN TARANTULA. *Isopoda sp.* NATURAL SIZE.

W. Saville-Kent, Photo.

CHAPTER X.

VEGETABLE VAGARIES.

HAD Virgil emigrated to Australia "*sub tegmine boabi*" might have stood in place of that allusion to the more homely beech which constitutes the prelude to his familiar pastoral. The dense and welcome shade that the Boab or Baobab is capable of affording the weary traveller, and his possibly wearier steed, is well exemplified by the figure of a full-foliaged tree, reproduced in Chapter I., Plate V. As indicated, indeed, in the accompanying illustration, the up-to-date Phyllis and Adonis may not disdain to foregather beneath its ample shelter and to dispense thereunder sweet ambrosia in its more modern garb, bohea, to themselves and to the stranger who may be tarrying in their midst.

Verily, among Vegetable Vagaries the same Baobab or Bottle-tree, *Adansonia rupestris*, is of

"SUB TEGMINE BOABI."
W. Saville-Kent, Photo.

the most extravagantly droll. The half a dozen or so individual examples reproduced in the Plate overleaf, Plate XLVI., represents but a few selections from upwards of fifty photographs of this tree taken by the writer in the neighbourhoods of Derby and Wyndham in Western Australia, and every one of these possesses a most distinctly marked individuality. With many of them the resemblance of their trunks to smooth, symmetrically-shaped radishes or rough-rinded mangold-wurzels of Brobdingnagian proportions, is ludicrous. The contour of such an one as is portrayed by No. 3 in the above-quoted Plate is particularly suggestive of the first-named salad adjunct, while in No. 5 the two types indicated are growing in close proximity. One feels sorely tempted at Derby, Western Australia, when Christmastide comes round, to sally forth with a huge pot of crimson or vermilion paint, and to add that thin veneer of colour that is alone wanting to transform these bizarre trunks into "property" vegetables that would be the envy and admiration of the chief pantomime artist at Drury Lane.

In some instances, as illustrated by No. 1 of the series in Plate XLVI., the main trunk is almost spherical in shape, and with its crown of bare radiating branches bears a by no means inconsiderable resemblance to a colossal octopus. In the background of this picture other trees of the same description may be noted, and also a large termite mound of the Kimberley type, closely corresponding in shape and dimensions with the examples figured in Plate XVI. It often happens that double or twin trunks are produced by the Baobab tree. Three such instances are yielded in the Plate illustrative of this type, and a fourth on page 269. In fig. 2 of the Plate quoted the growth plan has varied to the extent of forming a three-shafted column. In its extreme youth, the Baobab presents little or nothing to distinguish it from the ordinary run of trees. A very juvenile sapling is included in the foreground, and a little to the right of the central figure in No. 3. of Plate XLVI. It in all ways resembles a small detached branchlet from an adult tree.

As is clearly indicated by the many specimens here figured, the Baobab is essentially deciduous in its habits, the majority of the examples portrayed representing the trees in the dry or winter season, when they are for the most part completely divested of foliage and show the singularities of their form to the greatest advantage. There is even, however, in this matter of leaf-shedding, as with the formula of growth, a marked individuality among the trees, and a reluctance to conform to any hard and fast rule of periodicity. There are some trees to be found in full foliage at any season of the year. Fig. 4. in Plate XLVI. represents a by no means unusual example that was photographed in full leaf, at the time when the great majority of its companions exhibited only bare poles.

The aspect of the handsome, dark, glossy palmate leaves of the Baobab are clearly indicated in the photograph of a full-foliaged tree reproduced in Chapter I., Plate V. The flowers are also notable in size and appearance. They are of a creamy white hue, and somewhat thick waxy consistence, five-petalled with a slightly prolonged tubular calyx, an elongate pistil, and a surrounding sheaf of fine long stalked stamens and golden anthers, resembling those of the ordinary Myrtaceæ. At the time of the writer's visits to the North West, the flowering season for the Baobab trees had long since passed. By dint, however, of carefully scouring the country round, one solitary bloom was, with the kindly assistance of Dr. Ernest Black, ultimately secured. A pencil sketch made of it is reproduced below.

The large ovate fruits or nuts of the Boab tree may be observed scattered among the branches of many of the examples figured, but are especially conspicuous in the one that forms the subject of the illustration on the next page. A characteristic representation of a single nut, illustrating the more normal bluntly ovate contour, is also afforded by the engraved example included with the carved shell ornaments reproduced on page 10 of Chapter I. This figure delineates the nut on a scale of one-fifth of its natural size. Boab nuts are found, however, to differ to a remarkable degree both in shape and size even on the same tree. Normally they are obtusely ovate, five to seven or eight inches long, and covered externally with a thin green felt-like pile. They may vary, however, from the foregoing normal contour in the one direction to an elongate acuminate and in the other to an almost spheroidal contour, while the matured nuts may

W. Saville-Kent, del.
FLOWERS OF BAOBAB TREE, *Adansonia rupestris*. HALF NATURAL SIZE.

PLATE XLVI

not exceed a greater length than two inches. When ripe, the interior of the thin shell is filled with a friable, flour-like, slightly acid powder, in which the somewhat small bean-like seeds are embedded.

In both their matured condition and in an earlier state the boab nuts are esteemed as food by the natives and have found favour with Europeans. Grey, in his " Expeditions of Discovery," Vol. 1, 1841, describing

BAOBAB TREE, NUT-BEARING EXAMPLE WITH DOUBLE TRUNK.

it as the " gouty stem-tree," remarks : " The foliage of this tree was slight but graceful, and it was loaded with fruit of an elliptical form as large as a cocoa-nut. This fruit was enclosed in a rind resembling that of an almond, and inside the rind was a shell containing a soft white pulp in which were placed a species of almond very palatable to the taste, and arranged in this pulp much in the manner in which the seeds are placed in the pomegranate." The bark of the tree, when cut, is attested to by the same explorer as yielding a nutritious white gum which,

PROSTRATE BAOBAB, WITH REJUVENESCENT TRUNKS, p. 270.

both in appearance and taste, resembles macaroni; upon the bark being soaked in hot water an agreeable mucilaginous drink was also produced. This and other earlier references to this tree, conjoined with the fact that the corresponding fruit of the allied African species, *Adansonia digitata*, is also highly appreciated by the natives, and has at times, as in the case of Major Pedley when searching for Mungo Park, sufficed as the sole nutriment to Europeans for many days, would seem to justify the anticipation that the Western Australian Baobab is a tree abounding in latent possibilities, and well worthy the attention of specialists in economic botany.

The dimensions of the Baobab or Bottle-tree, combined with its exceedingly slow rate of growth, indicate that it must attain to a very considerable antiquity. King has attested to an individual tree which measured no less than 85 feet in circumference, and it has been recorded of an example raised under favourable conditions at Kew that it took eight years to grow to a height of $4\frac{1}{2}$ feet. The celebrated Dragon-tree of Orotava, Teneriffe, belonging to the same family group of the Sterculariaceæ, which was finally destroyed by a hurricane in the year 1867, had an age of no less than 5,000 or 6,000 years ascribed to it by Humboldt. Up to the date of its destruction it was pronounced by Sir John Herschel and other authorities to be probably the oldest tree in the world.

The tenacity of life of the *Adansonia rupestris* is well evidenced by examples that were hewn down many years since during the process of road-making around the township of Derby, Western Australia. The trunks, though literally denuded by the axe of every root and branch, have been by no means deprived of their vitality, and are in many instances putting out fresh branches which, time permitting, will constitute new trees. In the adjacent bush or so-called "pindan" where the tree abounds, it is, indeed, by no means an uncommon sight to see a trunk prostrated, probably centuries ago by some abnormal storm, out of which a fresh tree has reared itself phœnix-like with renewed youth and vigour.

An example of the rejuvenescent properties possessed by this tree is afforded by the preceding photographic reproduction, wherein three vigorous independent trees are, as it were, growing up from the erstwhile apparently wrecked and prostrate trunk. The adage "As the tree falls so shall it lie" does not at all events apply to the Western Australian Baobab, or it is at any rate one of the brilliant exceptions that seemingly exist to prove the rule in most mundane matters. It is a rare thing to see an actually dead Baobab. The only veritable occasion of such mortality present to the author's mind is the example reproduced in the accompanying illustration. This, however, is no case of ordinarily inherited decrepitude and decay, but, as an ancient British jury would have it, it is the outcome of the direct act of Providence. It is, in fact, what must have been the towering monarch of the surrounding forest, shattered in the full plenitude of its might by a lightning flash. There is no resurrecting, phœnix-wise, as in the previously illustrated instance, from this cataclasmic overthrow. Destruction is thorough and complete. And yet, as though even now the whole soul of the living organism had not abandoned it, one

LIGHTNING-SHATTERED BAOBAB TREE, NEAR DERBY, KING'S SOUND, WESTERN AUSTRALIA.

upstanding angle of the blasted trunk is fashioned into the similitude of a weird monstrous bird that has mounted guard and broods like a disembodied spirit over the desolation wrought.

Among the most characteristic and bizarre of the Australian indigenous vegetable products must be included the so-called "Grass-trees," referable to the genera Xanthorrhæa and Kingia. The more popular title for these singular plants is that of "Blackboys," the aspect of the usually more or less incinerated trunks of the arborescent species, as seen from a little distance, being to a considerable extent suggestive of the aboriginal representatives of humanity. The group is most essentially Australian, limited in its distribution to the temperate districts, and apparently the surviving relic of an age in which they and other colossal members of the Sedge and Rush tribe, Juncaceæ, were the paramount forms of vegetation.

The most familiar and widely spread representative of this tribe is the arborescent Grass-tree, or Blackboy, *Xanthorrhœa arborea*, characteristically portrayed by several growth phases in Plate XLVII. As it more commonly occurs, this tree rarely exceeds eight or ten feet in height, and there is only a single or but sparsely divided crown of long narrow grass-like leaves at the summit of the black, cylindrical stem. In situations specially favourable to its growth, however, it attains to a much more luxuriant development. An example that is reputed to be the finest of its species extant in Western Australia is represented in the lower figure of the Plate above quoted. This remarkably fine example is growing, and haply preserved from destruction, in the grounds of the hostelry that constitutes the half-way house between Donybrook and Bridgetown, in the above-named Colony. As indicated by the lad standing near it, its altitude is little short of twenty feet. The tree has not produced many of its characteristic flower-spikes within recent years, though one may be observed emerging from the crowns of foliage towards the right.

Drakesbrook, on the Bunbury line in the South-Western district of Western Australia, furnished the subject of the photograph reproduced in the upper moiety of the Plate under notice. It embodies not only fine examples of the same species of Blackboy, but a luxuriant growth of other vegetation that imparts to the scene an almost tropical appearance. Most conspicuous among these secondary growths are the pinnate, palm-like, leaves of *Macrozamia Fraseri*, one of the Cycadaceæ. Space unfortunately forbids the reproduction of the separate photographs taken of finer isolated examples of this plant, in which the large central pine-apple-like fruit are clearly depicted. Mingled with these Cycads may be observed an abundant growth of a fern that is accounted by botanical authorities to present little, if any, distinction from the familiar English bracken, *Pteris aquilina*.

The Xanthorrhœas figured in this illustration have produced an abundant crop of flower-spikes. They are for the most part, however, of the previous season's growth, and are in consequence dark in hue and much weathered and dilapidated. A correct idea of the pristine aspect of these flower-spikes, when their contour was symmetrical, may be formed by a reference to the figure of an allied species,

GRASS-TREES, CYCADS AND FERNS, DRAKESBROOK, WESTERN AUSTRALIA, p. 272.

ARBORESCENT GRASS-TREE OR "BLACKBOY." *Xanthorrhœa arborea*, p. 272.

X. hastilis, reproduced on the left-hand side of Plate XLVIII. This photograph was taken when the flower-spikes were most fully developed. At that season their contour is perfectly symmetrical and their cylindrical surfaces completely studded with tiny cream-white starlike florets. The name of the Underground Blackboy, or Grass-tree, has been popularly applied to this species with reference to the circumstance that it produces no conspicuous stem or trunk like the other species, the crown of leaves springing direct from an underground rhizome or root-stock. Very abundant growths of this Blackboy occur on the borders of the railway track between Perth and Fremantle, Western Australia. The view here reproduced was taken in this locality.

The most striking representative of the Grass-tree tribe is probably the one sharing with the species last referred to a portion of Plate XLVIII. This is the so-called Drumstick Grass-tree or Blackboy, *Kingia australis*, which is limited in its distribution to the Southern districts of Western Australia. The popular name applied to this Grass-tree bears a very obvious relationship to the contour of the flower-spikes. A number of these are developed from the terminal crown, and in place of being abnormally elongate and spear-like are short and capitate. With this species of Grass-tree the trunk is never subdivided as in the case of the arborescent Xanthorrhœas, the finest plants being represented by a single erect cylindrical column with its crown of wiry grass-like leaves and capitate flower stalks. The average height to which these Drumstick Grass-trees grow is ten or twelve feet, though they occasionally exceed twice this altitude. A remarkably fine example of this plant, presented to the nation by Sir Malcolm Fraser, K.C.M.G., the Agent General for the Colony of Western Australia, has been transported to and set up in the Botanical Galleries of the British, Natural History, Museum. It is of no less a height than thirty feet.

The development of the leafy crown of Kingia is subject to considerable variation. In the example figured it is rather stunted and short, the selection having been made with reference more to the height of the tree and to the abundance of the flower stalks. In some of the specimens to the rear of the chief individual figured it may be observed that the fibrous leaves occupy a much larger relative mass. The locality which yielded this picture is in the vicinity of Pinjarrah, Western Australia, and consists of enclosed, partly cleared, pasture ground that was previously thickly grown over with this and the ordinary arborescent Grass-tree. Examples of the last-named species are observable beneath the shadow of the larger trees on both the right and left-hand sides of the photograph reproduced. The small light-coloured irregularly shaped hillocks dotted about throughout the ground area in this view represent, it may be mentioned, clay tenements of the Termites or White Ants that are feeding on the decaying Grass-tree roots and stumps. This form of the White Ant nest has been already referred to in Chapter IV., and well illustrates the relatively small dimensions of the edifices raised by these insect architects outside the limits of the tropics.

The higher bushy trees, with large distinctly serrated leaves, that overshadow the Xanthorrhœas on either side in Plate XLVIII., typify a very characteristic Australian group, as exemplified by the genus Banksia of the family of the Proteaceæ. The species here depicted, *Banksia grandis*, is one of the handsomest of its tribe. Its large dark glossy leaves and bright yellow flower-spikes are often a foot or more in length, and constitute quite a conspicuous feature in those districts, where it forms extensive plantations. No less than thirty-four species of Banksia have been described, the majority of which are indigenous to the southern, temperate regions of the Australian continent and the adjacent island of Tasmania. Throughout the Australian Colonies the more common popular name for these trees is that of "Honeysuckle." This title has been conferred upon them with reference to the large quantity of honey the abundant blossoms produce, which afford an almost continual feast for a vast number of exclusively honey-eating birds. The matured flower-spikes of the species now under notice literally drip with the sweet secretions, and are the object of assiduous attention, not only from birds, but also from the Western Australian aboriginals, whose dusky countenances during the honeysuckle season are for the nonce transfigured beyond recognition by an adherent mixture of Banksia honey and yellow pollen.

The large seed cones, commonly a foot long and five or six inches thick, of *Banksia grandis* are in great favour as an article of firewood in those localities where the species represents one of the principal forest growths, the abundance of resin contained in them creating as cheerful a blaze as the cedar or pine cones esteemed for a like purpose in northern latitudes. Although most extensively known as "Honeysuckle," the commoner name applied to the tree, and more especially with reference to the cones vended for kindling purposes in the neighbourhoods of Perth and Fremantle, Western Australia, is the "Banshee." It was difficult, at first, to determine how this seemingly incongruous title had become affixed to it. Apparently, however, it has been evolved as an illiterate corruption of the generic name of Banksia.

Towards the vagaries of the vegetable world the somewhat heterogeneous group of the Mangroves contributes a conspicuous instalment. Although by no means restricted to Australia, being abundant in innumerable varieties throughout all tropical coasts and estuaries, the northern sea-board of this Island-Continent is especially fertile in both specific types and individual developments. Mangroves, as a concrete group, do not pertain to any special family, or even order, of the vegetable kingdom. They have been correlated under a common title with reference only to the circumstance that their habitats and environments have induced their collective assumption of corresponding or closely parallel structural adaptations. All Mangroves, in the popularly accepted sense of the term, agree with one another in being inhabitants of the foreshores of coast-lines and river estuaries under such conditions that they are subject to being, alternately, partially submerged and completely exposed to atmospheric influences with the rise and fall of the tide. Two of the most abundantly distributed representatives of the mangrove tribe are represented in the lower figure of Plate XLIX.

The lighter-barked tree in the centre is the so-called White Mangrove, *Avicennia officinalis*, belonging to the botanical order of the Verbenaceæ, a group that includes the handsome cultivated Lantanas. Flanking each side of this tree are characteristic growths of the Red or Orange Mangrove, *Rhizophora mucronata*, belonging to the order of the Rhizophoreæ, which, in this instance, includes several other genera, *e.g.*, Ceriops and Bruguiera, that participate in corresponding Mangrove habits. The essential features in each of the two species figured is the peculiar modification of their root elements. In Avicennia, the central form, the radiating course of the horizontal slightly submerged roots from the central trunk is distinctly traceable to a considerable distance by the outgrowth therefrom of vertical, thickly crowded, shoot-like developments, popularly known as "Cobbler's pegs," which, penetrating the superincumbent soil, attain to the height of from about one foot to eighteen inches above it. To the uninitiated, it would scarcely occur that this mass of vertical twigs was other than a luxuriant development of the suckers or root shoots by which so many plants ordinarily propagate and extend their borders.

The interpretation of the true function of these vertical growths would appear to have been first arrived at by the late Dr. James Bancroft, of Brisbane, Queensland, who contributed the account of some interesting investigations he made to the 1888 Meeting of the Australian Association for the Advancement of Science. One day, when examining the shore in the vicinity of Avicennia growths at half tide,

Dr. Bancroft observed a white powder floating on the water and, on nearer investigation, discovered that it was being thrown off the surface of the vertical root shoots of the Mangrove. Examined more minutely, it was found that the cortex of these shoots were thickly riddled with pores which much resembled the so-called "lenticels" of ordinary plants. These pores communicated with an extensive system of air-passages which permeated the substance of the roots, and in the process of their enlargement threw off as broken down cellular tissue the powder observed floating on the surface of the water. These shoots, or aerial roots as they are more correctly termed, were thus demonstrated to be virtually a system of air-shafts which, projecting above the surface of the mud, secured for the roots abundant aeration and ventilation with every fall of the tide. It was owing to this efficient system of air-passages that the roots were preserved from becoming water-logged and rotten, as would happen to those of ordinary forest trees if submitted to similar conditions. A corresponding exaggerated lenticel-like aeration system was found by Dr. Bancroft to appertain to all of the peculiar root or trunk modifications possessed by the other species of Mangroves he examined. These included, in addition to the true aerial roots of Rhizophora next described, the nodular or knee-shaped woody protuberances developed from the ordinary horizontal roots in the genus Bruguiera and the greatly elongated vertical offshoots, much resembling those of Avicennia, possessed by *Sonneratia acida*, but which, in some instances, were elevated to a height of no less than six feet above the surface of the soil.

The phenomenon of aerial, ventilating roots is most conspicuously manifested in the Red or Orange Mangrove, *Rhizophora mucronata*, which occupies the greater portion of the photographic view reproduced in Plate XLIX. Arched, repeatedly bifurcating offshoots are in this type abundantly produced from the main trunk, while adventitious shoots that hang down and bifurcate in a similar manner are developed from the larger branches. Together they form a labyrinth of reticulations that in large plantations is often many acres or even miles in extent, and of such a height as to present an almost impenetrable barrier to the explorer. In most instances, moreover, the muddy ooze in which this species delights to grow, is of such depth and so soft and treacherous a character as to effectually debar all progress, except by recourse to the expedient of clambering upon the surface of the roots themselves. In locations, however, where the ground is firmer and the trees not so thickly planted, the naturalist may reap an abundant harvest and endless recreation within and without the precincts of a Mangrove thicket. On the outer, land-side, margin the specific

A. MANGROVE-FREQUENTING FRUIT BAT, *Pteropus conspicillatus*, p. 278.

B. ORANGE AND WHITE MANGROVES, *Rhizophora mucronata*, and *Avicennia officinalis*, BROOME CREEK, WESTERN AUSTRALIA, p. 275.

varieties of Mangroves are the more numerous. Ceriops, Ægiceras, Excœcaria, Bruguiera and other generic types are mingled with Rhizophora and Avicennia. The majority of these possess large handsome glossy foliage as ornamental in character as that of cultivated Laurels, Euonymi, or Rhododendrons. Rhizophora, in fact, when forming as it does on the outskirts compact symmetrical growths, might be readily mistaken for a rhododendron bush, and the same might be said of *Bruguiera Rheedii*. Many of the Mangrove species, moreover, produce an abundance of flowers which, if not conspicuous for size and beauty, are highly scented and load the surrounding atmosphere with sweet perfume. *Ægiceras majus* is one of these, the white flowers growing in profuse bunches at the extremities of the branches and emitting a scent almost as powerful as and somewhat resembling that of the Garden Syringa. *Rhizophora mucronata*, again, produces star-shaped white woolly flowers that are, though more diminutive, to a considerable extent suggestive of those of the Fringed Violet, Thysanotis, hereafter described.

The brilliant colours and quaint habits of the singular crabs belonging to the genus Gelasimus that occur abundantly among the Mangroves have been referred to in a previous Chapter. Birds of many varieties abound in the Mangrove thickets. They include Honey-eaters attracted by the flowers, flycatchers, king-fishers, waders, and on the sea margin, the fish-eating hawks. Not unfrequently the Fruit-eating Bats, or Flying Foxes of the colonists, belonging to the genus Pteropus, take up their abode in the denser, rarely invaded, depths of the Mangrove forests. In these secure retreats, or rookeries as they are somewhat inappropriately termed—"battery" would seem to be the correct word—the animals assemble in hundreds or even thousands and pass the whole day hanging head downwards asleep, or in a semi-torpid state, from the Mangrove branches. As soon, however, as the shades of evening fall, they awake to activity and sally forth in long streams to every point of the compass in quest of food. The somewhat bean-like fruit of the White Mangrove, *Avicennia officinalis*, would appear to yield these bats an abundant repast at some seasons of the year, while the flowers of the many species of Eucalyptus, of which some kinds are almost always in bloom, represent a yet more permanent diet. Cultivated fruits such as peaches, grapes, mangoes or bananas, are, as dwellers in bat countries well know, a terrible temptation to these night marauders, who will travel incredible distances from their mangrove or scrub fastnesses to take their toll, often a most heavy one, from the crop in season. An individual out of a dense crowd that was surprised and captured in the Mangrove thickets in Roebuck Bay, Western Australia, has furnished the subject for the top illustration in Plate XLIX. The species is apparently identical with the Northern Fruit-eating Bat, *Pteropus conspicillatus*.

What would at first sight appear to be a floral product closely resembling a rose or camelia in size, shape, and structure, is depicted in the accompanying illustration. It is as a matter of fact no flower at all, but a vagary of vegetation induced through the interference of a gall-insect, the plant thus distinguished being the ordinary White Mangrove, *Avicennia officinalis*. Several of the bushes in the vicinity of the one from which the examples figured were taken, were literally laden with these 'rosette-shaped galls, one small spray gathered (which was also photographed) bearing as many as twenty of these singular products. The colour of these "gall flowers" approximated to that of the leaves of the tree that bore them. From an æsthetic point of view, they were undoubtedly as worthy of admiration as the green roses which are assiduously cultivated and held in high repute in the gardens of rose specialists. Contemplating these Mangrove "roses" in conjunction with the verdant variety of *Rosa hortensis* that happens to be flourishing in the writer's garden and thus available for comparison while penning these lines, the mind is exercised with the thought as to whether gall-insects have not had something to do with the

ROSETTE GALLS OF WHITE MANGROVE.

origin of some of our most highly prized horticultural triumphs. If, in fact, the green rose is the archetype of the exquisite blooms that through high cultivation and under another colour smell so sweet, have we not in this mangrove gall created straightaway the prototype of a floral gem that simply needs bleaching and the addition of a drop of scent to produce a buttonhole camelia that would satisfy the most fastidious.

There is yet another noteworthy point concerning this Mangrove vagary. That "big fleas have lesser fleas, &c." is a time-worn truism that might scarcely be suspected to be capable of practical demonstration in the case of a leaf gall. A very interesting verification of the adage, however, is afforded by our Mangrove "Camelia." On examining the single "buttonhole" example with three diverging normal leaves in the photograph reproduced, three hemispherical elevations may be seen on the edge of one of the petal-like sub-divisions which show white against the dark background of the left-hand leaf. A fourth is also visible more to the right. It might be imagined from their smooth spheroidal contour that they were adherent drops of water. In place of this they are, however, the product of a second species of gall-insect which has inoculated with its ovipositor the leaves already metamorphosed by its predecessor.

A vegetable freak that will be tolerably familiar to all who have visited the remote Nor' West is depicted in the accompanying figure. It portrays a branchlet with attached blossoms of a member of the pea tribe, *Crotularia Cunninghami*, whose flowers bear a most grotesque resemblance to little green brown-striped birds. The bush producing them averages a height of from three to four or five feet, and is particularly abundant on the sand hills close to the sea-shore in the vicinity of Broome, Roebuck Bay, Western Australia. One such bush is in fact included in the foreground of the Chapter-heading illustration of Broome Creek reproduced on

W. Saville-Kent, Photo.

BIRD-PEA, ROEBUCK BAY, W.A.

280   THE NATURALIST IN AUSTRALIA.

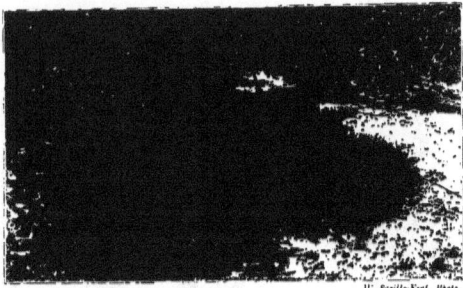

Salicornaria sp., CYGNET BAY, WESTERN AUSTRALIA.   *W. Saville-Kent, Photo.*

page 195. To the ancient writers, by whom the long-stalked Barnacles, *Lepas anitifera*, were declared to be the direct progenitors of living Liliputian geese, this bird-plant of North-West Australia would have proved a veritable windfall.

Two very characteristic plant growths dominant in the neighbourhood of the Western Australian Mangrove swamps are photographically represented on this page. The upper of these illustrations depicts a species of the Chænipodiaceous tribe belonging to the genus Salicornaria. This, and several allied forms, are notably abundant on the salt flats near Carnarvon, at the mouth of the Gascoigne river. The particularly symmetrical growth here selected for illustration was photographed, however, at Cygnet Bay, which

"ROLEY-POLEY" GRASS, *Spinifex longifolius*, SHARK'S BAY, WESTERN AUSTRALIA.   *W. Saville-Kent, Photo.*

is nearly midway between Broome and King's Sound, further north. A band of low-growing Mangroves constitutes, as will be recognised, the background to this picture.

On the sandy ridges in close proximity to the outskirts of the Mangrove belt in Roebuck Bay, the very characteristic grass, *Spinifex longifolius*, represented by the lower of the two figures on page 280 will scarcely escape notice. With the colonists it is popularly known as Roley-Poley grass. This title has been given to it with reference to the spherical contour of the seed heads, and the unique conditions under which their distribution is accomplished. On arriving at maturity these globular seed-heads become detached *en masse* from their supporting stalk, and roll before the wind, it may be for miles, along the surface of the sand, before they break up and the seeds are distributed. The most luxuriant development of this grass is attained in those districts of the sub-tropical Australian coast-line where huge drifting sand-hills constitute a leading feature. In such locations, this Spinifex fulfils a very useful rôle. Independently of the above-recorded method of dispersing its seed over widely extended areas, the grass itself throws out vigorous runners of great length which, penetrating among the sand, to a large measure counteract its natural tendency to drift. Fresh Water Camp, Shark's Bay, in Western Australia, already referred to in conjunction with the Chapter dealing with Pearls and Pearl-oysters, is a locality where this grass attains to a remarkably luxuriant state of development, and represents the scene of the photograph reproduced.

From among the wealth of Wild Flowers for which Australia in general, and Western Australia in particular, is so justly celebrated, the limits of our nearly exhausted space permit only of brief reference to the accompanying illustration of a single noteworthy example. The type selected for this marked distinction shares the patronymic of a modest but most highly-prized British flower, being popularly known throughout the area of its recurrence as the "Fringed Violet." Technically, the plant is in no way related to the Viola tribe, being more nearly allied to the Iris family or Iridaceæ. The precise specific appellation of the form here figured is *Thysanotis dichotoma*. There are many known representatives of the same genus, all indigenous to Australia, and exhibiting among themselves a considerable range of variation with regard to both their contour and growth habits. All, however, agree with one another in their tri-petallate structure, the pellucid texture and lilac tints of their fragile blossoms, and in the exquisitely delicate fringe of filaments which borders each separate petal. To a talented floral artist the production of a replica of these strikingly characteristic flowers should be a comparatively facile task. A sheeny lilac silk cut to the required contour, and with the edges symmetrically frayed out, would reproduce all the essential features, to which the addition of the remaining structural elements would involve but little labour.

The species here figured is one of the finest of its kind. It forms somewhat irregular loosely growing grass-like tufts, with flower-bearing panicles that attain to two or three feet in height. In another form, *T. multiflorus*, common in the same district, the plant forms more compact hemispherical masses a foot or so in height, with much more abundant but smaller flowers. In a third species again, *Thysanotis Patersoni*, the plant takes the form of a climber, its exquisite lilac stars bespangling all manner of adventitious vegetation, while its own slender wire-like repent stem is almost invisible. As might be anticipated, the frail beauty of these Fringed Violets, like that of our native Harebells, is essentially evanescent, departing within a short interval of their being plucked. To successfully photograph them it is desirable to take the camera to their native haunts, and to portray them *in situ* or in freshly-gathered groups. The examples figured were thus culled from that veritable mine of wild-flower wealth, the railway embankment between Perth and Fremantle.

The tempting field of indigenous vegetation has to be abandoned here to

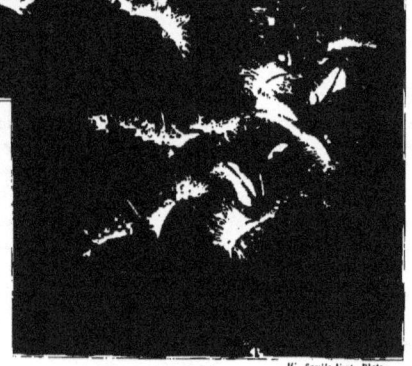

W. Saville-Kent, Photo.
FRINGED VIOLET, *Thysanotis dichotoma*.

permit of a brief record of Australian potentialities for the entertainment of distinguished immigrants from other climes. The dry atmosphere, wealth of sunshine, and congenial temperature predominating in the extra-tropical metropolitan centres has proved to be singularly adapted to the acclimatisation, among other exotic plants, of many of the most handsome representatives of the genus Cereus, belonging to the tropical American Cactus tribe. A few excerpts from a considerable number of photographs taken by the writer, reproduced in these pages, will suffice to establish the correctness of this assertion. The Cacti of the genus Cereus are notable for the circumstances that their handsome blossoms are, in the majority of instances, purely white, varied perhaps by tinges of delicate pink or sulphur yellow, and, more especially, that they only condescend to display their floral charms in the night season, closing their petals and hastening to decay with the rising sun. To see these plants, and, above all, to photograph them in their glory, one must therefore—unless employing magnesium light, which would be scarcely practical in the case of several of the larger examples here illustrated—be an early riser. The witching hour to secure a favourable "sitting" is at daybreak, just before the sun appears above the horizon, when with a prolonged exposure there is sufficient actinic illumination to impress the plate.

The finest collection of Cerei or other Cacti growing upon Australian territory is undoubtedly that contained in the Adelaide Botanical Gardens. The examples selected for the border illustrations of pages 284 and 285 were derived from this source. The Mexican species *Cereus chalybæus*, represented on the side of page 285, formed, as may be recognised by its comparison with the rail-fence close beside it, a sheaf of polygonal flower-bearing columns over thirty feet high. A couple of flowers of the same type, taken from a nearer standpoint, are portrayed on the opposite side border. The outer petals in this species were slightly tinged with pink. The two basement border illustrations to the same pages represent another form, *Cereus nitens*, that was established close beside the tall columnar species, but differed in its growth habits to the extent of forming procumbent straggling masses, whose highest stalks were elevated to no greater heights than from one to two feet above the surface of the ground. The expanded blossoms in this variety are of a translucent snowy whiteness about six inches in diameter, and, as seen in clusters of eight or nine individuals, presented a most fascinating spectacle.

The most strikingly remarkable example of Cactus blooms here portrayed is undoubtedly that of *Cereus grandiflorus*, represented on page 286. The expanded goblet-shaped flowers of this superb species measure as much as nine inches or, with

CEREUS CHALYBÆUS

the outer bracts, a foot in diameter. The inner petals are pure white and the outer ones have an inclination to pale lemon yellow, which becomes more decidedly emphasized in the radiating bracts. The species is essentially repent in its habits, and has been extensively introduced as an ornamental creeper on verandahs and trellises throughout the suburbs of Perth and Fremantle, Western Australia. The remarkably fine blossoming example represented by the upper of the two figures here reproduced is one growing in the grounds of the Perth Government Gardens, which, after scaling a high brick wall on the further side, hangs in flowering festoons over the one presented to view. That this grand spectacular display should be enacted but one night each year, and then for the delectation only of the bats and moths or phenomenally early birds,

CEREUS NATENS.

appears to be a sad, almost sinful, waste of one of Flora's most marvellous creations. Doubtless, however, it is the moths only that are required in this instance to assist Dame Nature in the accomplishment of the essential purpose of all floral structures. To these insects the glistening sheen of the white cactus lamps in the pale moonlight must prove an unerring beacon to the honied banquet that awaits them. As they flit from cup to cup the large protruding pistil and stellate stigma, distinctly shown in the lower of the two figures, forms a sort of landing stage or front door-mat upon which the moths must first alight. "Please wipe your shoes before entering" is the *mot d'ordre*, unwritten but obeyed, at every golden portal. Con-

CEREUS CHALYBÆUS.

CEREUS NITENS.

W. Saville-Kent, Photo.

forming to this behest, the winged guests detach from their downy feet a more or less considerable portion of the yellow, powder-like pollen adventitiously brought away from the anthers of the bloom last visited. And thus the seed germs are fertilised.

Many and anxious were the tentative daybreak visits paid to this splendid Cactus in order not to miss the coveted opportunity of immortalising its snowy charms. For several days previous the swelling buds appeared to be on the very brink of bursting open, but on as many occasions disappointed anticipations. At length, on the very night preceding the writer's departure by steamer for the farther North, and England, the coy beauty, from her

CEREUS GRANDIFLORUS. *H. Vedhoken, Photo.*

VEGETABLE VAGARIES.

coign of vantage "over the garden wall," relentingly unbent to her constant wooer, and furnished the subject of our picture.

The photographic replica forming the lower of the two floral subjects included in Plate L. is a highly interesting instance of felicitous acclimatisation somewhat accidentally brought about. The location, as in the case of two of the Cacti previously referred to, is the Botanical Gardens, Adelaide, and the aquatic plant figured is a Brazilian species, *Pontideria crassipes*, locally termed the "Water Hyacinth." For many years this Pontideria was tenderly nursed and coddled in the warm water, *Victoria regia*, tanks of the adjacent hothouses,—but without its condescending to produce a single bloom. The decree at length went forth from the long-suffering but patience-exhausted Curator to

cast the unprofitable servant into the outer lake, where, lo and behold, the following year, it clothed its mirrored surface with a sheet of exquisite amethystine blossoms. Such, as the Poet Moralist has written, are the "sweet uses of adversity."

Queensland is the scene of the plant figured in the upper sub-division of Plate L. The species, *Beaumontia grandiflora*, is one of the many favorite creepers acclimatised in the Brisbane Gardens. The flowers are trumpet-shaped, as white and as large as virgin lilies, while the plant forms festoons or a network of rope-like stems and dark glossy leaves that may cover, as here shown, yards of space or traverse the whole length of a paddock fence. They and innumerable other garden beauties, many of them photographically recorded, and including such forms as huge bushes of Frangipani (Plumiera), Francisces, Bougainvilleas, Hibiscus, Pomegranate, Bauhinia, Allamandas, Stephanotis, Antigonon, and a host of other English hothouse types, all growing out of doors in rampant luxuriance, beside English Roses, Camelias, and all the ordinary garden plants, are to the writer among the happy memories of the Queensland bungalow and garden paradise he occupied there a few years since. Such an Eden had necessarily also its quota of luscious fruits in equally bewildering variety. Pineapples, Mangoes, Bananas, Persimmons, Passion-fruit, Granidella, Grapes, Peaches, the sweetmeat-like *Monstera deliciosa*, and, to crown all, English Strawberries, yielded a dessert menu scarcely to be despised. Is it to be wondered at that emigrants of, it may be, but a few years' standing only to Australia's prolific soil and sunny clime, find it difficult to rehabilitate themselves contentedly amidst the grudgingly responsive fallows, predominating fogs, and murky skies of their native land?

The concluding border illustration to this Chapter affords a glimpse of the manner in which tree-ferns are wont to grow in Australia. The upper figure depicts a sheltered nook in the Brisbane, Queensland, Botanic Gardens, wherein they have been artificially introduced in company with a mingled variety of tropical vegetation. A closely identical picture, except for the presence of the banana-like Strelitzia and the artificial foreground, is presented by a photograph taken by the writer in the natural scrub in the vicinity of the Barron Falls, North Queensland. Tree-ferns in their fullest state of development must undoubtedly, however, be sought for in the southern island-colony of Tasmania, where, more especially on the moisture-laiden flanks of Mount Wellington, they grow in the greatest luxuriance. A fern-grown area in this notable locality is portrayed by the lower of the two photographs reproduced. The typical Tasmanian Tree-fern, *Dicsonia antarctica*, occupies the greater portion of the field, growing so thickly that their tree-like trunks are for the most part concealed from

PLATE L.

A. TRUMPET CREEPER, *Beaumontia Grandiflora*, ON A QUEENSLAND VERANDAH, p. 288.

B. WATER HYACINTH, *Pontederia Crassipes*, ADELAIDE BOTANIC GARDENS, p. 287.

view. In the immediate foreground, however, a small space is monopolised by another indigenous Tasmanian fern, apparently a species of Lomaria, notable for its production of diverse barren and fertile stalks, and for the characteristic coppery or even crimson tinge of the young, immature, plants and newly unrolled fronds.

The snap-shot photograph selected as the concluding tail-piece to these Chapters represents a clash of incoming and resurgent breakers taken by the author at the "Causeway" on the outer or ocean beach at Bunbury, Western Australia.

"BREAK — BREAK — BREAK."

APPENDIX A.

At Page 11, Chapter I., the almost unique character of the Australian boomerang as a native weapon was reservedly referred to. The reservation, which may be here explained, was made with reference to some highly interesting data concerning this instrument that were communicated to the author by Mr. Henry Balfour, the Curator of the Ethnographical Department of the Oxford University Museum, though a little too late for incorporation in the Chapter indicated.

As pointed out by the above-named ethnological authority, the boomerang is of higher antiquity and its existing use is more widely diffused than is commonly supposed. According to Mr. Balfour, two forms of this weapon are peculiar to India. One of these, of a simple curved shape and made of wood, is possessed by the Koli tribe, belonging to the district of Guzerat. A second Indian form belongs to the Marawas of Madura, and differs in shape from both the above and the Australian type. The contour of this Marawan boomerang is almost crescentic, perfectly flat, but much broader at the more remote or distal extremity of the instrument as held in the hand. The narrower proximal or handle end is, moreover, fashioned into the form of a conveniently prehensible knob, which is usually roughly carved. Although commonly made of wood, it is not unfrequently constructed of steel or even of ivory. An example recently in the writer's possession—since contributed by him to the Oxford Museum—while made of wood, has a terminal metal capping, and, as shown by the worn marking, had originally three ornamental metal studs on each side. An illustration of this boomerang is herewith given.

INDIAN (MARAWAN) BOOMERANG.

This description of boomerang has been proved by General Pitt Rivers to belong to that category of these weapons which will return to the thrower when dexterously manipulated. The ancient Egyptians, Mr. Balfour informs the writer, undoubtedly used boomerangs which, in some cases, were also returnable, as proved by the successful trial of carefully reconstructed models. There are a few African tribes that still use flat, curved throwing sticks, closely resembling boomerangs, as also do the natives of Arizona and New Mexico. Mr. Balfour further testifies to having recognised in certain illustrations inscribed on ancient Etruscan vases what appear to be undoubted representations of boomerangs being thrown at game.

From the multiplicity of evidence here recorded, the boomerang must evidently be regarded as a weapon that did not originate adventitiously with the Australian Aborigines, or at any rate upon Australian soil, but was in all probability brought there with the earliest immigrants from the Asiatic continent. The yet wider distribution of the characteristic Australian "Woomera," or spear-thrower, yields substantial testimony towards a corresponding conclusion with relation to that instrument.

INDEX.

A

	Page
Aboriginals, Australian	7
Aboriginal women	14
Abrolhos corals	139
,, Islands	132
,, ,, avifauna	136
,, ,, fish fauna	150
Abrus precatorius	13
Acanthias	192
,, vulgaris	154
Acclimatisation of English trout and salmon	182
Acclimatised cacti	283
Acrobates pygmæus	26
Acrocephalus australis	54
Acrozoanthus australiæ	224
Actinodendron alcyonidium	223
Actinopyga mauritiana	149
,, obesa	149
,, polymorpha	149
Adansonia digitata	37, 270
,, rupestris	37, 266
Adelaide Botanical Gardens cacti	283
Ægiceras majus	277
Albany Pass termite mounds	116
Alcyone pusilla	54
Alopecias vulpes	154
Alpheus avarus	203
Ameghino, Florentino, on Patagonian marsupials	20
Amphibolurus barbatus	81
Amphiprion bicinctus	220
,, Clarkii	220
,, melanopus	220
,, percula	220
Anchovy	155
Anemones, giant varieties	220
,, harbouring living fish	220
Angel-fish	154
Anguilla australis	159, 184
Antarctic Continent, pre-existing	2

	Page
Ant-hill Point, Albany Pass	116
Ants employed surgically	255
,, green	252
,, weaving habits	253
Aphanapteryx	6
Apteryx	5
Aracana aurita	187
,, ornata	187
Arachnura Higgensii	260
Arapaima gigas	4, 180
Arctogea	2
Argiope cophinaria	262
,, regalis	262
Argyrodes	259
Arhamphus	179
Arius thalassinus	176
Army crabs	241
Arripis georgianus	178
,, salar	178
Artamus	7
Artificial production of pearls	214
Ascidian-covered rocks, Roebuck Bay	216
Astacopsis Franklinii	157, 238
,, serratus	238
Asterina calcar	243
Atticora	7
Aulopus purpurissatus	150, 176
Australian aboriginals	7
,, badger	21
,, bear	26
,, Cyclodus	98
,, fish fauna	153
,, flora	37
,, house swallow	66
,, jungle fowl	31
,, kingfishers	54
,, magpie	52
,, monitor	94
,, song birds	54
,, whitings	173

	Page
Australian wrens	66
Avicennia officinalis	275
Avifauna, Houtman's Abrolhos	136

B

	Page
"Bahmeen"	169
Bancroft, Dr. James, on Mangrove respiration	275
Bandicoot	20
Balistes	189
Banksia grandis	274
"Banshee"	274
Baobab	37
„ or Bottle-tree	37, 266
„ rejuvenescent properties	270
"Barok" pearls	198
Barracouta	170
„ method of taking	171
Barracuda	172
Barramundi	160, 179
Barrier reef bêche-de-mer	149
Bastard trumpeter	163
"Batavia" wreck on Houtman's Abrolhos	134
Bearded lizard	81
Beaumontia grandiflora	288
Bêche-de-Mer fishery, Western Australia	196
„ of Houtman's Abrolhos	148
„ Roebuck Bay	217
„ "black fish"	149
„ "red fish"	149
„ "surf red"	149
„ "teat fish"	149
Belencia Clytie, its social chrysalides	257
Bell bird	66
Bellows fish	154
Bennett, Dr. George, on Ornithorhynchus	16
Berycidæ	167
Boryx affinis	167
„ Mulleri	167
Bird life on Murray River	36
„ -like pea-blossom	279
Bird's-foot sea-star	244
Birds frequenting Houtman's Abrolhos	136
„ of Paradise	34
"Birth of a coral island"	147
Blackboys, or Grass-trees	271
„ -edged pearl shell	212
„ fish, Tasmanian	156

	Page
Black-headed Gouldian finch	57
„ swan	36
Blister pearls	198
Blue crab	150, 238
„ fish	170
„ groper	174
„ -head parrot fish	174
„ shark	154, 193
„ -tongued lizard	98
Boab or Baobab	37, 266
Bob-tailed lizard	94
Bombay duck	176
Boobook owl	41
Boomerang, Australian	11
„ Indian	290
"Boomer" kangaroo	21
Boongarry	21
"Bottled-nosed" Snapper	161
Bottle-tree	37
„ or Baobab	37, 266
Bower birds	33
Broome, port of, Roebuck Bay	197
Bruguiera Rheedii	277
Brush Turkey	32
Buckland Fish Museum	164
Bull-dog ants	252
„ shark	193
Bummuloh	176
Burnet River salmon	179
Butcher-bird	51
Butter fish	166

C

	Page
Cacti acclimatised in Australia	283
Caldwell, W. H., Oviparous habits of Ornithorhynchus first demonstrated by	16
Callorhynchus antarcticus	155, 193
Calotermes flavicollis	110
Campbell, A. J., Avifauna of Houtman's Abrolhos	136
Carangidæ	169
Caranx trachurus	169
„ gallus	169
„ georgianus	170
„ radiatus	169
Carassius vulgaris	184
Carcharias glaucus	154, 193

INDEX.

	Page
Carp, sea	166
Cassowary	5
Casuarius	5
Cat-fish eel	176
,, fishes	176
Caulospongia verticillata	235
Celænia excavata	259
Centriscus scolopax	154
Ceratodus Forsteri	3, 179
Cereus chalybæus	283
,, grandiflorus	283
,, nitens	283
Ceriops	277
Ceriornis Lathami	33
Cestracion Phillipi	192
Chærops cyanodon	174
Chanos salmoneus	175
Chatham Islands	6
Chelone mydas	237
Chilodactylus Allporti	166
,, bizonarius	166
,, gibbosus	166
,, macropterus	166
,, nigrescens	166
,, nigricans	166
Chimæra monstrosa	155, 193
Chilomycterus jaculiferus	191
Chlamydera maculata	34
Chlamydosaurus Kingi	70
Clupea sprattus	154, 175
,, sagax	175
Clupeidæ	174
"Cobbler's pegs"	275
Cod family	155
Cœnolestes obscurus	3
Colochirus anceps	217
Colonial salmon	178
Colour metamorphoses of Hobart trumpeter	163
Commensal spiders	259, 263
Compass ant-hills	121
Compsognathus longipes	76
Condylactis	223
Conger eel	154
Copidoglanis tandanus	176
Coral beach, Houtman's Abrolhos	148
,, -boring sponge	219
,, island, initial growth	147
,, rock oysters	248
Corals, Australian	228

	Page
Corals, cup varieties	229
,, growth rate at Houtman's Abrolhos	146
,, of Houtman's Abrolhos	139
"Corroboree"	15
Corythophanes Hernandezii	80
Cossyphus unimaculatus	174
Cow-fishes	187
Crab, blue	238
,, commensal of sea anemone	220
,, mangrove	239
,, swimming	238
Crabs, soldier	240
Cracticus torquatus	51
Cradle, aboriginal	14
Cray-fish, Tasmanian	238
,, Murray river	238
Crepidogaster	155
Crotularia Cunninghami	279
Ctenolates ambiguus	158
Cuckoo wrass	187
Cucumber mullet	180
Culcita grex	243
Cultivation of mother-of-pearl shell	204
Cup corals	230
,, sponge	236
Cuscus	26
Cybium Commersoni	177
Cycadaceæ	272
Cyclodus	98
Cygnus atratus	36
,, olor	36
Cyprus pine	8

D

	Page
Dacelo gigas	53
,, Leachii	53
Dancing birds recorded by W. H. Hudson	62
,, habits of Gouldian finches	59
Dasyures	20
Dendrolagus Bennetianus	21
,, Lumholtzi	21
Dendrophyllia axifuga	231
Derwent smelt	5
Diamond sparrow	61
Dicsonia antarctica	288
Didelphyidæ	3
Dimona porrigens, its behatted larvæ	256

	Page
Dingo	15
Dinornis	6
Dinosaurian resemblances of frilled lizard	76
Diodon maculata	191
Dipnoi	3, 179
Diprotodon australis	28
Discosoma Haddoni	219
,, Kenti	219
Dodo	6
Doris imperialis	151
Dromaius	5
Dromornis	6
Drumstick Grass-tree	273
Duck-billed Platypus	15
Dwarf oysters	250

E

	Page
Eagle ray	154
Echidna aculeata	15, 18
,, feeding habits	19
Eels	184
,, Victorian	180
Egernia Cunninghami	94
,, depressa	92
,, Kingii	93
,, Stokesii	92
Elephant-fish	155, 163
Emu	5
,, wren	67
English perch	184
,, salmon, acclimatisation at the Antipodes	182
,, trout	180
Engraulis encrasicholus	155
,, antarcticus	155
Epimachus magnus	35
Estralda bella	61, 64
Eumaeus algeriensis	99
Eutermes arboreum	114
,, tenuis	111
Excoecaria	277

F

	Page
Fantastic rocks, Roebuck Bay	215
Festal dance or "Corroboree"	15
Fierasfer, fish living as a commensal within mother-of-pearl shells	201
Finch, Firetailed	61
,, Gouldian	58
,, Grass	55
,, White-headed	61
Fire-producing apparatus	13
,, tail	64
,, tailed finch	61, 64
,, ,, ,, playful habits	65
,, ,, ,, vocal powers	65
Fish acclimatisation, Western Australia	158
,, commensals of sea anemones	220
,, fauna of Australia	153
Fishes of Houtman's Abrolhos	150
Flathead	173
Flower and Lydekker, on Ornithorhynchus, kangaroos and phalangers	17, 22
Flying foxes	277
,, mouse	26
,, opossum	22
,, phalanger	22
,, squirrel	22
Folsche, Paul, Port Darwin ant-hills	123
Forbes, H. O., on former Antarctic Continent	6, 37
Formica smaragdina	254
,, viridis	252
Fox shark	154
Fremantle herring	178
Frenella robusta	8
Fresh-water cod	156
,, cray-fish	238
,, lobster	238
Frilled lizard	70
,, ,, bipedal locomotion	73
,, ,, feeding habits	76
Fringed violet	281
Froggatt, W. W., Termite investigations	109
Frost fish	154, 170
Fruit-eating bats	277

G

	Page
Gadidae	155
Gadopsis marmoratus	156, 238
Galaxias	180
,, attenuatus	4
Galaxiidae	180
Galeocerdo Rayneri	193
Galls of mangrove	278
Gar-fish	179

INDEX

	Page
Gastrotokeus biaculentus	187
Gelasimus coarctata	239
„ variatus	241
Giant anemones	220
„ herrings	175
„ mackerel	176
„ kingfishers	53
„ perch	160
Gigantic crane	7
Gippsland Lakes, resort of black swans	36
„ perch	160
Glass spear heads	12
Glaucosoma hebraicum	177
„ scapulare	177
Globe-fish	191
Gnow	33
Gobiesocidæ	155
Golden pearls of Western Australia	211
„ perch	158
"Gooanna"	94
Gouldian finches	55
„ „ breeding in England	63
„ „ colour distinctions	57
„ „ dancing habits	59
„ „ vocal talents	58
Grammatophora barbatus	81
Grass finches	55
„ trees, Xanthorrhœa	271
Grassi and Sandias, on European termites	130
Grayling	180
Great Lake trout	181
Green ants	252
„ lizard	98
Grey mullets	173
Gropers	159
Growth rate of corals	146
Grus australasianus	36, 54
Guano deposits, Houtman's Abrolhos	137
Gurnets	173
Gymnorhina tibicens	52

H

	Page
Hagen, on termites	129
Halcyon sanctus	124
Haplochiton Sealii	5
Harpodon nehereus	176
Hatted caterpillar of Dimona porrigens	256
Hatteria (Sphenodon) punctatus	6, 99

	Page
Haviland, Dr., Collection of termites	109
Heloderma horridum	98
„ suspectum	98
Hemirhamphus intermedius	179
Herring tribe	174
Heterotis	179
„ niloticus	4
Hippocampus abdominalis	185
„ antiquorum	185
Hippoglossus vulgaris	178
Hirundo neoxena	66
Hobart trumpeter	162
„ „ colour metamorphoses	163
Hodotermes Havilandi	128
Holothuria mammifera	149
Holothuridæ	148
Holoxenus cutaneus	173
"Honeysuckle," or Banksia	274
Horned toad	87
Horse mackerel	154
„ mackerels	169
House swallow of Western Australia	66
Houtman's Abrolhos	132
„ „ avifauna	136
„ „ bêche-de-mer	148
„ „ coral beach	148
„ „ „ reefs	139
„ „ fish fauna	150
„ „ guano deposits	137
„ „ tropical marine fauna	151
„ „ rocks	135
Howard River meridian ant-hills	123
Hudson W. H., on dancing birds	62
Hydrosaurus salvator	96
Hyracodon fuliginosus	3

I

	Page
Inchmen ants	252
Infusorial parasites of termites	127
Initial growth of coral island	147
Isinglass, yielded by Polynemidæ	169
Isopoda	265

J

	Page
Jabiru	7
Jacana	7
Jelly fish, King's Sound	237

	Page
Jersey lizard	98
Jew-fish	154
,, ,, (W.A.)	177
,, ,, (Q. & N.S.W.)	177
,, lizard	81
John Dory	154
Julis	150
Jungle fowl	31

K

Kangaroo	21
Kea	6
Kimberley termite mounds	119
King-fish, Abrolhos	170
,, ,, Australian varieties	177
,, ,, Queensland	169
,, ,, Tasmanian	171
Kingfishers	54
King Snapper	167
Kingia australis	273
Kiwi	5
Koala	20
Koalemus	28

L

Labrax lupus	160
Labricthys ceruleus	174
Labrus mixtus	187
Lace lizard	94
Lacerta ocellata	98
,, viridis	98
Lagoon, Pelsart Island	143
Lamna cornubica	154
Lanioperca	172
Lankester, E. Ray, Plaster casts of Sphenodon punctatus	165
Lates calcarifer	160, 169
,, colonorum	160
Latris bilineata	166
,, ciliaris	165
,, hecateia	162
,, inornata	166
,, Mortoni	166
Laughing jackass	51, 53
Laura Valley meridian ant-hills	122
Leather jackets	189
,, mouth	160

	Page
Leipoa ocellata	33
Lepidogaster	155
Lepidopus caudatus	154, 170
Lepidosiren paradoxus	4, 179
Leseur's water lizard	97
Lichen-bedecked larva of Phoraxlesma	256
Linckia laevigata	243
Lingah, mother-of-pearl shell	212
Lizard, bearded	81
,, blue-tongued	98
,, bob-tailed	94
,, frilled	70
,, green	98
,, Jersey	98
,, Jew	81
,, lace	94
,, moloch	84
,, monitor	94
,, New Zealand	99
,, ocellated	98
,, poisonous species	98
,, sleepy	89
,, spine-tailed	93
,, spinous	83
,, stump-tailed	89
,, Tuatara	99
,, water	97
Lizards as pets	98
Lotella callarias	156
,, marginata	156
Lung fish	179
Lydekker and Flower, on Ornithorhynchus, kangaroos and phalangers	17, 22
Lyre bird	35

M

Mackerel	154
Macquaria australasica	158
Macquarie's perch	158
Macropodidæ	21
Macrozamia Fraseri	272
Madrepora corymbosa	140
,, hebes	140, 145
,, proteiformis	142
,, pulchra	140
,, reparative powers	144
,, sarmentosa	141

	Page
Madrepora syringoides	140
„ violet-tinted	143
Magnetic ant-hills	121
Magpie	52
„ perch	166
Maigre	154, 177
Malurus	66
Mangrove blossoms	277
„ crab	239
„ galls	278
„ oysters	249
Mangroves, respiratory structures	276
Maorie jig	171
Marching termites	111, 115
Market prices of mother-of-pearl shell	213
Mary River salmon	179
Mauritius avifauna	6
McCook, Dr. H. C., on American spiders	262, 265
Megalania prisca	30
Megalops cyprinoides	175
„ thryssoides	175
Megapodidæ	31
Megapodium tumulus	31
Meleagrina Cumingii	212
„ fimbriata	209
„ fucata	212
„ imbricata	138, 209
„ irradians	209
„ lacunata	209
„ margaritifera	138, 196
Menura Alberti	36
„ superba	35
Meridian ant-hills, Howard River	123
„ „ Port Darwin	123
„ „ Laura Valley	121
Merlucius gayi	155
Mesoprion Johni	150
Metapteryx bifrons	6
Miolania Oweni	31, 84
Mirage-elevated breakers	142
Moa	5
Moloch horridus	31, 83
„ „ feeding habits	84
„ lizard	84
Monacanthus Ayraudi	190
„ Browni	190
„ hippocrepis	190
„ megalurus	190
Monitor lizard	94

	Page
Monotremata	15
Monster crabs	238
Montipora circinata	146
Mope-hawk	40
"More-pork"	40
Morwonga	166
Mother-of-pearl shell, cultivation of	204
„ „ fisheries, W. Australia	197
„ „ its current value	213
„ „ Lingah variety	212
„ „ Queensland	195
„ „ Shark's Bay	209
Mound-building birds	33
Mountain devil	83
„ parrot	6
Mud oysters	246
Mugilidæ	173
Mulloway	177
Murray cod	158
„ cod, acclimatised in W. Australia	158
„ River, abundant avifauna	36
Mycteria australis	7
Mycteris longicarpus	241
„ marching habits	241
„ platycheles	241
Myliobatis aquila	154
Myrmicobius fasciatus	20
"Myrmidon," H.M.S., survey of Cambridge Gulf	168
Myzanthus melanophrys	66

N

	Page
Nair fish	160, 169
Nannygai	167
Native cat	20
„ companion	36, 54
„ pheasant	33
„ trout	180
Nephila fuscipes	259, 263
Neptune's cup sponge	236
Neptunus pelagicus	150, 238
Nereis, commensal with sea anemone	227
New Zealand lizard	99
Ninox boobook	40
Northern chimæra	155
„ salmon	168
Notogea	2

INDEX.

	Page
Notoryctes typhlops	28
Nudibranchiate mollusc, remarkable specimen	150
Nycticebus	26

O

	Page
Old Man Snapper	161
Oligorus gigas	159
„ macquariensis	158
Orange mangrove	275
Ord River phenomenal fishes	166
Oreaster nodosus	243
Ornithorhynchus parndoxus	15
Osprey	67
„ nesting habits	67
Osteoglossum bicirrhosum	4, 179
„ formosum	4, 179
„ Jardinei	4, 179
„ Leichardti	4, 179
Ostracion aurita	187
„ cornutum	189
„ ornata	187
Ostrea edulis	245
„ glomerata	138, 245
„ mordax	247
„ ordensis	250
Ostrich	6
Ox-eye herring	175
Oyster-crusher	192
„ grounds, Western Australia	246
„ spat collectors	247
Oysters, artificial cultivation	247
„ coral rock	248
„ dwarf variety	250
„ elongate variety	248
„ mud	246
„ rock	246
„ zonal growth habit	247

P

	Page
Pagrus major	150, 161
„ unicolor	161
Palmipes membranaceus	244
Palinurus vulgaris	238
Pamban salmon	169
Pandion leucocephalus	67, 136
Paradiseidæ	34
Parasites or commensals of mother-of-pearl shells	201

	Page
Parasitic sponge	219
Parra	7
Passiflora	38
Patagonian marsupials	20
Pearl blisters	198
„ resembling infant's head	202
„ shell cultivation, Abrolhos Islands	208
„ „ „ Broome Creek	205
„ „ „ Shark's Bay	206
„ „ „ Thursday Island	204
„ shelling station, Shark's Bay	210
„ , "Southern Cross"	199
Pearls, artificial production of	214
„ fantastic shapes	199
„ methods of abstraction	212
„ nature of	199
Pelecanus conspicillatus	67
Pelicans, young	67
Pelsart Island Lagoon	143
Pentagonaster australis	245
Perameles	20
Perca fluviatilis	184
Perch family	158
Percolates colonorum	160
Perodicticus	27
Petauroides volans	22
Petaurus breviceps	23
Phalangista fuliginosa	22
Phascolarctos cinereus	26
Phascolomys	21
Phascolonus gigas	21
Pheasant, native	33
Phenomenal trout	181
Phorodesma, lichen-bedecked larva of	256
Photographing night-blooming cacti	283
„ swimming turtles	237
Phrynosoma cornutum	88
Phyllopteryx eques	185
„ foliatus	185
Physignathus Lesueuri	97
Physobrachia Douglasi	222
Pig-fish	174, 193
Piked dog-fish	192
Pilchard	175
Pinna shells, as fulchra of attachment for mother-of-pearl shells	209
Pinnotheres, crab living as a commensal within mother-of-pearl shells	203
Pipe-fish	186

INDEX.

	Page
Piping crow	52
Plagiodus ferox	176
Plagusia	176
Plaster coats of Tasmanian fish	164
Platax orbicularis	150
Platycephalus bassensis	173
Platychoerops Gouldii	174
Platypus	15
" inflicting wound	17
Plectognathi	187
Plectropoma Richardsoni	159
Pleuronectidæ	178
Plotosus	176
Podargus, attitude when alarmed	42
" Cuvieri	40
" favourite food	47
" rain-bath manœuvres	45
" strigoides	40
" vocal notes	49
" warning note	43
Poephila Gouldæ	55
" mirabilis	55
"Pogee"	212
Poisonous lizards	98
Polynemidæ	168
Polynemus indicus	177
" Sheridani	169
" tetradactylus	168
" Vorekeri	168
Pontideria crassipes	287
Porbeagle	154
Porcupine ant-eater	15, 18
" fish	191
Porites, as reef constructors	229
" astræoides	229
Port Darwin meridian ant-hills	123
" Jackson shark	192
Poterion patera	236
Potto	27
Pouched mole	28
Psettodes erumei	178
Pseudocarcinus gigas	238
Pseudochirus	26
Pseudophycis barbatus	155
Pseudoscarus	150
Pteromys	22
Pteropus conspicillatus	278
Ptilonorhynchus holosericeus	34
Ptiloris paradiseus	35

	Page
Proechidna Brugnii	15
Prothylacinus patagonicus	20
Protopterus annectens	4, 179
Prototroctes	4
" maræna	180
Prussian carp	184

Q

Quartz spear heads	12
Quatrefages, on Termites	110
Queensland, garden produce	288
" shrike	51

R

Rabbit-fish	193
Raft of King's Sound natives	9
Raquet-tailed kingfisher	54
Rat Island, coral growths on jetty	146
Red mangrove	275
Reed warblers	54
Reef corals, Houtman's Abrolhos	138
Reptile House at Zoological Gardens	99
Reptilian Society	99
Rhina squatina	154
"Rhinoceros Rock," Roebuck Bay	215
Rhizophora mucronata	275
Rhombsolea monopus	178
Rhoa	6
Rifle bird	34
Rock cods, Australian	159
" oysters	246
Roebuck Bay natives	15
"Roley-poley" grass	280
Roughy	178

S

Salicornaria	280
Salmo fario	180
" ferox	181
" salar	182
" trutta	180
Salmon, Colonial	178
" English	182
" trout	180
Samson fish	170
Sand-fly, attacking spider	264

	Page
Sarcophilus ursinus	20
Satin bower bird	34
Sauromarpus gaudichaudi	54
Savage, T. J., Termite investigations	113
Scabbard fish	154, 170
Scarlet-headed Gouldian finch	57
Sciæna antarctica	154, 177
„ aquila	154
Scroll-coral	146
Scomber antarcticus	154
„ scomber	155
Scrub Turkey	32
Scylla serrata	239
Sea anemones, stinging variety	223
„ breams	160
„ cucumbers	148
„ dragon	186
„ eagle	67
„ horse	185
„ perch	166
„ pikes	172
„ trout	182
Seer fish	177
"Sergeant Baker"	176
Seriola gigas	150, 170
„ grandis	170
„ Lalandii	170, 178
Sewin	181
Sharks	192
„ abundance off Houtman's Abrolhos	141
Shark's Bay corals	230, 232
„ „ mother-of-pearl shell	209
„ „ pearl shelling station	210
„ „ sponge	235
Shell aprons	10
„ beach, Pelsart Island	148
Sillago	173
„ ciliata	150
Silver eel	159
„ eels	184
„ eyes	54
„ perch	158
„ trumpeter	163
Siluridæ	176
Skipjack	154
Sleepy lizard	89
Smeathman, Henry, Termite investigations	111
Snapper	161
Snook	170

	Page
Snouted termites	130
Social chrysalides of Belenois Clytie	257
Solaster papposa	243
Soldier crabs	240
Solenognathus spinosissimus	186
„ Hardwickii	186
Sonneratia acida	276
Sooty terns	136
South Australian sea-dragon	186
Southern chimæra	193
"Southern Cross" pearl	199
Sparidæ	160
Spear heads	12
Sphenodon punctatus	6, 99
Sphyræna langsar	172
„ obtusata	172
Sphyrænidæ	172
Spider cocoons	258
„ eccentricities	257
Spiders, commensal species	259, 263
„ diminutive males	263
Spinifex grass	12
„ gum	12
„ longifolius	281
Spine-tailed lizard	93
Spinous lizard	83
Spiny ant-eater	15
„ dog-fish	154
„ tailed monitor	97
Spoggodes, Alcyonarian coral	217
Sponge destructive of corals	219
„ , Neptune's cup	236
Spotted bower bird	34
Sprat	154
Spur of platypus	17
Squirrel, flying	23
„ Sugar	23
Stag's Horn corals, Houtman's Abrolhos	140
Stalactite-like concretions, Sweer's Island	250
Starfish, decorative adaptability	244
„ Tasmanian	243
Starfishes, Great Barrier Reef	243
St. Helena, termite depredations	111
Steganophora guttata	61
Stinging sea anemones	223
Stipiturus malichurus	67
Stirling, Dr. E. C., on Notoryctes	28
Struthio camelus	6
Stump-tailed lizard	89

INDEX.

	Page
Sugar-squirrel	23
"Surf-red" bêche-de-mer	149
Sweer's Island, stalactite-like concretions	250
Swimming crabs	238
Synapta Beselli	149
Synaptura	178
Syngnathus	186

T

	Page
Tailed-spider, Arachnura Higgensii	260
Tailor-fish	170
Talegalla Lathami	32
Tanysiptera sylvia	54
Tarantula	265
Tarpon, American	175
,, Australian representative	175
Tasmanian black-fish	238
,, carp	166
,, cow-fish	188
,, devil	20
,, fish, coloured plaster casts	164
,, starfishes	243
,, salmon	182
,, tiger	20
,, tree-ferns	288
,, trumpeter	162
,, wolf	20
Tassel fishes	168
Temnodon saltator	154, 170
Tench	184
Termes armiger	129
,, atrox	114
,, bellicosus	112
,, flavipes	104
,, lucifugus	102, 110
,, mordax	114
,, taprobanes	108
,, viarum	115, 128
Termite depredations at St. Helena	111
,, investigations, T. J. Savage	113
,, ,, W. W. Froggatt	109
,, mounds, Albany Pass	116
,, ,, "Columnar"	124
,, ,, "Kimberley" type	119
,, ,, "meridian" type	121
,, ,, "pyramidal"	117
,, ,, utilitarian value	125
Termites, as food	105

	Page
Termites, attacked by ordinary ants	126
,, collection by Dr. Haviland	109
,, eroding metal and glass	107
,, grass-eating species	108
,, infusorial parasites	127
,, marching variety	111
,, reconstructed mounds	120
,, snouted varieties	130
,, specific characters	129
,, structural modifications	129
,, swarming habits	104
,, (white ants)	102
,, wood-eating propensities	106
Tetrodon Hamiltoni	191
Therapon Richardsoni	158
Theridium riparium	261
Throwing stick	13
Thursday Island, as Queensland central pearl-shelling station	197
Thylacaleo	28
Thylacinus cynocephalus	20
Thynnus thynnus	154
Thyrsites atun	170
,, Solandri	172, 177
Thysanotis dichotoma	281
,, multiflorus	282
,, Patersoni	282
Tiger shark	193
Tiliqua scincoides	98
Tinca vulgaris	184
Toad-fishes	191
Trachinidæ	172
Trachurus trachurus	154
Trachysaurus rugosus	89
Tree-ferns, Tasmania and Queensland	288
,, kangaroo	21
Trepang, Houtman's Abrolhos	148
"Triantelope"	265
Trichiuridæ	170
Trichonympha Leidyi	127
Trigger fishes	189
Triodia irritans	12
Tropical marine fauna of Houtman's Abrolhos	151
Trout, English	180
,, native	180
Trumpeter	162
Tuatara lizard	99
Tunny	154
Turbinaria bifrons	235

	Page
Turbinaria conspicua	230
" its abundance on the Australian coasts	229
Turbinaria peltata	230
" revoluta	235
Turbinariæ collection and transport of massive specimens	232
" early growths	231
" expanded polyps of	231
" of abnormal size	230
Twelve-wired bird of paradise	35

U

Underground Grass-tree	273

V

Varanus acanthurus	97
" giganteus	96
" salvator	30, 96
" varius	94
Velvet-fish	173
Violet-tinted madrepora	143

W

Wallace, Dr. A. R., Island life	132
" " on racial affinities of Australian aboriginals	8
Water hyacinth	287
" lizard	97
Weaving habits of green ants	253

	Page
Western Australian golden pearls	211
White ants	101
" " individual types	102
" eyes	54
" -headed finch	61
" " osprey	136
" mangrove	275
"Wideawake" gulls	136
Wombat	21
Wooden cradle	14
Wood hen	6
Woomera	13
Worm, commensal with sea anemone	227
Wrass	187
Wrens, Australian	66

X

Xanthorrhœa arborea	272
" hastilis	273
"X-ray" spider	262

Y

Yellow-tail	178
" tails	170
Young turtles, photographed while swimming	237
York devil	83

Z

Zeus faber	154
Zoological Gardens, Reptile House	99
Zosterops dorsalis	55

THE GREAT BARRIER REEF OF AUSTRALIA.

ILLUSTRATION FROM "THE GREAT BARRIER REEF OF AUSTRALIA."   *W. Saville-Kent, Photo.*

EXTRACTS FROM OPINIONS OF THE PRESS
ON
MR. SAVILLE-KENT'S RECENT BOOK
"THE GREAT BARRIER REEF OF AUSTRALIA"
(PUBLISHED BY MESSRS. W. H. ALLEN & CO., LIMITED, LONDON.)

Super-royal Quarto (13½ × 10), containing 16 Chromo and 48 Whole-page Plates in Photo-mezzotype.   Net Price **£4. 4s.**

TIMES.

The sumptuous volume entitled, "The Great Barrier Reef of Australia: Its Products and Potentialities," by W. Saville-Kent, F.L.S., F.R.S., &c., will be interesting primarily to zoologists and naturalists in general, but it is not without attractions for those who concern themselves with the commercial interests of Australia. It contains an exhaustive "account with copious coloured and photographic illustrations of the corals and coral reefs, pearl and pearl shell, bêche-de-mer, and other fishing industries and the marine fauna of the Australian Great Barrier region." The illustrations are very skilfully executed and very interesting in themselves, and the letterpress consists of a series of elaborate monographs on the natural features and products of this wonderful region. . . . Mr. Saville-Kent's chapter on the commercial potentialities of the reef is a veritable romance of the sea, and his whole work is a labour of love and enthusiasm.

SATURDAY REVIEW.

This is a sumptuous book : a large quarto volume, illustrated by no less than forty-eight plates in photo-mezzotype and sixteen in chromo-lithography. Such a complete study of a coral reef has never before been published. It deals not only with the natural history of the Great Barrier Reef, but also with the marine industries of that region, which are of no small importance to the colony of Queensland. . . . Mr. Saville-Kent's photographs and descriptions give a wonderfully vivid idea of these strange "toilers of the sea" in every respect but colour, and that the chromo-lithographs enable us to imagine. . . . Among these delightful pictures it is difficult to single out any for special praise. . . . The book is so full of curious and interesting matter that it is hard to know where to stop and when to put it down. Mr. Saville-Kent has brought a coral reef and its wonders nearer to naturalists who cannot wander far from the shores of colder regions than anyone hitherto has succeeded in doing.

NATURE.

Coral and coral reefs are likely to become additionally popular from the publication of a really magnificent book entitled "The Great Barrier Reef of Australia: its Products and Potentialities." This work . . . presents us with what is emphatically an *édition de luxe*. Of large size, its pages teem with most beautiful coloured illustrations of the life of the reef, and with photographic reproductions of its scenery. Nothing finer in the way of book-illustration has come under our notice, and the illustrations will be all the more welcome to naturalists, in that they reproduce the characteristics of the Great Barrier with absolute fidelity, to which the word-painting of a Ruskin would be wholly unequal. . . . Mr. Saville-Kent's book contains a series of nature-pictures of the corals such as has never before been submitted to the scientific world, and a glance at his illustrations does more to familiarise one with the phases and aspects of the reef and its life than pages of written description.

MORNING POST.

In thus foreshadowing possible sources of wealth, and in presenting this luxuriously fashioned account of the Great Barrier Reef and its products, Mr. Saville-Kent has rendered eminent service alike to the province of Queensland and the cause of scientific progress and knowledge.

SCOTSMAN.

It is certain that since the appearance of Mr. Darwin's monograph on Coral Reefs no contribution of such importance has been made to the literature of this interesting department of physical science. . . . It is certain that by bringing his researches and collections so fully within reach of students as Mr. Saville-Kent has done by the production of this magnificently appointed volume he has rendered natural science a service which it would be difficult to over-estimate. The work will always be a first authority on its subject, and as indispensable book of reference for all who wish to have views of their own upon coral reefs.

[Continued over Leaf.

EXTRACTS FROM OPINIONS OF THE PRESS (Continued).

CAMBRIDGE REVIEW.

The most striking feature of Mr. Saville-Kent's magnificent monograph on "The Great Barrier Reef of Australia" is the wonderful series of photographic plates which illustrate in a way never before attempted the extraordinary variety of shape and beauty of form which corals present. It almost takes away our breath to be suddenly shown one of these plates; we feel that we are looking at the thing itself, and we are lost in admiration at the skill of the photographer and the care of the publisher which have combined to produce these results.

STANDARD.

The old naturalist probably never dreamt of the world having to welcome so sumptuous a monograph as that which Mr. Saville-Kent has issued. For not only is it all that its title claims—an exhaustive account of the great coral reef which stretches along the Queensland coast—but in addition it forms a very admirable history of the marine resources of the Colony with which the author was so honourably connected as Commissioner of Fisheries. Nothing seems to be omitted, and everything is illustrated in the most beautiful manner by forty-eight plates, in which photo-mezzotype appears at its best, and sixteen chromo-lithographs from the pencil of the accomplished zoologist whose services to science have been so long and so varied. Altogether, a work more satisfactory, from an artistic and a scientific point of view has seldom come before us.

WEST AUSTRALIAN.

Mr. Saville-Kent's book on the Great Barrier Reef of Australia is truly a monumental work, and is an important contribution to both science and art. . . . The book, which is a veritable *édition de luxe*, and inspires admiration on the part of the most casual individual who may pick it up, compels the interested attention of the reader. It is as superb a specimen of the printer's, photographer's and publisher's art as it has been our good fortune to see, and we can well believe that it succeeded in eliciting the special commendation of Her Majesty the Queen, who was graciously pleased to accept a copy from its author.

DAILY NEWS.

It is to this marvellous feature of the ocean on the eastern coast of the Australian continent that Mr. W. Saville-Kent has devoted the magnificent quarto volume with its numerous coloured and photographic illustrations. . . . Mr. Saville-Kent is, in the first place, a naturalist, and none who are acquainted with his great "Manual of the Infusoria," will need to be told that he has gone about his herculean task with inexhaustible industry and zeal, and has produced a work unique of its kind and little likely to be superseded unless it be by future editions embodying further researches by the same indefatigable explorer and student of nature. It would be impossible to convey to a reader who has not examined this massive volume an adequate notion of the matter of its chapters, or of the singular beauty and interest of its plates, coloured and otherwise, after photographs and drawings.

FIELD.

One of the most magnificent that was ever published. Too much praise can hardly be bestowed upon the illustrations, which are mainly original photographs of the largest quarto size, displaying the beauties of the corals and other animals constituting these marvellous structures with a degree of accuracy which has never been even attempted.

THE AUSTRALASIAN.

A great work on a great subject. . . . Only the perfection to which the photographic and chromo-lithographic arts have been brought could have rendered possible the production of such a really superb book as Mr. Saville-Kent's "Great Barrier Reef of Australia," the scientific value of which is so largely enhanced by the number and beauty of its illustrations. . . . The book is one which whether viewed as a scientific treatise on a fascinating subject, or as a contribution to our knowledge of the economic resources of that great colony, or as a work of typographical and illustrative art, is entitled to unqualified praise.

BOOKSELLER.

One of the most striking publications of the hour, if not the most imposing of all.

DAILY CHRONICLE.

No praise could be too high for this magnificent work. . . . The text is extremely interesting, and written throughout in a fresh and lively style, which is too often not the case with works containing a similar amount of solid information.

LA NATURE.

M. Saville-Kent qui, pendant près de huit années, a occupé le poste d'inspecteur des pêcheries de la Grande-Barrière, a fait sur celle-ci une série d'observations qu'il a réunies dans une magnifique publication. Ce livres, qui paraîtra sous peu, est orné de photogravures et de planches coloriées qui en font un véritable objet d'art; hélas! quand verrons-nous de pareils livres en France? A l'étranger, quand il s'agit de science, on trouve toujours les bourses largement ouvertes. Que les choses sont différentes chez nous!

NOTTS GUARDIAN.

This magnificently illustrated and finely got-up volume, though treating its subject on scientific lines, is written in so lucid a style that it can be read with pleasure and appreciation by any ordinarily well-educated reader. . . . It is impossible to convey in words an adequate idea of the beauty and delicacy of these photo-mezzotype plates, and of the while field which they cover. The subjects range from photographs of coral specimens and coral growths to views of reefs, islets and islands, from an illustration of a cultivated oyster tank to hurricane-stranded coral masses and wrecked ships, from groups of pearls to representations of the various species of bêche-de-mer and of the many and odd kinds of fishes which swarm in the waters of the Great Reef. Mr. Saville-Kent has, indeed, by the aid of the camera, placed at the disposal of scientists an invaluable mass of observations, and enabled the stay-at-home naturalist almost to realise what Mr. Saville-Kent tells his readers was a day-dream of his own, namely, to see these wonderful coral organisms growing in their native seas.

PHOTOGRAPHY.

The Great Barrier Coral Reef of Australia is one of the wonders of the world, and the work under our notice is in every way worthy of such a subject. The naturalist will find in its pages a wealth of scientific fact gathered by a master-hand, and the ordinary individual under its guidance may wander in scenes of beauty and wonder hitherto unknown to him. Such a work, from the pen of such a scientist as the author, is an important contribution to British literature.

BY THE SAME AUTHOR.

A MANUAL OF THE INFUSORIA,

A descriptive Monograph of the FLAGELLATE, CILIATE and TENTACULIFEROUS PROTOZOA.

Royal 8vo, Vols. I to III, over 900 pages of text, and 51 Plates containing upwards of 2,000 figures.

LONDON: W. H. ALLEN & Co., LIMITED. Price £4. 4s. net.

Waterlow & Sons Limited, Printers, London Wall, London.

www.ingramcontent.com/pod-product-compliance
Lightning Source LLC
Chambersburg PA
CBHW022103290426
44112CB00008B/530